Free Radicals in Exercise and Aging

Zsolt Radák, PhD
Editor

Laboratory of Exercise and Physiology,
Semmelweis University Budapest
School of Sport Sciences
Budapest, Hungary

Human Kinetics

Library of Congress Cataloging-in-Publication Data

Free Radicals in Exercise and Aging/Zsolt Radák, editor
    p. cm.
Includes bibliographical references and index
ISBN 0-88011-881-4
1. Exercise--Molecular aspects.   2. Aging--Molecular aspects.
3. Free radicals (Chemistry)--Physiological effect.   4. Antioxidants--Physiological effect.
I. Radák, Zsolt, 1961-
               00-44938
               CIP

ISBN: 0-88011-881-4

Acquisitions Editor: Michael S. Bahrke; **Managing Editor:** Melinda Graham; **Assistant Editor:** John Wentworth; **Copyeditor:** Julie Anderson; **Proofreader:** Erin Cler; **Indexer:** Mary E. Coe; **Permissions Manager:** Cheri Banks; **Graphic Designer:** Nancy Rasmus; **Graphic Artist:** Kathleen Boudreau-Fuoss; **Cover Designer:** Jack W. Davis; **Illustrators:** Sharon Smith and Kathleen Boudreau-Fuoss; **Printer:** Versa Press

Printed in the United States of America   10 9 8 7 6 5 4 3 2 1

Human Kinetics
Web site: www.humankinetics.com/

United States: Human Kinetics, P.O. Box 5076, Champaign, IL 61825-5076
800-747-4457
e-mail: humank@hkusa.com

Canada: Human Kinetics, 475 Devonshire Road Unit 100, Windsor, ON N8Y 2L5
800-465-7301 (in Canada only)
e-mail: humank@hkcanada.com

Europe: Human Kinetics, P.O. Box IW14, Leeds LS16 6TR, United Kingdom
+44 (0)113-278 1708
e-mail: humank@hkeurope.com

Australia: Human Kinetics, 57A Price Avenue, Lower Mitcham, South Australia 5062
(08) 82771555
e-mail: liahka@senet.com.au

New Zealand: Human Kinetics, P.O. Box 105-231, Auckland Central
09-309-1890
e-mail: humank@hknewz.com

To my mother, Erzsébet
and to my late father, Lajos.

Z.R.

Contents

Contributors

Francisco H. Andrade, Department of Neurology, Case Western Reserve University and University Hospitals of Cleveland, Cleveland, Ohio

Eli Carmeli, Physical Therapy Program, Sackler Faculty of Medicine, Tel-Aviv University, Ramat Aviv, Israel

Peter Ferdinandy, Department of Biochemistry, University of Szeged, Dóm tér 9, Szeged, H-6720 Hungary

Sataro Goto, Department of Biochemistry, School of Pharmaceutical Sciences, Toho University, Miyama, Funabashi, Japan

Shukoh Haga, Institute of Health and Sport Sciences, University of Tsukuba, Tsukuba, Japan

Jay W. Heinecke, Department of Medicine and Department of Molecular Biology and Pharmacology, Washington University School of Medicine, St Louis, USA

John Hollander, Department of Kinesiology and Interdisciplinary Graduate Program of Nutritional Science, University of Wisconsin, Madison, 2000 Observatory Drive, Madison, WI 53706

Li Li Ji, Department of Kinesiology and Interdisciplinary Graduate Program of Nutritional Science, University of Wisconsin–Madison, 2000 Observatory Drive, Madison, WI 53706

Takao Kaneko, Tokyo Metropolitan Institute of Gerontology, Tokyo, Japan

Takako Kizaki, Department of Hygiene, Kyorin University, School of Medicine, Mitaka, Japan

Gila Lavian, Laboratory of Clinical Neurophysiology, Rambam Medical Center and Technion, Faculty of Medicine, Haifa, Israel

Mitsuyoshi Matsuo, Department of Biology, Faculty of Science and High Technology Research Center, Konan University, Kobe, Japan

Hiromi Miyazaki, Institute of Health and Sport Sciences, University of Tsukuba, Tsukuba, Japan

Shuji Oh-ishi, 5th Department of Internal Medicine, Tokyo Medical University, Inashiki 300-0395, Japan

Hideki Ohno, Department of Hygiene, Kyorin University, School of Medicine, Mitaka, Japan

Tomomi Ookawara, Department of Biochemistry, Hyogo College of Medicine, Nishinomiya, Japan

Zsolt Radák, Laboratory of Exercise and Phyisology, School of Sport Sciences, Semmelweis University Budapest, Budapest, Hungary

Abraham Z. Reznick, Department of Anatomy and Cell Biology, The Bruce Rappaport Faculty of Medicine, Technion, Haifa, Israel

Zoltán Szilvássy, Department of Pharmacology, University of Debrecen, Debrecen, Hungary

Arpad Tosaki, Department of Pharmacology, University of Debrecen, Debrecen, Hungary

Foreword

Biological oxidation processes "burn" substrate and provide us with the energy to survive. The "combustion" is not, however, a direct reaction of molecular oxygen with the substrate but rather a transfer of electrons mediated by several enzyme systems, in which oxygen is the final electron acceptor. Energy thus produced, when trapped and stored chemically as pyrophosphate bonds, fuels work processes such as cell signal transduction, muscle tone, circulation, respiration, digestion, secretions, activities of the nervous system, and synthesis of cell constituents. Unlike most other elements, reduction of or electron acceptance by oxygen is complicated by the fact that the oxygen molecule has two parallel spinning unpaired electrons in its outermost orbital. So, to be reduced in one step, an oxygen molecule would need two electrons, spinning in the opposite direction with respect to the valence electrons of oxygen but spinning in the same direction with respect to each other, to enter its outermost atomic orbital simultaneously. This is not possible according to the Pauli exclusion principle. Thus, the only solution is the univalent reduction of oxygen, that is, the reduction of the oxygen molecule one electron at a time. Intermediates in the oxygen reduction process are *free radicals*—molecules containing an unpaired electron (radical) that are capable of independent existence (free). For example, when a single electron reduces the oxygen molecule, the resultant species, the superoxide anion radical, still contains one unpaired electron. Under resting conditions, 2 to 5% of the total oxygen consumed by tissues may contribute to the development of reactive oxygen species.

That oxygen may be harmful to human health was postulated by Paul Bert in 1878, just a century after Joseph Priestly discovered the gas. Oxygen toxicity, studied in those early days, usually referred to the toxic effects of oxygen at high pressure such as during diving, hyperbaric O_2 therapy, and aerospace travel. In his classic work *La Pression Barometrique,* Bert described the incidence of convulsions in various animal species exposed to high-pressure oxygen. During those days, the periodic table (Mendeleef and Meyer, 1869) and tetrahedral carbon (Van't Hoff and Le Bell, 1874) had just been introduced. The concept of ions in solution (Arrhenius, 1884) was not fully accepted until Thompson (1897) proved the existence of electrons. Given the scientific backdrop, Bert's conception of oxygen toxicity was a prodigy. Today's concept of oxygen toxicity not only is restricted to hyperbaric oxygen but includes a very important element—oxidative stress under normobaric conditions in our daily lives. As early as 1956, Denham Harman suggested that free radicals are likely to be involved in aging. This free radical theory of aging has gained considerable momentum in recent years. A late-breaking aspect of oxidant action that has drawn the limelight of current biomedical research is the ability of these reactive species to modulate a number of intracellular signal transduction processes that are critically linked to widespread biological processes.

In exercise physiology, a common approach to measuring physical fitness is based on the individual's ability to use atmospheric oxygen in a given interval of time per kilogram of body weight (i.e., the aerobic capacity). Therefore, athletes aim to boost their aerobic capacities to the highest possible limit. Under resting conditions, oxygen content in arterial and venous blood of the skeletal muscle tissue is 20 and 15 ml per 100 ml blood, respectively. Physical exercise may increase skeletal muscle arteriovenous oxygen difference 3-fold and blood flow through the tissue 30-fold. As a result, exercise may cause up to a 100-fold increase in oxygen flux through the active skeletal muscles.

So, is exercising bad? Certainly not. A physically active lifestyle coupled with well-balanced nutrition is necessary to improve quality of life and ensure healthy aging. Depending on the duration and type of work, antioxidant defenses of active tissues may be overwhelmed by excess reactive oxygen species generated during exercise. Understanding the possible mechanisms that contribute to exercise-induced oxygen toxicity and designing appropriate measures to circumvent or minimize such toxicity are fundamental for three reasons: to enhance the efficacy of physical exercise as a preventive and therapeutic tool, to control exercise-induced oxidative tissue damage and augmentation of health risks, and possibly to enhance endurance performance capacity in sports. In this volume, our current understanding of some of these issues is discussed.

Chandan K. Sen, PhD, FACSM
Lawrence Berkeley National Laboratory
University of California, Berkeley and
University of Kuopio, Finland

Preface

Molecular oxygen has been present in the biosphere for the last two billion years as a result of photosynthetic activity of microorganisms. Oxygen is one of the most abundant elements on our planet, and its ability to accept electrons makes it vital for a variety of physiological processes. Aerobic organisms learned to produce energy by reducing molecular oxygen to water. However, during this process, by coincidence or by physiological means, the electron transport chain releases reactive oxygen intermediates. These reactive oxygen species continuously challenge the integrity of cell components. The negative charge of these reactive oxygen species makes them extremely active in interactions with a variety of molecules (e.g., the diffusion distance of hydroxyl radical is less than three molecular distance). Chapter 1 describes the manifestation of free radical species and discusses their reactivity, appearance, and site of generation.

Increased oxygen consumption during exercise might lead to the conclusion that exercise is dangerous because it generates free radicals and induces oxidative damage. However, let us look at this phenomenon from a different perspective. Can we increase our endurance or strength without feeling fatigued? Can we increase our bodies' resistance against foreign bodies and viruses without being exposed to them? Think about regular exercise as a special vaccine. We "infect" our bodies with small, but larger than so-called normal, levels of free radical species and let our bodies reply by up-regulating their antioxidant systems and tolerance to radical species. The second chapter of this book talks about the antioxidant defense mechanism that developed to protect our bodies against free radicals. The involvement of free radical species in aging was suggested more than 40 years ago, and this hypothesis has gained great support. Exercise and aging affect the antioxidant enzyme system in different ways. Chapter 3 shows how we can affect our antioxidant system by nutritional factors. Many of us take supplements to protect against reactive oxygen species, and this chapter provides some advice on supplementation. Chapters 4 and 5 discuss how free radical species in a certain concentration are key factors for a variety of physiological processes, such as preconditioning, fatigue, or cell signaling. These reactive oxygen species might up-regulate the resistance of the myocardium (a small amount of poison might be medicine!), modulate the force-generating ability of skeletal muscle (a large dose of antioxidant decreases the force generation; an extra amount of medicine might be a poison), and control vital redox processes. On the other hand, when the free radical concentration exceeds a certain level, and this level might depend on cell types, age, and history of free radical exposure, these reactive species can have deleterious effects. Chapters 6 and 7 provide the latest information on protein, DNA, and lipoprotein damage related to exercise and aging.

Aging generally is regarded as a process during which the ability of an organism to cope with internal and external effects decreases. This seems to involve the efficiency of the antioxidant and oxidative damage repair systems. Moreover, as a result of these defects, the amount of free radical generation might increase and age-associated oxidative damage might accumulate.

Exercise seems to be an easy way to delay or decrease the age-dependent loss of physiological function. If we put our bicycles in our closets for the winter and do not use them until spring, we might see some rust on them. On the other hand, a bicycle used throughout the year might become more resistant to the deleterious effects of oxygen. So get on your bike and ride, and do not worry about your age!

Acknowledgments

A number of my friends have kindly contributed their work and without their expertise, the book simply would not have been completed. The names of these exceptional individuals can be read on the first page of each chapter. I would like to express my thinks to Ferenc Katai for his encouragement and friendship during my career. I would also like to thank my friend Albert V. Taylor who continues to help me a lot. Thank you to Dr. Andras Monus because of his friendly support which was indispensable for this work. I also would like to express my thanks to Dr. Katsumi Asano for fostering my interest in exercise physiology. I am very thankful to my family for the patience and support which continue to be important and necessary for my scientific work.

A very special thanks to Melinda Graham, Dr. Rainer Martens, and the rest of the staff at Human Kinetics for their expertise and patience in bringing this book to publication.

Chapter 1

The Chemistry of Reactive Oxygen Species and Related Free Radicals

Mitsuyoshi Matsuo
Department of Biology, Faculty of Science and High Technology Research Center, Konan University, Kobe, Japan

Takao Kaneko
Tokyo Metropolitan Institute of Gerontology, Tokyo, Japan

Reactive oxygen species is a general term for molecular oxygen–derived molecules that are reactive species or that are converted easily to reactive species; some reactive oxygen species are free radicals. The uncontrolled generation of reactive oxygen species in vivo oxidizes biomolecules, such as nucleic acids, proteins, and lipids, which alters genetic information, denatures proteins, inactivates enzymes, and disorders biomembranes (figure 1.1). Thus, generation of reactive oxygen species causes oxidative stress, which is the harmful influence of in vivo oxidation due to a disturbance in the oxidant–antioxidant balance. Such reactive oxygen species damage organisms, leading to disease, poisoning, and presumably, aging (figure 1.1). Enhanced antioxidant defense against reactive oxygen species–induced oxidative stress is thought to be an important medical treatment because of scavenging them. In addition, lifestyle, including exercise, nutrition, smoking, and drinking, is being reconsidered for its role in oxidative stress. Research has been conducted on reactive oxygen species for about 90 years, but only recently have they been studied in the biological and medical sciences. Studies of reactive oxygen species and related free radicals in these areas are under way.

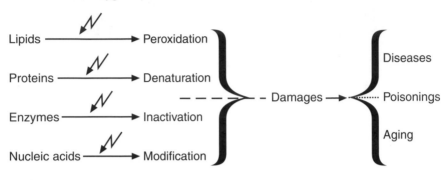

Figure 1.1 Damage of biomolecules due to reactive oxygen species.

Free Radicals

An atom consists of an atomic nucleus and electrons. The atomic nucleus is composed of protons and neutrons, and the electrons are located in atomic orbitals. When a molecule is formed by the bonding of atoms, usually its molecular orbitals are organized and each is occupied by a pair of electrons with opposite spin to each other. If an unpaired electron exists in the molecular orbital, the molecule is called a free radical (hereafter referred to as a radical). In general, a radical is very reactive chemically, because its unpaired electron seeks to pair with another electron to be stabilized. A radical (A•) reacts to accept an electron from another molecule (B), and then the electron-donated molecule (A) and another radical (B•) are produced:

$$A\bullet + B \rightarrow A + B\bullet$$

The newly produced radical is usually labile and can also react with a third molecule (C) to produce a third radical (C•):

$$B\bullet + C \rightarrow B + C\bullet$$

If the third radical reacts with the initially electron-donated molecule to produce the original radical, a radical chain reaction may start:

$$C\bullet + A \rightarrow C + A\bullet$$

The radical chain reaction continues until a termination reaction takes place. Thus, if the radical chain reaction proceeds in vivo without regulation, it may damage biological systems. The termination reaction occurs when radicals (R•) react with each other to produce nonradical species:

$$R\bullet + R\bullet \rightarrow RR$$

where R represents A, B, or C. The radical chain reaction is also terminated with antioxidants, such as vitamin E and butylated hydroxytoluene, which react with radicals to give nonradical species, as will be discussed later.

Molecular Oxygen and Reactive Oxygen Species

In the atmosphere, the content of molecular oxygen (hereafter referred to as oxygen) is nearly 21% by volume. Oxygen is soluble at a concentration of 1.3 mM in water at 25° C under 1 atm (Dean, 1973). Aerobic organisms have adapted very well to the atmosphere, evolving to use atmospheric oxygen by respiration and to obtain energy efficiently. Respiration can release up to 2870 kJ from the complete oxidation of 1 mol glucose to carbon dioxide and water, whereas fermentation releases only a small fraction, 197 kJ, from the conversion of 1 mol glucose to 2 mol lactate. The highly efficient energy gain is based on the high oxidizing potential of oxygen:

$$\tfrac{1}{2}\,O_2 + 2\,e^- + 2\,H^+ \rightarrow H_2O \qquad E^{o'} = 0.815\ V$$

where $E^{o'}$ represents the standard reduction potential in neutral solution. A poor oxidizing agent would make less efficient use of substrates. In biological systems, the energy is accumulated as adenosine $5'$-triphosphate (ATP) and released by its hydrolysis. Aerobic organisms, including humans, maintain high activity by respiration and cannot exist without oxygen.

Oxygen itself is a radical, because its two unpaired electrons are located in different antibonding π^* orbitals (figure 1.2); such a radical is called a diradical. These two electrons have the same spin (i.e., paralleled spin) in a triplet state. This is the most stable state, the so-called ground state, of oxygen ($^3\Sigma_g^- \, O_2$ or $^3O_2$) (Ingraham and Meyer, 1985).

Although oxygen has a high oxidizing potential, it does not rapidly oxidize ordinary organic compounds. Oxygen does not react as a normal double bond between the two oxygen atoms but rather acts as a diradical because of the triplet state. To maintain spin conservation during the reaction, oxygen must either react with another molecule with an unpaired electron or produce a triplet state product. Stable triplet states are unusual, so this kinetically restricts oxygen reactions. Electron transfer requires that the first electron be placed in

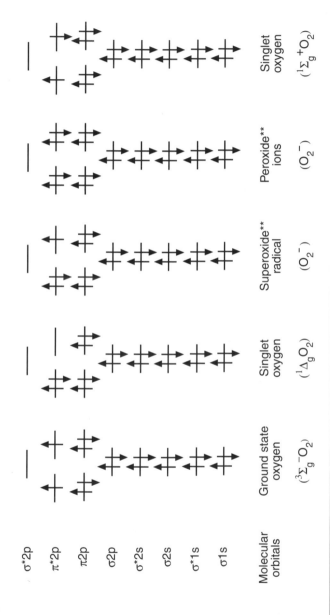

Figure 1.2 Spin states of electrons in diatomic oxygens. An arrow represents an electron, and its direction means the positive and negative signs of spin quantum number. *See Table 1.1, "RR' CO*", **See Table 1.1, "Superoxide Radical."

a partially filled antibonding π^* orbital. This also creates a barrier to electron transfer reactions. That is, oxygen has a kinetic barrier to oxidation of ordinary organic compounds.

Most biomolecules are organic compounds and nonradicals and, in accordance with Pauli's principle, have paired electrons with opposite spin, i.e., in singlet states. Hence, reaction of oxygen with biomolecules is restricted and extremely slow. Because of the high oxidizing potential of oxygen, the human body, which is composed of biomolecules, should burn in air. By virtue of the kinetic barrier, fortunately, organisms can exist in air and can even use oxygen to obtain energy.

Reactive oxygen species relevant to biological systems are listed in table 1.1; bear in mind that reactive oxygen species cannot be rigidly defined in technical terms. Among them, the typical ones are singlet oxygen ($^1\Delta_g O_2$), the superoxide radical (O_2^-), hydrogen peroxide (H_2O_2), and the hydroxyl radical ($HO\bullet$) (the superoxide radical and the hydroxyl radical are radicals also). In addition, nitrogen monoxide (NO, also referred to as nitric oxide) is an important radical species.

In respiration, oxygen undergoes a series of one-electron reductions with cytochrome c oxidase in complex IV of the mitochondrial electron transport chain to produce the superoxide radical, hydrogen peroxide, the hydroxyl radical, and water:

$$O_2 \xrightarrow{e^-} O_2^- \xrightarrow{e^-} H_2O_2 \xrightarrow{e^-} HO\bullet \xrightarrow{e^-} H_2O$$

Although oxygen is reduced very efficiently, there is a risk that the superoxide radical and hydrogen peroxide may leak from complex IV. From the generation of the superoxide radical and hydrogen peroxide in vitro in the mitochondria and microsomes from rat lung, Fridovich and Freeman (1986) estimated the formation rate of reactive oxygen species to be about 50 nmol • g-tissue^{-1} • min^{-1} or about 10^{11} radicals • cell^{-1} • day^{-1}. In exchange for having high activity, aerobic organisms are always exposed to oxidative stress due to oxygen and related species.

Singlet Oxygen

Ground state oxygen is physically or chemically excited to give singlet oxygen with an additional energy of 94 kJ • mol^{-1} ($^1\Delta_g O_2$) or 157 kJ • mol^{-1} ($^1\Sigma_g^+ O_2$) (Ingraham and Meyer, 1985). The lower energy $^1\Delta_g$ or higher energy $^1\Sigma_g^+$ state has two electrons with opposite spin in an antibonding π^* orbital or separate antibonding π^* orbitals, respectively (figure 1.2).

The lifetime of $^1\Sigma_g^+$ singlet oxygen is only 10 ps, and it usually decays to the $^1\Delta_g$ state before it reacts with anything. Hence, $^1\Sigma_g^+$ singlet oxygen is thought to have no biological significance.

Table 1.1 Reactive Oxygen Species Relevant to Biological Systems

Reactive oxygen species	Molecular formula	Formation
Free radical species		
Superoxide radical	O_2^-	One-electron reduction of ground state molecular oxygen
Hydroperoxyl radical	HOO•	Protonation of the superoxide radical
Hydroxyl radical	HO•	One-electron reduction of hydrogen peroxide and three-electron reduction of ground state molecular oxygen
Nitrogen monoxide	NO	One-electron reduction of nitrite
Alkoxyl radical	RO•	One-electron reduction of hydroperoxide
Peroxyl radical	ROO•	One-electron oxidation of hydroperoxide
Nonradical species		
Singlet oxygen	$^1\Delta_g O_2$ ($^1O_2$)	Excitation of ground state molecular oxygen
Hydrogen peroxide	H_2O_2	Two-electron reduction of ground state molecular oxygen, followed by protonation, and protonation of the peroxide ion
Peroxynitrite	ONOO$^-$	Reaction of nitrogen monoxide with the superoxide radical
Hydroperoxide	ROOH	Autoxidation and singlet oxygen oxygenation of unsaturated compounds
Ozone	O_3	Oxidation of ground state molecular oxygen with atomic oxygen formed by photolysis of ground state molecular oxygen
Hypochlorous acid	HClO	Hydrolysis of molecular chlorine
Excited carbonyl	RR′CO*	Cleavage of dioxetane and self-decomposition of peroxyl radicals

On the other hand, the half-life of $^1\Delta_g$ singlet oxygen is 2 μs in water, which is ample time for biological reactions. The half-life is highly solvent-dependent and, surprisingly, increases to 20 μs in deuterated water (Kearns, 1979). Thus, $^1\Delta_g$ singlet oxygen–mediated reactions may be enhanced in deuterated solvents. Furthermore, having no unpaired electrons, $^1\Delta_g$ singlet oxygen is not a radical and hence can react with ordinary organic compounds without kinetic restriction. This means that $^1\Delta_g$ singlet oxygen can react with most biomolecules at high reaction rates; that is, $^1\Delta_g$ singlet oxygen is a real reactive oxygen species. Hereafter, $^1\Delta_g$ singlet oxygen is referred to as singlet oxygen ($^1O_2$) (table 1.1).

Carotenoids are generally the most efficient quenchers of singlet oxygen, with rates near the diffusion limit; for example, the rate constant of β-carotene is 1.2×10^{10} $M^{-1} \cdot s^{-1}$ (Foote, 1979). However, the quenching mechanism of carotenoids is still not understood.

Although excitation of oxygen to the singlet states proceeds photochemically, fortunately it cannot be achieved directly by light irradiation. This is another important factor in the low reactivity of oxygen that is extremely beneficial to life. For experimental use, singlet oxygen is commonly generated from ground state oxygen by photosensitization:

$$^1Sens + \xrightarrow{h\nu} {}^1Sens^* \rightarrow {}^3Sens^*$$

$$^3Sens^* + {}^3O_2 \rightarrow {}^1Sens + {}^1O_2$$

In the so-called Type II reaction, a triplet sensitizer ($^3Sens^*$) formed from a photochemically ($h\nu$) excited singlet sensitizer ($^1Sens^*$) interacts with ground state oxygen to produce singlet oxygen. Dyes, including methylene blue, rose bengal, and protoporphylin, are used as sensitizers (^1Sens). Singlet oxygen is thus likely to be formed in pigmented organs, including the eye lens and skin, when exposed to light. Singlet oxygen damages proteins and polyunsaturated lipids.

Singlet oxygen can also be chemically generated from a mixture of hydrogen peroxide and the hypochlorite ion (ClO^-) (Murray, 1979), which is a powerful microbicidal bleaching agent:

$$ClO^- + H_2O_2 \rightarrow Cl^- + H_2O + {}^1O_2$$

This reaction might be biologically relevant, because the hypochlorite ion can be formed by myeloperoxidase during phagocytosis as follows (Sawyer, 1991):

$$Cl^- + H_2O_2 \xrightarrow{H^+} ClO^- + H_2O$$

In the presence of hydrogen peroxide and chloride or iodide, myeloperoxidase can kill a number of bacteria and fungi in vitro. It is thought to play an important role in microbial killing by oxidizing the chloride ion into hypochlorous acid (HClO) and consequently by generating singlet oxygen from the hypochlorite ion and hydrogen peroxide.

Singlet oxygen acts as an electrophile. Unsaturated compounds are peroxidized by singlet oxygen oxygenation via two types of addition reaction—an ene reaction and cycloaddition:

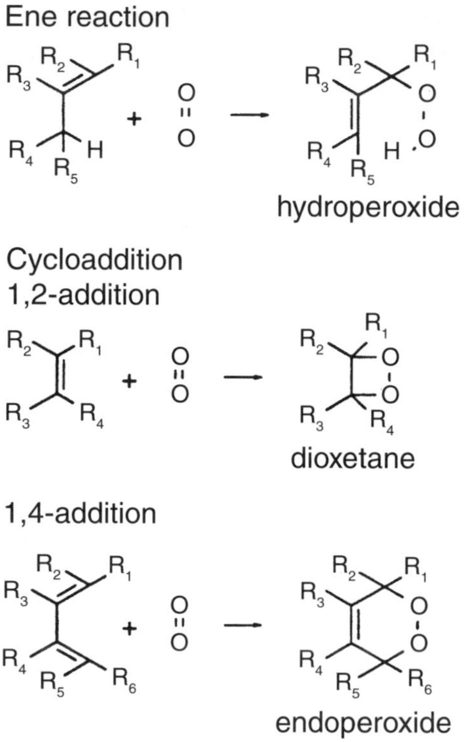

Ene reaction

hydroperoxide

Cycloaddition
1,2-addition

dioxetane

1,4-addition

endoperoxide

The rates of these reactions vary depending on the structures of substrates. For example, the rate constants of ene reactions and cycloadditions are $<5 \times 10^7$ (Gollnick and Kuhn, 1979) and $<5 \times 10^8$ $M^{-1} \cdot s^{-1}$ (Schaap and Zaklika, 1979), respectively.

Hydroperoxides are formed from both the auto-oxidation (see Hydroxyl Radical, p. 13) and singlet oxygen oxygenation of unsaturated compounds. These two reactions can be differentiated, because the double bond of each substrate migrates to the next carbon atom during the formation of hydroperoxide due to singlet oxygen oxygenation, as shown previously. The cooxidation of cholesterol provides an example of this differentiation. When hydroperoxides are formed from unsaturated compounds in the presence of cholesterol, the formation of cholesterol 5α-hydroperoxide provides evidence for singlet oxygen oxygenation, whereas the formation of cholesterol 7α- and 7β-hydroperoxides indicates auto-oxidation (Teng and Smith, 1973).

There are many reports that singlet oxygen is formed by dismutation of superoxide radicals and during the respiratory burst of neurophils. However, singlet oxygen formation must be carefully confirmed in biological systems. Frequently, singlet oxygen formation is proved only by light emission from the system (luminescence) or by inhibition of observed phenomena with so-called singlet oxygen scavengers such as 1,4-diazabicyclo[2.2.2]octane (DABCO), diphenylisobenzofuran, histidine, or azide (Halliwell and Gutteridge, 1989). There are, however, no specific scavengers of singlet oxygen. Although there are five light emission bands for singlet oxygen at 1268, 762, 634, 476, and 381 nm (Kasha and Brabham, 1979), observation of these bands is practically impossible because of a large number of contaminants that have fluorescence in the range.

Superoxide Radical

The superoxide radical is produced from one electron reduction of oxygen, as shown previously. It is negatively charged at the pH of neutral solution, because it has a pK_a of 4.69 (Bielski, 1978) and can pass through biomembranes only on an anion channel (Lynch and Fridovich, 1978). This is often written as $O_2 \overline{\cdot}$, where a dot represents an unpaired electron (i.e., a radical species; figure 1.2), and a small bar represents a negative charge (i.e., an anion species). Thus, the superoxide radical is precisely an anion radical. Because its spin density is delocalized between the two oxygen atoms, it has a limited radical character. Because of its negative charge, it is a weak Brønsted base in water and a powerful nucleophile, as well as a Brønsted base, in aprotic solvents.

In aqueous solution, the superoxide radical is extensively hydrated to give the hydroperoxyl radical (HOO•), which acts as a reducing agent and weak oxidizing agent. For example, the hydroperoxyl radical will reduce cytochrome c or nitroblue tetrazolium and will oxidize ascorbic acid. It also undergoes a dismutation reaction, which can be written overall as:

$$2\ O_2^- + 2\ H^+ \rightarrow H_2O_2 + O_2$$

Theoretically, however, this dismutation should be very slow because superoxide radicals electrostatically repel each other; in fact, the rate constant of this dismutation is less than 0.3 $M^{-1} \cdot s^{-1}$ (Bielski and Allen, 1977). In neutral to acidic solution, however, decomposition is largely due to bimolecular dismutation of either two protonated molecules or one protonated and one unprotonated molecule:

$$2\ HOO• \rightarrow H_2O_2 + O_2$$

$$HOO• + O_2^- + H^+ \rightarrow H_2O_2 + O_2$$

The elementary rate constants of these upper and lower reactions are 8.6×10^5 $M^{-1} \cdot s^{-1}$ and $1.0 \times 10^8\ M^{-1} \cdot s^{-1}$, respectively (Bielski, 1978). The maximum rate occurs at a pH that is equivalent to the pK_a for the hydroperoxyl radical, and the rate decreases with a decrease in the hydrogen ion concentration. In neutral solution, the overall rate of the preceding dismutation is estimated to be about 5×10^5 M^{-1} • s^{-1}. For a steady-state flux of 30 μM of the superoxide radical/hydroperoxyl radical at pH 5, the approximate half-life is about 30 ms (Sawyer, 1991). Any reaction that the superoxide radical undergoes in aqueous solution will compete with this dismutation reaction.

The superoxide radical can be produced chemically, physicochemically, and enzymatically. It is simply obtained from a solution of tetramethylammonium superoxide [(CH$_3$)$_4$ NO$_2$] in an aprotic polar solvent such as dimethyl sulfoxide and acetonitrile; unfortunately, commercially available potassium superoxide (KO$_2$) is only slightly soluble in such a solvent. Furthermore, its convenient source is by electrolysis of an oxygenated aprotic solvent containing an electrolyte. When only trace amounts are needed, it is commonly generated in solution by the xanthine oxidase–xanthine reaction or by pulse radiolysis of oxygenated, aqueous solutions containing formate ions. There are several superoxide radical-forming enzymes besides xanthine oxidase, such as nitropropane dioxygenase, indoleamine dioxygenase, tryptophan dioxygenase, and aldehyde oxidase.

The superoxide radical is formed in almost all aerobic cells (Fridovich, 1983). A major source is leakage from the oxygen reduction pathways in the electron transport chains of mitochondria and endoplasmic reticulum. The amount of leakage, that is, the production rate of the superoxide radical, increases as atmospheric oxygen concentration increases (Freeman and Crapo, 1981). The superoxide radical is also produced during the accelerated oxygen uptake, that is, the respiratory burst of phagocytic cells, such as neutrophils, macrophages, and lymphocytes. The oxygen uptake is due to the activation of a nicotinamide adenine dinucleotide phosphate (reduced form) (NADPH) oxidase complex bonded to the plasma membrane of these cells (Babior, 1978). The complex catalyzes the reduction of oxygen to the superoxide radical and the oxidation of NADPH into its oxidized form (NADP$^+$). The NADPH is provided by the pentose phosphate pathway in the cytosol. The superoxide radical plays a key role in the bactericidal action of these cells.

The reactivity of the superoxide radical is rather low in aqueous solution; thus, it is not a "reactive" oxygen species. However, biological damage by aqueous superoxide radical–generating systems has frequently been observed (Fridovich, 1983; Halliwell and Gutteridge, 1984). The xanthine oxidase–xanthine (or hypoxanthine) system damages biomolecules, including deoxyribonucleic acid (DNA), hyaluronic acid, collagen, and ornithine decarboxylase, in addition to, organelles, cells, and organs. The superoxide radical is dangerous to biological systems, because it is easily converted to real reactive oxygen species. As discussed later, the superoxide radical is dismutated to hydrogen peroxide and oxygen, and the hydrogen peroxide produces the hydroxyl radical in the presence of reduced transition metal ions by the Fenton reaction (see Fenton Reaction, p. 19). Furthermore, the superoxide radical reacts with nitrogen monoxide to give peroxynitrite (see Nitrogen Monoxide, p. 14). These products are very reactive.

Much evidence indicates that the superoxide radical is removed in vivo by superoxide dismutase (SOD) (see Antioxidant Defense, p. 21). Superoxide dismutases dismutate the superoxide radical under near diffusion control with a rate constant of $3 \times 10^9 \bullet M^{-1} s^{-1}$ and an approximate half-life of about 10 μs (Sawyer, 1991). Superoxide dismutases are present in all aerobes and are important antioxidant enzymes in conjunction with hydrogen peroxide–removing enzymes such as catalase and glutathione peroxidase.

Because superoxide dismutase is specific for the superoxide radical as a catalytic substrate, it is often assumed that, when superoxide dismutase inhibits a reaction, the superoxide radical is required for that reaction to proceed. Before making such an assumption, we should examine the inhibitory effect of the heat-denatured protein and the apoenzyme on the reaction. The superoxide dismutase protein acts as a good scavenger of singlet oxygen and the hydroxyl radical in chemical reactions, as do the apoenzyme and the holoenzyme (Matheson et al., 1975).

In biological systems, many quinones, such as ubiquinones and vitamin K, react reversibly with the superoxide radical:

$$\text{quinone} + O_2^- \leftrightarrow \text{semiquinone} + O_2$$

As the superoxide radical is scavenged by addition of superoxide dismutase, the conversion of quinone to semiquinone will decrease. So semiquinone-induced reactions might be erroneously attributed to the superoxide radical, because such reactions would be inhibited.

Hydrogen Peroxide

Having no unpaired electrons (figure 1.2), hydrogen peroxide is not a radical. Hydrogen peroxide is a relatively stable compound and is commercially available. It is not very reactive and acts as either a mild oxidizing or weak reducing agent. This means that in biological systems, it may be long-lived and may travel a long distance. Furthermore, hydrogen peroxide can readily pass through biological membranes, whereas the superoxide radical cannot do so without the aid of an anion channel.

The reactivity of hydrogen peroxide primarily depends on the unique bond energies of oxygen–oxygen and hydrogen–oxygen bonds, which are 213 and 376 kJ $\bullet$ mol^{-1}, respectively (Sawyer, 1991). This allows low-energy rearrangements to produce the extremely reactive hydroxyl radical. Although hydrogen peroxide itself is not so reactive, it is a potent reactive species to be converted to the hydroxyl radical. When hydrogen peroxide enters the cell, it may react with iron (II) ions or possibly copper (I) ions to give the hydroxyl radical (see Fenton Reaction, p. 19).

Hydrogen peroxide can inactivate a few enzymes, usually by oxidation of essential thiol groups. For example, glyceraldehyde-3-phosphate dehydrogenase in the glycolytic pathway is inactivated by hydrogen peroxide (Brodie and Reed, 1987). Thus, exposure of cells to large doses of hydrogen peroxide can deplete ATP by inhibiting glycolysis.

Hydrogen peroxide is often used as a disinfectant, and some bacteria are very sensitive to it. Some animal cells can be damaged in culture media at micromolar concentrations of hydrogen peroxide. The intensity of hydrogen peroxide toxicity to cells and organisms has been reported to be inconsistent. This can be accounted for by the differences both in the activity of hydrogen peroxide–scavenging enzymes and in the rate of conversion of hydrogen peroxide to more reactive species.

Any biological system that generates the superoxide radical will produce hydrogen peroxide by nonenzymatic or superoxide dismutase–catalyzed dismutation. Hydrogen peroxide, probably mainly via the superoxide radical, is produced from microsomes, mitochondria, and phagocytic cells. Furthermore, several enzymes, such as monoamine oxidase and galactose oxidase, produce hydrogen peroxide directly.

Microsomes from animal tissues have been shown to produce the superoxide radical and hydrogen peroxide at high rates in the presence of NADPH. These species largely arise from the NADPH-cytochrome P-450 reductase-cytochrome P-450 system (Terelius and Ingelman-Sundberg, 1988). However, the production rate of hydrogen peroxide in the perfused rat liver is much lower than that expected from the production rate of hydrogen peroxide in microsomes (Oshino et al., 1975). In the rat liver, the steady-state concentration of hydrogen peroxide is estimated to be $1\text{-}10^2$ nM. The liver has effective mechanisms for scavenging hydrogen peroxide. In addition, microsomes are an artifact of subcellular fractionation, and disruption of the endoplasmic reticulum and plasma membranes may increase hydrogen peroxide production.

Many bacteria and mycoplasma release hydrogen peroxide. Some animal cells, such as spermatozoa, also release hydrogen peroxide into the surrounding medium. Hydrogen peroxide is contained in human eye humors (Rose et al., 1998) and exhaled breath (Jobsis et al., 1998). Hydrogen peroxide and the superoxide radical are produced by photochemical reactions in sea water; incidentally, the former is present at less than micromolar concentrations in natural water (Zafiriou, 1987).

Hydroxyl Radical

The hydroxyl radical is produced by radiolysis of water under high-energy ionizing radiation and is thought to be produced mainly in biological systems by reductive cleavage of hydrogen peroxide in the Fenton reaction (see Fenton Reaction, p. 19):

$$Fe^{2+} + H_2O_2 \rightarrow Fe^{3+} + HO\bullet + HO^-$$

Another pathway to generate the hydroxyl radical may involve the one-electron reduction of hypochlorite by the superoxide radical or an iron (II) complex (Candeias et al., 1993, 1994):

$$HClO + O_2^- \rightarrow Cl^- + O_2 + HO\bullet$$

$$HClO + Fe^{2+} + \rightarrow Cl^- + Fe^{3+} + HO\bullet$$

The hydroxyl radical has a very high oxidation potential (Sawyer, 1991) and is highly reactive:

$$HO\bullet + e^- + H^+ \rightarrow H_2O \qquad E^{o\cdot} = 2.31 \text{ V}$$

The rate constants of hydroxyl radical reactions with most biomolecules are $10^8 - 10^9$ M$^{-1} \bullet$ s^{-1} (e.g., 5.1×10^9 M$^{-1} \bullet$ s^{-1} for methionine; Anber and Neta, 1967). Thus, the hydroxyl radical will react nearly under diffusion control by either addition to or hydrogen abstraction from molecules, so that another radical species is formed:

$$SH + HO\bullet \rightarrow HO(H)S\bullet$$

$$SH + HO\bullet \rightarrow S\bullet + H_2O$$

where SH represents a substrate. In either case, a radical chain reaction may take place. Such chain reactions are particularly important to lipid peroxidation, whereby polyunsaturated fatty acid moieties of membrane phospholipids may be destructively oxidized.

Once formed in vivo, the hydroxyl radical may react rapidly and indiscriminately with almost any biomolecule that is near its formation site. However, because the hydroxyl radical has extremely high reactivity and hence must be short-lived in vivo, it may be unable to react with any molecules far from the formation site. Thus, the damaging effect of the hydroxyl radical should be site-specific. Accordingly, the type of damage would depend on its formation site; for example, generation of the hydroxyl radical close to DNA could modify its bases or cause strand breakage (Mello Filho et al., 1984).

As previously described, reaction of the hydroxyl radical with a biomolecule will produce other radicals, the reactivity of which is usually lower than that of the hydroxyl radical. Less reactive radicals can sometimes diffuse away from the formation site and attack specific biomolecules. The importance of such secondarily formed radicals in vivo is shown by the ability of the hydroxyl radical to initiate lipid peroxidation through hydrogen abstraction and to produce the carbon-centered and peroxyl radicals (see Miscellaneous, p. 16; also equations 1.1-1.3).

Nitrogen Monoxide

Nitrogen monoxide has an unpaired electron in an antibonding π^* orbital and is a rather stable radical. Unlike many radicals, such as the hydroxyl radical, it does not react rapidly with most biomolecules. In contrast, it reacts rapidly with free and bound transition metals or some radicals, including oxygen. Its lifetime in vivo is thought to be less than 10 s (Moncada et al., 1991).

The fate and effect of nitrogen monoxide on biological systems also depend on its physical properties. Nitrogen monoxide is a colorless gas at room temperature. It is soluble at a concentration of 1.9 mM in aqueous solution under 1 atm (Dean, 1973). These properties are similar to those of oxygen.

Nitrogen monoxide gas is commercially available. It can be synthesized by several convenient methods, for example, the reduction of nitrite with iodide ions in aqueous solution:

$$2\ HNO_2 + 2\ I^- + 2\ H^+ \rightarrow NO + I_2 + 2\ H_2O$$

Nitrogen monoxide can be stored at room temperature after purification. It is formed in vivo from the conversion of arginine to citrulline with nitrogen monoxide synthases (Knowles and Moncada, 1994).

Examining reactions of nitrogen monoxide with metal complexes is important to understanding its bioregulatory behavior, including biological signaling. It reacts with some transition metal complexes to give metal-nitrosyl adducts. The rate of these reactions, as well as the stability of the metal-nitrosyl adducts, depends on the forms of metals and ligands in the complexes. Reaction of nitrogen monoxide with the iron (II) ion in acidic solutions produces a simple iron-nitrosyl adduct (Epstein et al., 1980). In contrast, the iron (III) ion does not produce any stable nitrosyl adduct. Furthermore, heme proteins are intimately associated with the biological action of nitrogen monoxide. Nitrogen monoxide binds to the heme moiety of guanylate cylase, thereby stimulating the conversion of guanosine 5'-triphosphate (GTP) to cyclic guanosine 3',5'-monophosphate (cGMP), a second messenger for relaxing blood vessels

(Wink et al., 1996). Thus, nitrogen monoxide released from endothelium is called the endothelium-derived relaxing factor (EDRF).

Nitrogen monoxide reacts with the hydroxyl radical at an almost diffusion-controlled rate in aqueous solution:

$$NO + HO\bullet \rightarrow HNO_2$$

where the rate constant is 4.5×10^9 $M^{-1} \bullet s^{-1}$ (Anber and Neta, 1967). It also reacts with nitrogen dioxide (NO_2) at a near diffusion-controlled rate to form dinitrogen trioxide (N_2O_3), which reverts to nitrogen dioxide in the presence of water. Nitrogen monoxide reacts rapidly with alkylperoxyl radicals as well as alkyl and alkoxyl radicals:

$$NO + ROO\bullet \rightarrow ROONO$$

where R represents an alkyl group and the rate constant is $1\text{-}3 \times 10^9$ $M^{-1} \bullet s^{-1}$ (Padmaja and Huie, 1993). This may be relevant to the effect of nitrogen monoxide on lipid peroxidation.

Reaction of nitrogen monoxide with the superoxide radical gives peroxynitrite ($ONOO^-$), which has received considerable attention both as a potentially deleterious reaction and as a detoxification mechanism for reactive oxygen species. The reaction between them is nearly diffusion controlled:

$$NO + O_2^- \rightarrow ONOO^-$$

where the rate constant is 6.7×10^9 $M^{-1} \bullet s^{-1}$ (Huie and Padmaja, 1993). Indeed, this reaction is three times faster than superoxide dismutase–catalyzed dismutation of the superoxide radical. Nitrogen monoxide does not directly react with hydrogen peroxide in neutral solution, but it seems to react with the hydroperoxyl anion (HOO^-) to form peroxynitrite (Koppenol, 1997).

Peroxynitrite is a relatively long-lived, toxic compound. It reacts with thiol compounds to give disulfides; the rate constant for reaction with cysteine is 5.9×10^3 $M^{-1} \bullet s^{-1}$ (Radi et al., 1991). Protonation of peroxynitrite ($pK_a = 6.8$) (Ramezanian et al., 1996) gives peroxynitrous acid (HOONO), which can oxidize various biological molecules, including nucleic acids, proteins, and phospholipids, in competition with isomerization to nitrate (NO_3^-). The reactive intermediates formed after protonation were first thought to be nitrogen dioxide and the hydroxyl radical. It has been suggested, however, that the intermediate responsible for oxidation of various biomolecules is an isomer of peroxynitrous acid (Pryor and Squadrito, 1995). Another potentially important reaction of peroxynitrite is nitration of tyrosine in the presence of transition metal ions (Ischiropoulos et al., 1992), which may inhibit and alter some enzymes and proteins.

Reaction of nitrogen monoxide with oxygen (i.e., its auto-oxidation) is very important. In hydrophobic media, auto-oxidation produces nitrogen dioxide, dinitrogen trioxide, and dinitrogen tetraoxide (N_2O_4), which are commonly associated with air pollution. In aqueous solution, however, there appears to be a different reactive species (NO_x) (Wink and Ford, 1995), which is rapidly hydrolyzed to nitrite. This species is capable of both oxidation of redox-active compounds and nitrosation of amino and thiol compounds. It reacts with proteins and peptides with thiol groups (RSH), such as glutathione, to give S-nitrosothiol adducts (RSNO), although nitrogen monoxide itself does not react directly with thiol groups to form S-nitrosothiol adducts (Wink et al., 1994);

$$NO_X + RSH \rightarrow NO_2^- + RSNO$$

Where R represents an alkyl group, SH a thiol group, and SNO an S-nitrosothiol group.

These adducts have been shown to form endogenously in the cardiovascular and pulmonary systems. Because they release nitrogen monoxide over a period of time and activate guanylate cyclase in vivo, they have been proposed as alternative endothelium-derived relaxing factors (Myers et al., 1990).

Miscellaneous

Hydroperoxide (ROOH) is produced by auto-oxidation and singlet oxygen oxygenation of unsaturated compounds (Matsuo and Kaneko, 1998). In particular, it is important as the primary product of lipid peroxidation. For lipid peroxidation, either an unsaturated lipid or oxygen must be activated. For example, an unsaturated lipid is activated to the carbon-centered unsaturated lipid radical for lipid peroxidation by an auto-oxidation mechanism (see following explanation), and oxygen is activated to singlet oxygen for lipid peroxidation by a singlet oxygen oxygenation mechanism (see Singlet Oxygen, p. 5). In addition, typical enzymatic lipid peroxidation catalyzed by lipoxygenase proceeds by a reaction mechanism similar to an auto-oxidation mechanism.

The auto-oxidation mechanism for lipid peroxidation has been well established. Lipid peroxidation consists of the following three processes:

Initiation

1.1   RH + I • Æ R• + IH

Propagation

1.2 $R• + O_2 \rightarrow ROO•$

1.3 $ROO\bullet + RH \rightarrow ROOH + R\bullet$

Termination

1.4 $R\bullet + R\bullet \rightarrow RR$

1.5 $R\bullet + ROO\bullet \rightarrow ROOR$

1.6 $ROO\bullet + ROO\bullet \rightarrow ROOR + O_2$

where RH represents an unsaturated lipid; $I\bullet$, a radical such as the hydroxyl radical derived from an initiator, $R\bullet$ a carbon-centered lipid radical; $ROO\bullet$, the lipid peroxyl radical; ROOH, lipid hydroperoxide; RR, a lipid dimer; and ROOR, lipid peroxide.

In initiation, a hydrogen atom is abstracted from a lipid molecule to yield the carbon-centered lipid radical (equation 1.1). In propagation, oxygen is added to the radical to give the lipid peroxyl radical (equation 1.2), which abstracts a hydrogen atom from another lipid molecule to produce both lipid hydroperoxide and the carbon-centered lipid radical (equation 1.3). Another oxygen is added to the resulting carbon-centered lipid radical (equation 1.2), and a chain reaction starts. The chain reaction can continue until termination reactions arise (equations 1.4-1.6).

In addition, accumulation of lipid hydroperoxide leads to a series of decomposition, by which a variety of products, including aldehydes, are formed. A decomposition mechanism (i.e., a self-reaction) is known as the Russell mechanism (Howard, 1973; Russell, 1957), where a combination of two peroxyl radicals forms a tetraoxide intermediate (ROOOOR):

$$2\ ROO\bullet \rightarrow [ROOOOR] \rightarrow R'R''\ {}^3CO^* + ROH + {}^3O_2$$

$$[R'R''\ {}^3CO^*, {}^3O_2]_{cage} \rightarrow R'R''\ CO + {}^1O_2$$

Where R represents R'R''CH. The tetraoxide decays to form singlet and triplet excited-state products, detected as singlet oxygen and triplet excited carbonyls ($R'R''\ {}^3CO^*$), end products of which are ketones ($R'R''\ CO$) and alcohols (ROH). Quenching of the triplet carbonyl by oxygen in a solvent cage ($[R'R''\ {}^3CO^*, {}^3O_2]$) is thought to give singlet oxygen.

Lipid hydroperoxide, similar to hydrogen peroxide, is reduced with an iron (II) complex, and its oxygen–oxygen bond is cleaved to give the alkoxyl radical ($RO\bullet$):

$$ROOH + Fe^{2+} \rightarrow RO\bullet + HO^- + Fe^{3+}$$

An iron (III) complex can produce the peroxyl radical (ROO•) from hydroperoxide, although the rate of this reaction is much slower than the rate of the iron (II)–mediated reaction:

$$ROOH + Fe^{3+} \rightarrow ROO\bullet + H^+ + Fe^{2+}$$

Inevitably, the iron (II) complex reacts with hydroperoxide to give the alkoxyl radical as shown above.

Ozone (O_3) is formed from ground state oxygen by solar radiation in the upper atmosphere:

$$3\,O_2 + \overset{h\nu}{\rightarrow} 2\,O_3$$

which provides a protective shield against short wavelength ultraviolet radiation.

Ozone adds a double bond in an unsaturated compound to give an ozonide, which decomposes to a ketone, according to the Criegee mechanism (Bailey, 1978):

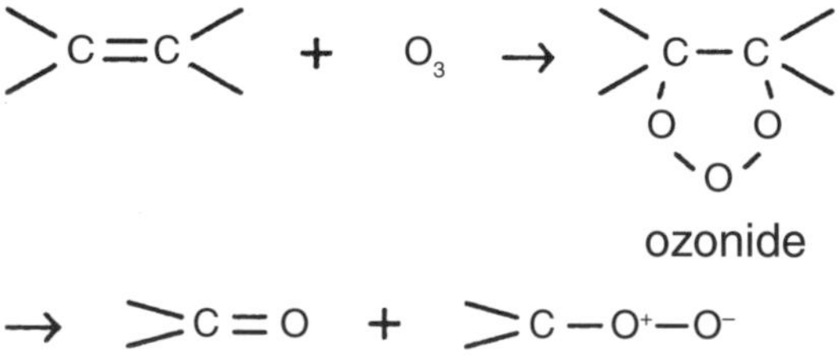

ozonide

Although ozone itself is a nonradical species, it can react with various unsaturated compounds to produce radical species and can stimulate lipid peroxidation. Ozone exposure seriously damages the lungs of animals and humans.

Hypochlorous acid (HClO) is produced, for example, by distillation following acidification of molecular chlorine dissolved into a potassium hydroxide solution. It is in equilibrium with molecular chlorine in water:

$$Cl_2 + H_2O \leftrightarrow HClO + HCl$$

The hypochlorite ion (ClO^-) is a bleaching agent and also a powerful microbicidal agent. As described before, it is produced from halide and hydrogen peroxide by action of myeloperoxidase and is related to the generation of singlet oxygen in the presence of hydrogen peroxide (see Singlet Oxygen, p. 5). Furthermore, hypochlorous acid is reduced with the superoxide radical or an iron (II) complex to give the hydroxyl radical (see Hydroxyl Radical, p. 13).

Excited carbonyls (>C=O*) are produced from cleavage of 1,2-dioxetane and α-peroxylactone, as well as self-reaction of peroxyl radicals as shown previously:

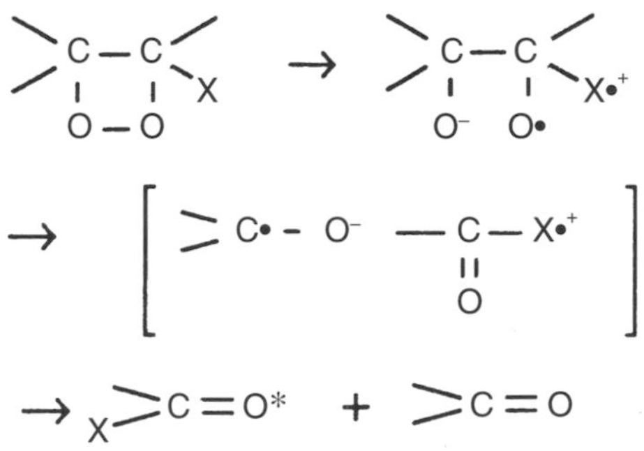

Interestingly, in chemiluminescence and bioluminescence, light is emitted when excited carbonyls are quenched (Adam, 1983).

Fenton Reaction

Most hydroxyl radicals in vivo are believed to generate from the reductive decomposition of hydrogen peroxide with reduced transition metal ions, which is called the Fenton reaction (Fenton, 1894):

$$Mt^{n+} + H_2O_2 \rightarrow Mt^{(n+1)+} + HO\bullet + HO^-$$

where Mt^{n+} or $Mt^{(n+1)+}$ represents a reduced or oxidized transition metal ion, respectively. In most cases, the metal ion is presumed to be the iron (II) or copper (I) ion. The rate constant of the iron (II)- or copper (I)-catalyzed reaction is reported to be 76 $M^{-1} \bullet s^{-1}$ (Walling, 1975) or $4.7 \times 10^3\,M^{-1} \bullet s^{-1}$ (Halliwell and Gutteridge, 1990), respectively, the latter being much greater than the former.

There is controversy about whether the Fenton reaction can be performed in vivo. In particular, we should discuss the claims for the in vivo occurrence of the Fenton reaction, because its rate constant is too small and the concentration of Fenton reaction-reactive transition metal ions is too low.

Obviously the rate constant of the Fenton reaction is low. However, the generation rate of the hydroxyl radical in a cell can be estimated as follows: When 1 μM each of hydrogen peroxide and iron (II) ion are mixed, the generation rate of the hydroxyl radical is $7.6 \times 10^{-11}\,M^{-1} \bullet s^{-1}$, using a rate constant of 76 $M^{-1} \bullet s^{-1}$ for the Fenton reaction; if the average volume of a cell is about 10^{-12} L (Metzler, 1977), then the generation rate of the hydroxyl radical in the cell is about 4×10^6 radicals $\bullet$ day^{-1}. Thus, it is presumed that a considerable amount of hydroxyl radicals is generated for the Fenton reaction in cells.

Because both substrates of the iron (II) ion and hydrogen peroxide will be consumed as the reaction proceeds, they need to be continuously supplied for hydroxyl radical generation. The supply of iron ions may limit hydroxyl radical generation and enhance tissue injury and cell damage. It is known that mechanical injury of the brain releases iron ions to stimulate lipid peroxidation. In addition, oxidative stress can provide iron ions for the Fenton reaction, for example, by releasing iron ions from ferritin or heme proteins.

Iron ions are present in biological systems as an essential part of proteins, such as hemoglobin, ferritin, and cytochromes, but only at very low concentrations as Fenton reaction-active low molecular weight complexes. If hydrogen peroxide reacts with iron ions in a protein to give the hydroxyl radical, the protein will be damaged in specific amino acid residues, in particular histidine, lysine, proline, and arginine residues, which either are very close to iron-binding sites or themselves act as iron-binding sites (Stadtman, 1990). As mentioned (see Hydroxyl Radical, p. 13), the site specificity of molecular damages is an important aspect of hydroxyl radical–inducing oxidative damage to nucleic acids, proteins, and lipids. Presumably, the site specificity of molecular damages depends on specific binding structures between iron ions and/or copper ions and macromolecules. In addition, the metal ions in the specific binding sites may be inaccessible to hydroxyl radical scavengers added.

Metal-Catalyzed Haber-Weiss Reaction

Several types of metal ion catalysts play important roles in hydroxyl radical generation in biological systems (Halliwell and Gutteridge, 1984, 1990). Iron ions are the most likely candidates as stimulating agents for generating the hydroxyl radical and other oxygen-centered radicals in vivo.

Hydroxyl radical generation in superoxide radical–generating systems is usually inhibited with catalase or superoxide dismutase. It is quite reasonable that catalase inhibits hydroxyl radical generation, because the hydroxyl radical arises from hydrogen peroxide by the Fenton reaction, and because catalase dismutates hydrogen peroxide to water and oxygen (see Antioxidant Defense, p. 21). The inhibitory action of superoxide dismutase clearly shows that the superoxide radical is also involved in hydroxyl radical generation.

The Haber-Weiss reaction occurs when the superoxide radical reacts with hydrogen peroxide to give the hydroxyl radical (Haber and Weiss, 1934):

1.7 $O_2^- + H_2O_2 \rightarrow O_2 + HO\bullet + HO^-$

If this reaction is performed in vivo, the superoxide radical is directly associated with generation of the hydroxyl radical. However, the rate constant of this reaction in neutral solution is about 2 $M^{-1} \cdot s^{-1}$ (Ferradini et al., 1978). Thus, the Haber-Weiss reaction hardly occurs in vivo.

An alternative role of the superoxide radical in generating the hydroxyl radical has been proposed. Because the superoxide radical acts as a reducing agent, it reduces oxidized forms of transition metal ions to reduced forms, which enhance the Fenton reaction. The superoxide radical will reduce the iron (III) ion to the iron (II) ion:

1.8 $Fe^{3+} + O_2^- \rightarrow Fe^{2+} + O_2$

The iron (II) ion will react with hydrogen peroxide to form the hydroxyl radical:

1.9 $Fe^{2+} + H_2O_2 \rightarrow Fe^{3+} + HO\bullet + HO^-$

The net of the preceding two reactions (equation 1.8 plus equation 1.9) is the Haber-Weiss reaction (equation 1.7). This series of reactions is often called the metal-catalyzed Haber-Weiss reaction or the superoxide radical-driven Fenton reaction (Koppenol et al., 1978; McCord and Day, 1978). Because iron ions act catalytically, they are needed only in very small amounts as a Fenton reaction-active low molecular weight complex.

The superoxide radical can be interpreted merely as a reducing agent, but it does explain a number of experimental observations. Other reducing agents stimulate the generation of the hydroxyl radical in the presence of the superoxide radical, hydrogen peroxide, and iron ions. Both NAD(P)H (Rowley and Halliwell, 1982) and thiol compounds (Saez et al., 1982), such as reduced glutathione and cysteine, can increase the metal ion-dependent hydroxyl radical generation from the superoxide radical and hydrogen peroxide. Another important biological reducing agent is ascorbic acid (vitamin C), which can replace the superoxide radical in several hydroxyl radical–generating systems with iron or copper ions (Rowley and Halliwell, 1983). When ascorbate acts as a reducing agent, superoxide dismutase does not prevent hydroxyl radical generation, whereas catalase still inhibits hydroxyl radical generation. If both the superoxide radical and ascorbate are available, the relative contributions of each to the hydroxyl radical generation depend on their concentrations.

Antioxidant Defense

It is well known that oxygen is toxic. For example, anaerobes, such as several *Clostridium* species, are killed in the presence of oxygen. Furthermore, when maintained under 100% oxygen, rats die within about 3 days. It is clear that organisms are always exposed to oxidative stress due to oxygen and reactive oxygen species.

Oxidative stress can lethally damage organisms. It is believed that, to avert damage, aerobic organisms have acquired both antioxidant defense mechanisms and oxidative damage-repairing mechanisms during evolution (table 1.2). Antioxidant defense is composed of both antioxidant enzymes and biological antioxidants.

Antioxidant enzymes include superoxide dismutase, catalase, glutathione peroxidase, and glutathione S-transferase (Halliwell and Gutteridge, 1989). Superoxide dismutase dismutates the superoxide radical to hydrogen peroxide and oxygen:

$$2\ O_2^- + 2\ H^+ \rightarrow H_2O_2 + O_2$$

Several forms of superoxide dismutase exist in biological systems. Copper- and zinc-containing superoxide dismutase (CuZn-SOD) is found in cytosol; manganese-containing superoxide dismutase (Mn-SOD) in mitochondria; and a small amount of extracellular superoxide dismutase (Ec-SOD), which is a form of copper- and zinc-containing superoxide dismutase, in extracellular fluid. Further, iron-containing superoxide dismutase (Fe-SOD) is in bacteria and plants. Catalase dismutates hydrogen peroxide to water and oxygen:

$$2\ H_2O_2 \rightarrow 2\ H_2O + O_2$$

Glutathione peroxidase reduces hydroperoxides (ROOH) to alcohols (ROH) and also reduces hydrogen peroxide to water at the expense of reducing glutathione (GSH) to oxidized glutathione (GSSG):

$$ROOH + 2\ GSH \rightarrow ROH + H_2O + GSSG$$

$$H_2O_2 + 2\ GSH \rightarrow 2\ H_2O + GSSG$$

Glutathione S-transferase also reduces hydroperoxides to alcohols using reduced glutathione as a reducing agent.

Superoxide dismutases dismutate the superoxide radical to hydrogen peroxide and oxygen. Thus, it is thought that, for antioxidant defense, the action of superoxide dismutases must be followed by the actions of catalase and/or glutathione peroxidase to scavenge hydrogen peroxide. However, an alternative role of superoxide dismutase has been proposed based on the view that an excess of superoxide dismutase should decrease the steady-state level of the superoxide radical without increasing, rather with decreasing, the endogeneous formation of hydrogen peroxide (Liochev and Fridovich, 1994). A more effi-

Table 1.2   Biological Defense Against Oxidative Stress

Antioxidant defense		Occurrence
Preventive antioxidant enzymes		
Superoxide dismutase	Cu,Zn-SOD	Cytosol
	Ec-SOD	Extracellular fluid
	Mn-SOD	Mitochondria
	Fe-SOD	Cytosol, chloroplasts
Catalase		Peroxisomes
Glutathione peroxidase		Cytosol, extracellular fluid, biomembranes
Glutathione *S*-transferase		Cytosol
Antioxidants		
α-Tocopherol (vitamin E)		Biomembranes, lipoprotein
Ascorbic acid (vitamin C)		Cytosol, extracellular fluid
Ubiquinol		Biomembranes, lipoprotein
Carotenoids		Biomembranes, lipoprotein
Reduced glutathione	GSH	Cytosol, extracellular fluid
Uric acid		Cytosol, extracellular fluid
Metal sequestration		
Transferrin		Extracellular fluid
Ferritin		Cytosol
Ceruloplasmin		Extracellular fluid

Oxidation damage repair	Occurrence
Repair enzymes	
Nuclease	Nuclei, cytosol
Glycosylase	Nuclei, cytosol
DNA polymerase	Cytosol
Protease	Cytosol
Phospholipase	Biomembranes, lipoprotein

cient dismutation system of the superoxide radical would prevent the formation of higher stoichiometric levels of hydrogen peroxide by reactions other than dismutation, such as reduction of superoxide radical by [4Fe-4S]-

containing dehydratases. In fact, it has been reported that the superoxide dismutase–overexpressing clones of Chinese hamster cells present reduced steady-state levels of hydrogen peroxide (Teixeira et al., 1998). Superoxide dismutase might contribute to antioxidant defense by itself, without the aid of catalase or glutathione peroxidase.

Furthermore, the "radical sink" hypothesis suggests a concerted antioxidant interaction between reduced glutathione and superoxide dismutase (Winterboun, 1993). In this proposal, biologically generated radicals (R•) oxidize reduced glutathione to form its thiyl radical (GS•) (equation 1.10). This oxidizing thiyl radical is not biologically benign and can undergo other potentially harmful reactions. The reaction of the thiyl radical with the glutathione anion (GS⁻) produces the glutathione disulfide anion radical (GSSG⁻) (equation 1.11), which, in turn, reduces oxygen, forming the superoxide radical and glutathione disulfide (GSSG) (equation 1.12).

1.10 $R\bullet + GSH \leftrightarrow RH + GS\bullet$

1.11 $GS\bullet + GS^- \leftrightarrow GSSG^-$

1.12 $GSSG^- + O_2 \to GSSG + O_2^-$

In a concerted effort, superoxide dismutase will catalyze the dismutation of the superoxide radical, terminating biologically generated radicals.

Biological antioxidants are composed of both water-soluble and fat-soluble antioxidants (Halliwell and Gutteridge, 1989). Water-soluble antioxidants include reduced glutathione, ascorbic acid, and uric acid; and fat-soluble antioxidants include vitamin E (tocopherols, mainly α-tocopherol), ubiquinols, and carotenoids. Reduced glutathione, one of the most abundant biological reducing agents, acts as a thiol reagent converting disulfides to thiols and as a substrate for glutathione peroxidase and glutathione S-transferase. Ascorbic acid is another abundant biological reducing agent. Uric acid and carotenoids behave as singlet-oxygen quenchers and radical scavengers. Tocopherols and ubiquinols exist mainly in biomembranes and function as radical scavengers. α-Tocopherol localizes within the membrane bilayer at a concentration of about 1 molecule per 10^3 phospholipid molecules (Buttriss and Diplock, 1988).

Typical lipid peroxidation-inhibiting antioxidants are so-called chain-breaking antioxidants, like α-tocopherol (Burton and Ingold, 1981). If α-tocopherol (TOH) is added to a lipid peroxidation system (equations 1.1-1.3), then the resulting lipid peroxyl radical (ROO•) abstracts a hydrogen atom from α-tocopherol to give lipid hydroperoxide (ROOH) and the α-tocopheroxyl radical (TO•). The α-tocopheroxyl radical does not abstract a hydrogen atom from any lipid molecule and is converted to nonradical products, thus breaking the chain of lipid peroxidation.

$$ROO\bullet + TOH \rightarrow ROOH + TO\bullet$$

$$ROO\bullet + TO\bullet \rightarrow \text{nonradical products}$$

In the presence of ascorbate in the aqueous phase, the α-tocopheroxyl radical is reduced back to original α-tocopherol, and ascorbate is oxidized to the ascorbyl radical (Packer et al., 1979). The fate of the ascorbyl radical is obscure, but it may be dismutated to ascorbate and dehydroascorbate (Yamazaki and Piette, 1961) or may revert to ascorbate in the presence of thioredoxin reductase (May et al., 1998).

Organisms appear to have evolved mechanisms to protect themselves from metal-inducing damage, as well as mechanisms to use metal. For example, iron-binding proteins, such as transferrin and ferritin, and copper-binding proteins, such as ceruloplasmin, may minimize the pool size of dangerous transition metal ions in cells. Thus, transition metal sequestration may be a part of antioxidant defense.

Oxidative damage–repairing mechanisms also are considered necessary to maintain homeostasis in organisms. Oxidized DNA may be repaired by the action of enzymes such as nuclease and glycosylase. Oxidized proteins may be removed by proteases. Oxidized lipids may be reduced by glutathione peroxidase directly or after hydrolysis with phospholipase.

Oxidative Stress

Organisms are always exposed to oxidative stress and have antioxidant defenses against it. Oxidative stress occurs when the homeostatic balance between oxidant (C_o) and antioxidant (C_a) capacities in biological systems is disturbed and the redox state becomes more pro-oxidizing:

Oxidative stress: $C_o > C_a$

If the oxidant capacity is greater than the antioxidant capacity, a small amount of reactive oxygen species may escape from the antioxidant defense in cells. Thus, the ratio of reduced to oxidized glutathione concentration in biological systems can be used as an index of oxidative stress, because reduced glutathione is one of the most abundant antioxidants in vivo and, in addition, acts as the substrate of glutathione peroxidase and glutathione S-transferase. In addition, the ratio of NADPH concentration to $NADP^+$ concentration may also reflect oxidative stress, because NADPH is the electron donor for reduction of oxidized glutathione to reduced glutathione with glutathione reductase.

Under oxidative stress, biomolecules will be oxidized in tissues and organs. In fact, cells always contain oxidized products of nucleic acids, proteins, and lipids. This means, not only that biomolecules are oxidized in vivo, but also that their oxidized products are not completely removed. The observed levels of the oxidized products reflect an equilibrium between their formation rate and their removal and repair rates. Presumably, repair processes are more important to protect against the damage of DNA, and degradation processes are more important to protect against the damage of proteins and lipids.

Quantitative Analysis of Reactive Oxygen Species in Biological Systems

Generally, it is very difficult to directly detect reactive oxygen species generated in biological systems. This is because reactive oxygen species normally exist at extremely low concentrations, and because reactive oxygen species react almost immediately at their formation sites and hence do not accumulate to a great degree. Usually, the generation of reactive oxygen species is estimated either by measuring reactive oxygen species trapped with exogenous trapping agents or by measuring oxidized biomolecules.

Theoretically, all radicals can be measured by electron spin resonance (ESR) spectroscopy, because they have paramagnetism due to an unpaired electron. However, oxygen-centered radicals generated in biological systems cannot be directly detected by ESR because of their low concentration and high reactivity. Furthermore, ESR measurement of biological samples at room temperature is seriously constrained by the high dielectric constant of fluid water in the microwave region. The largest volume of an aqueous sample in the sensitive region of an ESR cavity is usually about 50 μL when an ordinary cavity and capillary cell are used, and about 200 μL when a special cavity and flat cell are used (Mason, 1996).

Such radicals may possibly be detected by spin trapping (Janzen, 1980). Spin traps, which are ESR-negative, can rapidly react with radicals to give unique and relatively stable spin adducts, which are ESR-positive radicals. In addition, even though the free radical formation rate is low, the spin adducts may accumulate with time and may be quantitated.

α-Phenyl-*tert*-butyl nitrone (PBN) or 5,5-dimethylpyrroline *N*-oxide (DMPO) is frequently used as a spin trap for radicals in biological systems (Janzen, 1980) (figure 1.3). A radical adds to its carbon-nitrogen double bond to give a spin adduct of nitroxide. Because the nitroxide is relatively stable, its ESR spectrum can be measured. The radical can be characterized on the basis

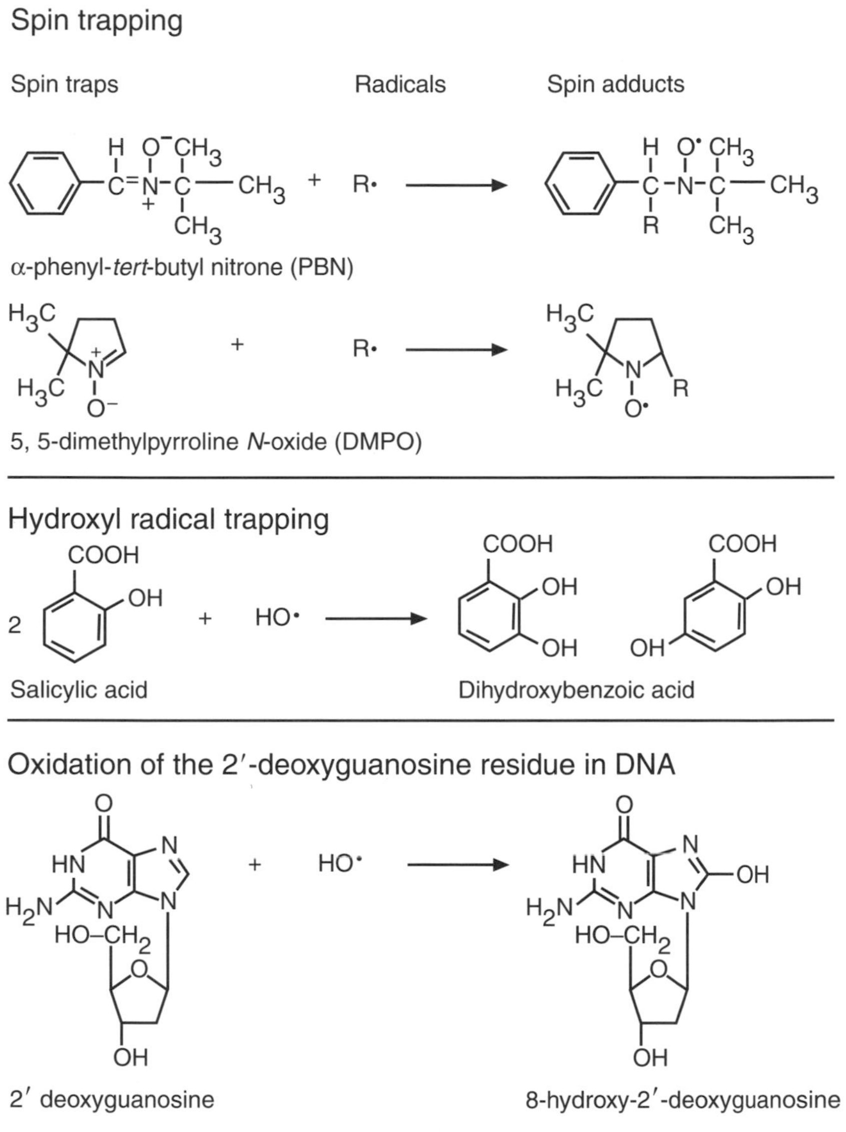

Spin trapping

Spin traps   Radicals   Spin adducts

α-phenyl-*tert*-butyl nitrone (PBN)

5, 5-dimethylpyrroline *N*-oxide (DMPO)

Hydroxyl radical trapping

Salicylic acid   Dihydroxybenzoic acid

Oxidation of the 2′-deoxyguanosine residue in DNA

2′ deoxyguanosine   8-hydroxy-2′-deoxyguanosine

Figure 1.3   Spin trapping, hydroxyl radical trapping, and oxidation of 2′-deoxyguanosine residue in DNA.

of the hyper-fine coupling constants of the spin adduct. At a concentration suitable for ESR spectroscopy, PBN is both water-soluble and lipid-soluble and permeates all tissues equivalently (Chen et al., 1990).

Salicylate appears to permeate tissues and reacts under diffusion control with hydroxyl radicals to form dihydroxybenzoic acid (figure 1.3) (Kaur and Halliwell, 1996). The ratio of dihydroxybenzoic acid to salicylate in tissues can be determined by high-performance liquid chromatography (HPLC) with electrochemical detection.

When biomolecules are damaged under oxidative stress, several methods are available to estimate oxidized products. For example, DNA is oxidized with reactive oxygen species. One of the indices of oxidized DNA is its ratio of 8-hydroxy-2′-deoxyguanosine (8-OHdG; figure 1.3) residues to 2′-deoxyguanosine (dG) residues. After hydrolysis of DNA, 8-OHdG is measured at very high sensitivity by HPLC with electrochemical detection (Floyd et al., 1986). Amounts of 8-OHdG as small as 20 fmol can be quantitated. In normal tissue DNA, the ratio of 8-OHdG residues to dG residues is $0.5\text{-}2{:}10^5$. Figure 1.4 shows age-dependent changes of the ratio of 8-OHdG residues to dG residues in nuclear DNA from livers, hearts, kidneys, and brains of male Fischer 344 rats (Kaneko et al., 1996).

An alternative approach is "fingerprinting" of oxidized products. The hydroxyl radical–induced damage of naturally occurring biomolecules can be identified by analyzing the specific pattern (i.e., "fingerprint") of oxidized products from biomolecules (Dizdaroglu, 1986). For example, a hydroxyl radical attack on DNA bases produces a multitude of products, which can be derivatized, separated by gas liquid chromatography, and identified by mass spectroscopy. This complicated pattern, which results from base modification, may be used as a fingerprint of DNA damage by the hydroxyl radical (Aruoma et al., 1989).

Conclusion

In this chapter, we outlined the reactions of reactive oxygen species—mainly singlet oxygen, the superoxide radical, hydrogen peroxide, the hydroxyl radical, and nitrogen monoxide—and related radicals, emphasizing reactions of biological relevance. We also discussed the Fenton reaction and metal-catalyzed Haber-Weiss reaction, which are important aspects of the harmful effects of reactive oxygen species. Antioxidant defense and oxidative stress were discussed as biological responses to these species. Finally, we briefly covered the quantitative analysis of these species in biological systems. The chemistry of reactive oxygen species is a huge area of study and is reviewed here only introductorily. Experimental data are available not only from biochemistry research but also from physical, organic, and inorganic chemistry researches. These data include rate constants and mechanisms for many reactions of reactive oxygen species with biomolecules and their model compounds.

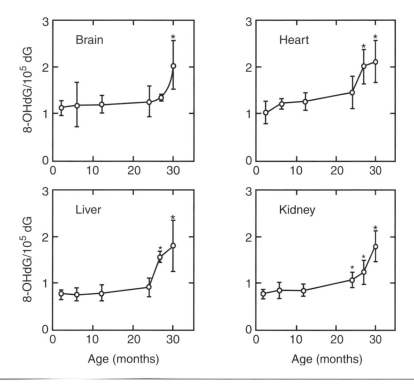

Figure 1.4 Ratio of 8-hydroxy-2'-deoxyguanosine (8-OHdG) residues to 2'-deoxyguanosine (dG) residues in nuclear DNA from organs of male Fischer 344 rats during aging. A value represents a mean with its standard deviation.
*Differences between the values of old and young rats are statistically significant at $p < .05$.
Data from Kaneko et al. 1996, with permission.

In biological systems, most reactive oxygen species are convertible to other reactive oxygen species, and some reactive oxygen species are highly reactive, as their names signify. Thus, it is extremely difficult to observe precisely the behavior of each species in biological systems. Consequently, the biological relevance of these species is not yet fully understood and should be further investigated.

References

Adam, W. 1983. Four-membered ring peroxide: 1,2-dioxetanes and α-peroxylactones. In: *The Chemistry of Peroxides*, S. Patai (ed.), pp. 829-920. Wiley, Chichester, UK.
Anber, M., and P. Neta. 1967. A compilation of specific bimolecular rate constants for the reactions of hydrated electrons, hydrogen atoms and hydroxyl radicals with inorganic

and organic compounds in aqueous solution. *International Journal of Applied Radiation and Isotopes* 18:493-523.

Aruoma, O.I., B. Halliwell, and M. Dizdaroglu. 1989. Iron ion-dependent modification of bases in DNA by the superoxide radical-generating system hypoxanthine/xanthine oxidase. *Journal of Biological Chemistry* 264:13024-13028.

Babior, B.M. 1978. Oxygen-dependent microbial killing by phagocytes. *New England Journal of Medicine* 298:659-668, 721-725.

Bailey, P.S. 1978. *Ozonation in Organic Chemistry, Vol. I.* Academic Press, New York.

Bielski, B.H. 1978. Reevaluation of the spectral and kinetic properties of HO_2 and O_2 free radicals. *Photochemistry and Photobiology* 28:645-649.

Bielski, B.H., and A.D. Allen. 1977. Mechanism of the disproportionation of superoxide radicals. *Journal of Physical Chemistry* 81:1048-1050.

Brodie, A.E., and D.J. Reed. 1987. Reversible oxidation of glyceraldehyde 3-phosphate dehydrogenase thiols in human lung carcinoma cells by hydrogen peroxide. *Biochemical and Biophysical Research Communications* 148:120-125.

Burton, G.W., and K.U. Ingold. 1981. Autoxidation of biological molecules: 1. The antioxidant activity of vitamin E and related chain-breaking phenolic antioxidants in vitro. *Journal of the American Chemical Society* 103:6472-6477.

Buttriss, J.L., and A.T. Diplock. 1988. The relationship between α-tocopherol and phospholipid fatty acids in rat liver subcellular membrane fractions. *Biochimica et Biophysica Acta* 962:81-90.

Candeias, L.P., K.B. Patel, M.R.L. Stratford, and P. Wardman. 1993. Free hydroxyl radicals are formed on reaction between the neutrophil-derived species superoxide anion and hypochlorous acid. *FEBS Letters* 333:151-153.

Candeias, L.P., M.R.L. Stratford, and P. Wardman. 1994. Formation of hydroxyl radicals on reaction of hypochlorous acid with ferrocyanide, a model iron (II) complex. *Free Radical Research* 20:241-249.

Chen, G.M., T.M. Bray, E.G. Janzen, and P.B. McCay. 1990. Excretion, metabolism and tissue distribution of a spin trapping agent, α-phenyl-*N-tert*-butyl-nitrone (PBN) in rats. *Free Radical Research Communications* 9:317-323.

Dean, J.A. (ed.) 1973. *Lange's Handbook of Chemistry, 11th ed.*, p. 10-7. McGraw-Hill, New York.

Dizdaroglu, M. 1986. Chemical characterization of ionizing radiation-induced damage to DNA. *BioTechniques* 4:536-546.

Epstein, I.R., K. Kustin, and L.J. Warshaw. 1980. A kinetic study of the oxidation of iron (II) by nitric acid. *Journal of the American Chemical Society* 102:3751-3758.

Fenton, H.J.H. 1894. Oxidation of tartaric acid in presence of iron. *Journal of the Chemical Society Transactions* 65:899-910.

Ferradini, C., J. Foos, C. Houee, and J. Pauchault. 1978. The reaction between superoxide anion and hydrogen peroxide. *Photochemistry and Photobiology* 28:697-700.

Floyd, R.A., J.J. Watson, J. Harris, M. West, and P.K. Wong. 1986. Formation of 8-hydroxydeoxyguanosine, hydroxyl free radical adduct of DNA in granulocytes exposed to the tumor promotor, tetradeconylphorbolacetate. *Biochemical and Biophysical Research Communications* 137:841-846.

Foote, C.S. 1979. Quenching of singlet oxygen. In: *Singlet Oxygen*, H.H. Wasserman and R.W. Murray (eds.), pp. 139-171. Academic Press, New York.

Freeman, B.A., and J.D. Crapo. 1981. Hyperoxia increases oxygen radical production in rat lungs and lung mitochondria. *Journal of Biological Chemistry* 256:10986-10992.

Fridovich, I. 1983. Superoxide radical: An endogeneous toxicant. *Annual Review of Pharmacology and Toxicology* 23:239-257.

Fridovich, I., and B.A. Freeman. 1986. Antioxidant defenses in the lung. *Annual Review of Physiology* 48:693-702.

Gollnick, K., and H.J. Kuhn. 1979. Ene-reactions with singlet oxygen. In: *Singlet Oxygen*, H.H. Wasserman and R.W. Murray (eds.), pp. 287-427. Academic Press, New York.

Haber, F., and J. Weiss. 1934. The catalytic decomposition of hydrogen peroxide by iron salts. *Proceedings of the Royal Society of London, Series A. Mathematical and Physical Sciences* 147:332-351.

Halliwell, B., and J.M.C. Gutteridge. 1984. Oxygen toxicity, oxygen radicals, transition metals and disease. *Biochemical Journal* 219:1-14.

Halliwell, B., and J.M.C. Gutteridge. 1989. *Free Radicals in Biology and Medicine, 2nd ed.* Oxford University Press (Clarendon), New York.

Halliwell, B., and J.M.C. Gutteridge. 1990. Role of free radicals and catalytic metal ions in human disease: An overview. *Methods in Enzymology* 186:1-85.

Howard, J.A. 1973. Homogeneous liquid-phase autoxidation. In: *Free Radicals, vol. II*, J.K. Kochi (ed.), pp. 3-62. Wiley, New York.

Huie, R.E., and S. Padmaja. 1993. The reaction of NO with superoxide. *Free Radical Research Communications* 18:195-199.

Ingraham, L.L., and D.L. Meyer. 1985. *Biochemistry of dioxygen.* Plenum Press, New York.

Ischiropoulos, H., L. Zhu, J. Chen, M. Tsai, J.C. Martin, C.D. Smith, and J.S. Beckman. 1992. Peroxynitrite-mediated tyrosine nitration catalyzed by superoxide dismutase. *Archives of Biochemistry and Biophysics* 298:431-437.

Janzen, E.G. 1980. A critical review of spin trapping in biological systems. In: *Free Radicals in Biology, vol. IV*, W.A. Pryor (ed.), pp. 115-154. Academic Press, New York.

Jobsis, Q., H.C. Raatgeep, S.L. Schellekens, W.C. Hop, P.W. Hermans, and J.C. de Jongste. 1998. Hydrogen peroxide in exhaled air of healthy children: Reference values. *European Respiratory Journal* 12:483-485.

Kaneko, T., S. Tahara, and M. Matsuo. 1996. Non-linear accumulation of 8-hydroxy-2'-deoxyguanosine, a marker of oxidized DNA damage, during aging. *Mutation Research* 316:277-285.

Kasha, M., and D.E. Brabham. 1979. Singlet oxygen electronic structure and photosensitization. In: *Singlet Oxygen*, H.H. Wasserman and R.W. Murray (eds.), pp. 1-33. Academic Press, New York.

Kaur, H., and B. Halliwell. 1996. Salicylic acid and phenylalanine as probes to detect hydroxyl radicals. In: *Free Radicals, A Practical Approach*, N.A. Punchard and F.J. Kelly (eds.), pp. 101-116. Oxford University Press (IRL Press), New York.

Kearns, D.R. 1979. Solvent and solvent isotope effects on the lifetime of singlet oxygen. In: *Singlet Oxygen*, H.H. Wasserman and R.W. Murray (eds.), pp. 115-137. Academic Press, New York.

Knowles, R.G., and S. Moncada. 1994. Nitric oxide synthases in mammals. *Biochemical Journal* 298:249-258.

Koppenol, W.H. 1997. The chemical reactivity of radicals. In: *Free Radical Toxicology*, K.B. Wallace (ed.), pp. 3-14. Taylor & Francis, Washington, DC.

Koppenol, W.H., J. Butler, and J.W. van Leeuwen. 1978. The Haber-Weiss cycle. *Photochemistry and Photobiology* 28:655-658.

Liochev, S.I., and I. Fridovich. 1994. The role of O_2 in the production of HO•: In vitro and in vivo. *Free Radical Biology and Medicine* 14:85-90.

Lynch, R.E., and I. Fridovich. 1978. Permeation of the erythrocyte stroma by superoxide radical. *Journal of Biological Chemistry* 253:4697-4699.

Matheson, I.B.C., R.D. Etheridge, N.R. Kratowich, and J. Lee. 1975. The quenching of singlet oxygen by amino acids and proteins. *Photochemistry and Photobiology* 21:165-171.

Mason, R.P. 1996. In vitro and in vivo detection of free radical metabolites with electron spin resonance. In: *Free Radicals, A Practical Approach*, N.A. Punchard and F.J. Kelly (eds.), pp. 11-24. Oxford University Press (IRL Press), New York.

Matsuo, M., and T. Kaneko. 1998. Lipid peroxidation. In: *Methods in Aging Research*, B.P. Yu (ed.), pp. 571-606. CRC Press, Boca Raton, FL.

May, J.M., C.E. Cobb, S. Mendiratta, K.E. Hill, and R.F. Burk. 1998. Reduction of the ascorbyl free radical to ascorbate by thioredoxin reductase. *Journal of Biological Chemistry* 273:23039-23045.

McCord, J.M., and E.D. Day. 1978. Superoxide dependent production of hydroxyl radical catalyzed by iron-EDTA complex. *FEBS Letters* 86:139-142.

Mello Filho, A.C., M.E. Hoffmann, and R. Meneghini. 1984. Cell killing and DNA damage by hydrogen peroxide are mediated by intracellular iron. *Biochemical Journal* 218:273-275.

Metzler, D.E. 1977. *Biochemistry, The Chemical Reactions of Living Cells*. Academic Press, New York.

Moncada, S., R.M.J. Palmer, and E.A. Higgs. 1991. Nitric oxide: Physiology, pathology, and pharmacology. *Pharmacological Reviews* 43:109-142.

Murray, R.W. 1979. Chemical sources of singlet oxygen. In: *Singlet Oxygen*, H.H. Wasserman and R.W. Murray (eds.), pp. 59-114. Academic Press, New York.

Myers, P.R., P.L. Minor Jr., R. Guerra Jr., J.N. Bates, and D.G. Harrison. 1990. Vasorelaxant properties of the endothelium-derived relaxing factor more closely resemble S-nitrosocysteine than nitric oxide. *Nature* 345:161-163.

Oshino, N., D. Jamieson, and B. Chance. 1975. The properties of hydrogen peroxide production under hyperoxic and hypoxic conditions of perfused rat liver. *Biochemical Journal* 146:53-65.

Padmaja, S., and R.E. Huie. 1993. The reaction of nitric oxide with organic peroxyl radicals. *Biochemical and Biophysical Research Communications* 195:539-544.

Packer, J.E., T.F. Slater, and R.L. Willson. 1979. Direct observation of a free radical interaction between vitamin E and vitamin C. *Nature* 278:737-738.

Pryor, W.A., and G.L. Squadrito. 1995. The chemistry of peroxynitrite: A product from the reaction of nitric oxide with superoxide. *American Journal of Physiology* 268:L699-L722.

Radi, R., J.S. Bechman, K.M. Bush, and B.A. Freeman. 1991. Peroxynitrite oxidation of sulfhydryls. The cytotoxic potential of superoxide and nitric oxide. *Journal of Biological Chemistry* 266:4244-4250.

Ramezanian, M.S., S. Padmaja, and W.H. Koppenol. 1996. Nitration and hydroxylation of phenolic compounds by peroxinitrite. *Methods in Enzymology* 269:195-201.

Rose, R.C., S.P. Richer, and A.M. Bode. 1998. Ocular oxidants and antioxidant protection. *Proceedings of the Society for Experimental Biology and Medicine* 217:397-407.

Rowley, D.A., and B. Halliwell. 1982. Superoxide-dependent formation of hydroxyl radicals from NADH and NADPH in the presence of iron salts. *FEBS Letters* 142:39-41.

Rowley, D.A., and B. Halliwell. 1983. Formation of hydroxyl radicals from hydrogen peroxide and iron salts by superoxide- and ascorbate-dependent mechanisms: Relevance to the pathology of rheumatoid disease. *Clinical Science* 64:649-653.

Russell, G.A. 1957. Deuterium-isotope effects in the autoxidation of aralkyl hydrocarbons. Mechanisms of the interaction of peroxyl radicals. *Journal of the American Chemical Society* 79:3871-3877.

Saez, G., P.J. Thornalley, H.A.O. Hill, R. Hems, and J.V. Bannister. 1982. The production of free radicals during the autoxidation of cysteine and their effect on isolated rat hepatocytes. *Biochimica et Biophysica Acta* 719:24-31.

Sawyer, D.T. 1991. *Oxygen Chemistry*. Oxford University Press, New York.

Schaap, A.P., and K.A. Zaklika. 1979. 1,2-Cycloaddition reactions of singlet oxygen. In: *Singlet Oxygen*, H.H. Wasserman and R.W. Murray (eds.), pp. 173-242. Academic Press, New York.

Stadtman, E.R. 1990. Metal ion catalyzed oxidation of proteins: Biochemical mechanism and biological consequences. *Free Radical Biology and Medicine* 9:315-325.

Teixeira, H.D., R.I. Schumacher, and R. Meneghni. 1998. Low intracellular hydrogen peroxide levels in cells overexpressing CuZn-superoxide dismutase. *Proceedings of the National Academy of Sciences of the USA* 95:7872-7875.

Teng, J.I., and L.L. Smith. 1973. Steroid metabolism. XXIV. On the unlikely participation of singlet molecular oxygen in several enzyme oxygenations. *Journal of the American Chemical Society* 95:4060-4061.

Terelius, Y., and M. Ingelman-Sundberg. 1988. Cytochrome P-450-dependent oxidase activity and hydroxyl radical production in micellar and membranous types of reconstituted systems. *Biochemical Pharmacology* 37:1383-1389.

Walling, C. 1975. Fenton's reagent revisited. *Accounts of Chemical Research* 8:125-131.

Wink, D.A., and P.C. Ford. 1995. Nitric oxide reactions important to biological systems: A survey of some kinetics investigations. *Methods: A Comparison to Methods in Enzymology* 7:14-20.

Wink, D.A., R.W. Nims, J.F. Darbyshire, D. Christodoulou, I. Hanbauer, G.W. Cox, F. Laval, J. Laval, J.A. Cook, M.C. Krishna, W. DeGraff, and J.B. Mitchell. 1994. Reaction kinetics for nitrosation of cysteine and glutathione in aerobic nitric oxide solution at neutral pH. Insights into the fate and physiological effects of intermediates generated in the NO/O_2 reaction. *Chemical Research in Toxicology* 7:519-525.

Wink, D.A, M.B. Grisham, J.B. Mitchell, and P.C. Ford. 1996. Direct and indirect effects of nitric oxide in chemical reactions relevant to biology. *Methods in Enzymology* 268:12-31.

Winterboun, C.C. 1993. Superoxide as an intracellular radical sink. *Free Radical Biology and Medicine* 14:85-90.

Yamazaki, I., and L.H. Piette. 1961. Mechanism of free radical formation and disappearance during the ascorbic acid oxidase and peroxidase reactions. *Biochimica et Biophysica Acta* 50:62-69.

Zafiriou, O.C. 1987. Is sea water a radical solution? *Nature* 325:481-482.

Chapter 2

Antioxidant Defense: Effects of Aging and Exercise

Li Li Ji and John Hollander

Department of Kinesiology and Interdisciplinary Graduate Program of Nutritional Science, University of Wisconsin–Madison

Physical exercise is an intimate part of animal life. To pursue food, escape predators, and ensure reproduction, animals have gained high mobility through evolution. The most prominent biological change that occurs during exercise is an increased metabolic rate, matched by an enhanced rate of oxygen consumption. A high rate of oxygen flux through the mitochondria may provoke increased electron "leakage" and impose an oxidative stress to components and organelles essential for cell function (Jenkins, 1993). The increased production of reactive oxygen species (ROS) and oxidative stress associated with physical exercise may contribute to aging itself. On the other hand, aging has been shown to increase ROS production and to alter cellular antioxidant defense systems (Yu, 1994). Aged animals and humans may have increased vulnerability to exercise-induced oxidative stress (Ji, 1998).

In this chapter, we will summarize the effects of exercise and aging on cellular antioxidant systems. We will focus on the skeletal muscle system for the following three reasons:

a. Aging and oxidative stress have been reviewed by many experts (Ames et al., 1993; Sohal and Weindruch, 1996; Yu, 1994), but relatively few reviews have been devoted to skeletal muscle oxidative stress and aging.

b. Deterioration of skeletal muscle function is an important issue in medical gerontology because of the critical role of muscle for mobility and normal life.

c. Skeletal muscle displays some unique characteristics during aging in terms of both free radical production and antioxidant systems.

Antioxidant Defense Mechanisms

Cellular antioxidant defenses are conventionally classified into two categories: enzymatic and nonenzymatic. Primary antioxidant enzymes include superoxide dismutase (SOD), glutathione peroxidase (GPX), and catalase (CAT) (Chance et al., 1979). A number of enzymes are involved in the supply of substrates and reducing power (nicotinamide adenine dinucleotide phosphate, reduced form; NADPH) for primary antioxidant enzymes, such as glutathione reductase (GR) and glucose 6-phosphate dehydrogenase (G6PDH), but they do not directly remove ROS. Glutathione sulfur-transferase (GST) conjugates glutathione (GSH) with xenobiotics and toxins that can potentially generate ROS. Nonenzymatic antioxidants such as vitamin E, vitamin C, and β-carotene directly scavenge O_2^- and $\cdot OH$, as well as singlet oxygen (Yu, 1994). GSH and other low molecular weight antioxidants play an important role in maintaining substrate levels for GPX and keeping vitamin E and vitamin C in the reduced state (Meister and Anderson, 1983). Another way to classify antioxidants is to consider whether they can be synthesized within the body and induced under oxidative stress; antioxidant enzymes and GSH belong to this category (table 2.1). The second category of antioxidants, represented by antioxidant vitamins, cannot be synthesized or induced and must be taken from the diet. Thus, nutrition significantly impacts cellular antioxidant systems.

Superoxide Dismutase (SOD)

SOD (EC 1.15.1.1), discovered by McCord and Fridovich (1969), represents a family of metalloenzymes that catalyzes a common one-electron dismutation of O_2^- to H_2O_2.

$$2O_2^- + 2H^+ \rightarrow H_2O_2 + O_2$$

This reaction occurs naturally with a rather slow rate ($t\frac{1}{2} = 7s$) and is also pH dependent. With 0.35 μM SOD present, the rate of $O_2\cdot$ dismutation is accelerated dramatically, with a $t\frac{1}{2} = 0.5$ ms regardless of pH (Chance et al., 1979).

Table 2.1 Antioxidant Enzyme Activity in Various Tissues

Tissues	SOD			GPX			CAT	GR	GST	G6PDH	CS
	Cu/Zn (unit/mg)	Mn (unit/mg)	Total (unit/g ww)	Cyto (unit/mg)	Mito (unit/mg)	Total (unit/g ww)	(unit/g ww)	(unit/g ww)	(unit/g ww)	(unit/g ww)	(unit/g ww)
Liver	500	50	14,400	550	430	85	670	40	940	8.0	18
Heart	65	21	2610	150	70	17	84	1.3	2.5	10.9	72
Soleus	48	7	1300	n.d.	n.d.	13	61	0.8	1.1	n.d.	40
DVL	21	8	1360	23	17	2	18	0.4	0.5	0.6	20
SVL	21	5	887	n.d.	n.d.	0.9	15	0.3	0.2	n.d.	6
Erythrocytes	n.a.	n.a.	8.8	n.a.	n.a.	25	10	35	1.0	2.3	n.d.

Note: Units of enzyme activity in all rat tissues are as specified under experimental conditions. Activity in erythrocytes: units per g Hb. SOD, superoxide dismutase; GPX, glutathione peroxidase; Cyto, cytosol; Mito, mitochondria; CAT, catalase; GR, glutathione reductase; GST, glutathione sulfer-transferase; G6PDH, glucose 6-phosphate dehydrogenase; CS, citrate synthase; DVL, deep vastus lateralis; SVL, superfacial vastus lateralis; n.a., not applicable; n.d., not determined. Sources: Ji, 1995; Leeuwenburgh et al., 1997.

There are three types of SOD, depending on the metal ion bound to its active site. Copper- and zinc-containing SOD (CuZn-SOD) is a highly stable enzyme found primarily in the cytosolic compartment of the eukaryotic cells such as yeast, plants, and animals but not generally found in the prokaryotes such as bacteria and algae. CuZn-SOD is a dimer (molecular weight [MW] = 32,000) and is sensitive to cyanide and H_2O_2 inhibition (Fridovich, 1995). Manganese-containing SOD (Mn-SOD) is a tetramer with a much larger MW of 88,000. Mn-SOD is present in the mitochondrial matrix of eukaryotes and is insensitive to cyanide and H_2O_2. However, it is not as stable as CuZn-SOD and can be inhibited by sodium dodecyl sulfate (SDS) and chloroform/ethanol treatments (Ohno et al., 1994). This cyanide sensitivity has been used to distinguish the two types of SOD in tissue extracts without separating mitochondria and cytosol. Bacteria contain a third type of SOD, which requires iron as a prosthetic group (Fe-SOD). Unlike most enzymes, SOD lacks a Michaelis constant (K_m), and its catalytic activity increases with increasing $O_2 \bullet^-$ concentration within a wide range (Chance et al., 1979). Because of the previously mentioned kinetic properties, assays of SOD are usually based on indirect methods that involve inhibiting a reaction in which $O_2 \bullet^-$ is generated (Fridovich, 1985). Therefore, it is not always meaningful to compare reported SOD activities between studies that use different assay methods. Mammalian species also contain extracellular SOD (Ec-SOD), located in the plasma and interstitial fluid. Although belonging to the CuZn-SOD family, Ec-SOD is a tetramer with a larger molecular weight and can be bound to heparin. Its primary function is to remove $O_2 \bullet^-$ generated outside cell membrane due to irradiation, inflammation, and ischemia-reperfusion (Fridovich, 1995; Ohno et al., 1994).

The two types of SOD have quite different characteristics in terms of protein turnover. Recombinant human SOD (rh-SOD) studies reveal that CuZn-SOD has a half-life ($t\frac{1}{2}$) of 6 to 10 min, whereas Mn-SOD has a much longer $t\frac{1}{2}$ of 5 to 6 h (Gorecki et al., 1991). This provides an important clue as to how SOD gene regulation is controlled. The relative abundance of CuZn-SOD messenger ribonucleic acid (mRNA) displays clear tissue-specific differences, with the liver possessing the highest levels followed by the heart, lung, and skeletal muscle (Ji, 1998). Among different muscle fibers, CuZn-SOD mRNA levels are the highest in type 1 muscle (soleus), followed by mixed muscle fiber type (plantaris), and then type 2 muscle vastus lateralis and gastrocnemius. Consistent with the mRNA levels, type 1 muscle also has a higher level of CuZn-SOD protein content than type 2 muscle (Ohno et al., 1994). These findings suggest that, at rest, CuZn-SOD gene expression may be regulated at the pretranslational level.

In general, Mn-SOD activity, content, and mRNA abundance follow the same order across tissues as displayed by CuZn-SOD. Mn-SOD in the eukaryotic cells is encoded by a nuclear gene. A large protoform of enzyme is synthesized in the cytosol and transported into the mitochondria (Zhang, 1996). Mn-SOD can be induced under oxidative stress, and the up-regulation of Mn-SOD gene expression is mediated, at least partially, by a transcriptional mechanism (Whitsett et al., 1992). A number of potential inducers of Mn-SOD have been identified, including tumor necrosis factor α (TNF-α), interleukin-1 (IL-1), and lipopolysaccharide (Visner et al., 1990). The Mn-SOD promoter contains both nuclear factor κB (NF-κB) and activator protein–1 binding sites. Transcriptional control of Mn-SOD may be mediated in part by activation of NF-κB binding (Das et al., 1995).

In mammals, the highest SOD activity is found in the liver, followed by kidney, brain, adrenal gland, and heart (Halliwell and Gutteridge, 1989). In the skeletal muscle, SOD activity is similar to that in the heart, and the differences between muscle fiber types seem small (Ji and Fu, 1992) (table 2.1). The distribution of SOD activity in the cell varies from tissue to tissue and also from species to species (Halliwell and Gutteridge, 1989). Thus, in rat liver about 90% of SOD activity is in the cytosol and 10% in the mitochondria, whereas in the myocardium and skeletal muscles, the contribution of Mn-SOD to total SOD activity is higher, ranging from 15 to 20% (Leeuwenburgh et al., 1997).

SOD provides the first line of enzymatic defense against intracellular free radical production by removing $O_2^{-\bullet}$. Although $O_2^{\bullet}$ itself is not highly toxic, it can extract an electron from biological membrane and other cell components, causing free radical chain reactions. Therefore, it is essential for the cell to keep superoxide anions in check. Administration of diethyldithiocarbamic acid (DDC), a SOD inhibitor, causes severe oxidative damage under a variety of experimental conditions (Halliwell and Gutteridge, 1989). The essential role of SOD is best illustrated by the phenotype of SOD mutants. In humans, mutagenesis of CuZn-SOD in Lou Gehrig's disease results in apoptosis of spinal neurons (amyotrophic lateral sclerosis) (Fridovich, 1995). However, unless both forms of SOD are defective, serious oxidative stress is rarely seen.

Glutathione Peroxidase (GPX)

GPX (EC 1.11.1.9) catalyzes the reduction of H_2O_2 and organic hydroperoxide to H_2O and alcohol, respectively, using GSH as the electron donor (Flohe, 1982):

$$2GSH + H_2O_2 \rightarrow GSSG + 2\ H_2O$$

or

$$2GSH + ROOH \rightarrow GSSG + ROH + H_2O$$

In mammalian cells, GPX catalyzes the reduction of both hydrogen and organic peroxides (e.g., lipid peroxide) by GSH, forming water and alcohol, respectively. By donating a pair of hydrogen ions, GSH is oxidized to glutathione disulfide (GSSG). Reduction of GSSG is catalyzed by GR, a flavin-containing enzyme, wherein NADPH is used as the reducing power. This reaction takes place simultaneously with GPX, thus providing a redox cycle for the regeneration of GSH from GSSG. NADPH is supplied by G6PDH, malic enzyme, and isocitrate dehydrogenase (Reed, 1986). GPX refers only to the Se-dependent enzyme (EC 1.11.1.9). The so-called Se-independent GPX is part of the GST (EC 2.5.1.18) family, which removes some species of organic peroxides (ROOH) (Habig et al., 1984). GPX is highly specific for its hydrogen donor GSH but has low specificity for hydroperoxide, ranging from H_2O_2 to complex organic hydroperoxides that include long-chain fatty acid hydroperoxides and nucleotide-derived hydroperoxides. This kinetic characteristic of GPX makes it a versatile hydroperoxide remover in the cell; thus, it plays an important role in inhibiting lipid peroxidation and preventing damage to deoxyribonucleic acid (DNA) and ribonucleic acid (RNA) (Flohe, 1982). Although GPX and CAT have an overlap of substrate H_2O_2, GPX (at least in mammals) has a much greater affinity for H_2O_2 at low concentrations ($K_m = 1$ μM) than CAT ($K_m = 1$ mM) (Sies, 1985).

GPX is a homotetramer with each 22 kDa subunit bound to a selenium atom existing as a selenocysteine. The expression of the GPX gene, hgpx1, occurs in a wide range of tissues controlled by different mechanisms in mammalian tissues, such as oxygen tension, metabolic rate (Flohe, 1982), growth and development (Moscow et al., 1992), and toxins and xenobiotics (Halliwell and Gutteridge, 1989).

GPX is located in both the cytosol and mitochondrial matrix of the cell, with a distribution ratio of ~2:1 (Chance et al., 1979). This allows GPX to reach a number of cellular sources of hydroperoxide generation. The activity of GPX is high in the liver and erythrocytes; moderate in the brain, kidney, and heart; and low in the skeletal muscle (table 2.1). However, the oxidative type 1 muscle (soleus) possesses a GPX activity close to the level in the heart.

Catalase (CAT)

The primary reaction that CAT (EC 1.11.1.6) catalyzes is the decomposition of H_2O_2 to H_2O, although CAT has other biological functions (Chance et al., 1979). It shares this function with GPX, but the substrate specificity and affinity as well as the cellular location of the two antioxidant enzymes are different, as discussed in the previous section.

$$2H_2O_2 \rightarrow 2H_2O + O_2$$

CAT is a tetramer with a relatively large MW of ~240,000. Heme (Fe^{3+}) is a ligand required to be bound to the enzyme's active site for its catalytic function. CAT resembles SOD in kinetic properties, such as the lack of an apparent K_m and Vmax, and its activity increases enormously with an increase in H_2O_2. Azide and cyanide are both inhibitors of CAT, an inhibition often used to partition CAT activity from GPX activity in enzyme assays of crude tissue extracts. CAT is also strongly inhibited by aminotriazole, a specific CAT inhibitor that can be fed to animals without causing gross metabolic defect (Halliwell and Gutteridge, 1989). Assays of CAT typically involve the addition of H_2O_2, and its removal is followed at 240 nm spectrophotometrically (Aebi, 1984). Caution should be taken with the reported catalytic activity, because it is determined not only by the enzyme protein present in the assay medium but also by the concentration of H_2O_2 used.

Murine CAT gene regulation has been studied extensively. A single gene (Cs) located on chromosome 2 is responsible for coding the primary structure of the enzyme (Holmes and Duley, 1975). Once synthesized, the polypeptide may be modified epigenetically in terms of sulfhydryl groups and carbohydrate or protein moieties. The normal (N) and epigenetically modified (E) polypeptidal subunits can produce five tetrameric isozymes, similar to the isozyme patterns of lactate dehydrogenase (Holmes and Master, 1970).

CAT is located primarily in the organelle called peroxisome (Aebi, 1984). However, mitochondria and other intracellular organelles such as endoplasmic reticular may also contain some CAT activity (Luhtala et al., 1994). There is considerable debate about whether the detected CAT activity in these sources results from contamination due to cell fractionation during assay preparation (Halliwell and Gutteridge, 1989). CAT activity among mammalian tissues follows the order of SOD, with liver being the highest and skeletal muscle the lowest (table 2.1). Among the various muscle types, type 1 muscle (soleus) displays the highest CAT activity, followed by type 2a muscle deep vastus lateralis (DVL). Type 2b muscle superficial vastus lateralis (SVL) has the lowest CAT activity.

The primary function of CAT is to remove H_2O_2 produced in the peroxisomes due to enzymes such as flavoprotein dehydrogenase in the β-oxidation of fatty acids, urate oxidase, and the metabolism of D-amino acids. When H_2O_2 concentration is high, the catalytic function of CAT prevails; that is, H_2O_2 is decomposed to O_2 and H_2O. Alternatively, CAT may exhibit a peroxidative activity when the H_2O_2 concentration is low ($<10^{-6}$ M) and in the presence of a suitable hydrogen donor, such as ethanol (Chance et al., 1979).

Antioxidant Vitamins

Vitamin E, vitamin C, and β-carotene play a critical role in protecting the cells from ROS-induced oxidative stress (Halliwell and Gutteridge, 1989; Packer, 1991; Yu, 1994). Because humans cannot synthesize these vital anti-oxidants, they depend exclusively on dietary intake. Recent research suggests that several other low molecular weight compounds, such as ubiquinone, uric acid, and α-lipoic acid, serve important antioxidant functions (Yu, 1994). Abundant evidence shows that tissue contents of certain antioxidant vitamins (e.g., vitamin E) decrease during aging and as a result of acute and chronic exercise, which narrows their protective margin against ROS. Ironically, aging and exercise tend to alter an individual's dietary habits, thus affecting antioxidant intake. There are no clear guidelines regarding optimal dietary intake of antioxidants for elderly people or physically active individuals beyond the recommended dietary allowances (RDAs).

Vitamin E

Vitamin E (α-tocopherol) is the most important fat-soluble chain-breaking antioxidant in the body. Although it is incorporated into virtually all cell membrane bilayers, a major portion of tissue vitamin E is concentrated in the inner mitochondrial membrane, where the electron transport chain is located (Gohil et al., 1986). However, vitamin E concentration in the cell membrane is rather small, about one in several thousand molecules of phospholipid (Packer, 1991). Vitamin E content is relatively constant (60-70 nmol/g) across several major body tissues, such as liver, heart, lung, and adipose tissue, but is most abundant in brown adipose tissue (Gohil et al., 1987). Skeletal muscle, however, has only 20 to 30 nmol/g of vitamin E, depending on the fiber type. These differences in vitamin E levels probably reflect the differences of mitochondrial content as well as the oxidative potential among the various tissues (Gohil et al., 1986). Despite the relatively low content of vitamin E, its concentration in tissue is very stable and it is difficult to deplete acutely. This is because, after vitamin E quenches an electron from a free radical species and is converted to a vitamin E radical, it can be reduced back to vitamin E by ascorbate (vitamin C) or GSH, either enzymatically or nonenzymatically (Beyer, 1994a). Vitamin E deficiency has been consistently linked to lipid peroxidation (Davies et al., 1982; Dillard et al., 1978). Animals fed a low vitamin E diet suffer from loss of cell membrane fluidity, reduction of mitochondrial respiratory coupling, skeletal muscle damage, and increased incidence of cardiomyopathy (Davies et al., 1982; Diplock, 1991; Jackson et al., 1983).

Vitamin C

Vitamin C (ascorbate) is a water-soluble vitamin present in the cytosolic compartment of the cell and the extracellular fluid (Mascio et al., 1991). Although vitamin C can interact directly with O_2^- and $\bullet OH$, thus functioning as an antioxidant, its better-known antioxidant function is closely related to that of vitamin E. Spatial arrangement allows vitamin C to have close contact with vitamin E radicals generated in the cell membrane phase (Halliwell and Gutteridge, 1989). After donating an electron to the vitamin E radical, ascorbate is oxidized to a semidehydroascorbate (SDA) radical, a less reactive compound. The SDA radical either is recycled by dihydrolipoate or goes through a disproportionation reaction to form dehydroascorbate (DHA). In the presence of GSH, the enzyme DHA reductase catalyzes the regeneration of ascorbate. In animals, SDA radicals also can be converted directly to ascorbate by the enzyme SDA reductase, using nicotinamide adenine dinucleotide (NADH) as the reducing power. Vitamin C is especially efficient in scavenging free radicals formed in the aqueous phase such as plasma, thus preventing damage to erythrocyte membrane (Beyer, 1994a).

β-Carotene

β-Carotene, a major carotenoid precursor of vitamin A, has recently received broad attention as an antioxidant (Burton and Ingold, 1984). Although its best-defined antioxidant function is to quench singlet oxygen (not a free radical), it may also be involved in other free radical reactions (Machlin and Bendich, 1987; Mascio et al., 1991). β-Carotene inhibits lipid peroxidation initiated by oxygen- or carbon-centered free radicals (Yu, 1994).

Glutathione and Other Antioxidants

GSH is a thiol-containing tripeptide found in high concentrations in virtually all animal and plant cells. The most important antioxidant function of GSH is to serve as a substrate for GPX to remove hydrogen and organic peroxides (e.g., lipid peroxide). By donating a pair of hydrogen ions, GSH is oxidized to GSSG. Reduction of GSSG is catalyzed by GR, a flavin-containing enzyme, wherein NADPH is used as the reducing power. This reaction takes place in conjunction with GPX, thus providing a redox cycle for the regeneration of GSH (Flohe, 1982). GSSG levels in most tissues are very low, and the intracellular ratio of GSH:GSSG recently has been found to be much higher than previously reported in the literature (Asuncion et al., 1996; Vina et al., 1995).

Recently, GSH has been shown to reduce a variety of antioxidants in the cell. For example, GSH has been postulated to reduce vitamin E (α-tocopheroxyl) radicals that are formed in the chain-breaking reactions with alkoxyl or lipid peroxyl radicals (Packer, 1991). GSH may also be used to reduce the SDA radical (vitamin C radical) derived in the recycling of vitamin E and to reduce α-lipoic acid to dihydrolipoate which recently have been hypothesized to play an important role in the recycling of ascorbic acid (Niki et al., 1995). Together, these reactions keep the limited resources of vitamin E and vitamin C in the reduced state at the expense of GSH, a relatively abundant reducing power in the cell.

GSH concentration in the cell is in the millimolar range for most tissues, but there is a great variability in GSH content in different organs, depending on their functions and oxidative capacities (table 2.2). With the exception of the eye lens, liver has the highest concentration of GSH (5-7 mM) in the body (Halliwell and Gutteridge, 1989). Other important organs such as the lung, kidney, and heart contain 2 to 3 mM of GSH. Red blood cells contain a high level of GSH (~2 mM) compared to plasma (<0.05 mM), primarily due to the protective role of GSH against oxidative damage to hemoglobin. Skeletal muscle GSH concentration varies, depending on muscle fiber type and animal species (Ji, 1995; Ji et al., 1992). In rats, type 1 fibers (e.g., soleus muscle) contain 6-fold higher GSH content (~3 mM) than type 2b fibers (e.g., white vastus lateralis). In contrast, the GSH:GSSG ratio appears remarkably consistent across various fiber types (Ji, 1995).

Intracellular GSH level is regulated both by GSH utilization and GSH synthesis. Although GSH is essential for normal cell function, most organs and tissues do not synthesize GSH de novo. Instead, GSH is taken up from extracellular sources and imported into the cell (Deneke and Fanburg, 1989). This process encompasses cross-membrane transport and resynthesis of GSH, known as the γ-glutamyl cycle. The membrane-borne γ-glutamyl transpeptidase (GGT) controls the cleavage of GSH, with subsequent translocation of amino acids across the cell membrane, whereas glutamylcysteine synthetase (GCS) catalyzes the formation of an initial peptide bond between cysteine and glutamate. This rate-limiting step of GSH synthesis is inhibited via negative feedback by GSH. The final step of GSH synthesis is catalyzed by GSH synthetase (GS).

The majority of the GSH pool in the body is synthesized de novo in the liver, which supplies approximately 90% of the circulating GSH under physiological conditions (Deneke and Fanburg, 1989; Meister and Anderson, 1983). Hepatic GSH synthesis is controlled by both substrate (amino acids) availability and hormonal regulation. Thus, fasting decreases liver GSH content, whereas refeeding quickly restores it (Leeuwenburgh and Ji, 1996). Insulin and glucocorticoids stimulate hepatic GSH synthesis via induction of GCS (Lu et al., 1992). In contrast, glucagon and several other adenosine 3',5'-cyclic monophosphate (cAMP)–stimulating agents down-regulate hepatic GSH

Table 2.2   Glutathione Content and Related Enzyme Activities in Various Tissues

	GSH (μmol/m)	GSSG (μmol/g ww)	GSH + GSSG (μmol/g ww)	GSH:GSSG (μmol/g ww)	GGT (unit/g ww)	GCS (unit/g ww)	Cysteine (nmol/g ww or ml)
Rat							
Plasma	0.01	0.002	0.014	6	n.a.	n.a.	5
SVL	0.5	0.03	0.5	17	n.d.	n.d.	n.d.
DVL	1.5	0.09	1.6	17	94	0.5	60
Soleus	2.2	0.18	2.4	18	27	0.3	50
Heart	1.4	0.11	2.4	13	38	0.3	40
Liver	5.4	0.29	5.9	19	30	3.5	1300
Mice							
Plasma	0.02	0.004	0.028	5	n.a.	n.a.	n.d.
Quadriceps	0.7	0.04	0.75	17	46	n.d.	n.d.
Heart	0.9	0.11	1.1	9	35	n.d.	n.d.
Kidney	3.1	0.17	3.4	20	44,000	n.d.	n.d.
Liver	6.2	0.28	6.8	22	46	4.2	n.d.

Note. Values are given as mean of no less than seven animals. GSH and GSSG were determined with high-performance liquid chromatography. GSH, glutathione; GSSG, glutathione disulfide; GGT, γ-glutamyl transpeptidase; GCS, glutamylcysteine synthetase; SVL, superficial vastus lateralis; DVL, deep vastus lateralis; n.a., not applicable; n.d., not determined. Sources: Leeuwenburgh & Ji, 1995, 1998; Leeuwenburgh et al., 1997.

synthesis by phosphorylating and inhibiting GCS (Lu et al. 1991). Release of GSH from the liver is promoted by catecholamines, glucagon, and vassopressin (Lu et al., 1992; Sies and Graf, 1985).

GSH turnover rate is high in most mammalian tissues, estimated to be 4.5, 2.7, and 1.6 μmol/h, for the liver, kidney, and skeletal muscle, respectively (Griffiths and Meister, 1979). Although GSH turnover rate is relatively low in the noncontracting skeletal muscle, muscle is an important GSH pool. Indeed, the large muscle mass of the body (~40% of body weight) and the relatively high GSH concentration in muscle may influence plasma GSH levels and GSH turnover under certain physiological and pathological conditions (Kretzschmar et al., 1992).

When oxidation of GSH to GSSG exceeds the reducing capacity of GR, skeletal muscle fibers, cardiac myocytes, and liver cells are all capable of exporting GSSG to maintain the GSH:GSSG ratio (Meister and Anderson, 1983). This process is important because high intracellular levels of GSSG may inactivate certain enzymes and cause protein cross-linkage and thus may cause damage to the cell (Halliwell and Gutteridge, 1989). However, GSSG export may also subject the tissues to net loss of GSH under oxidative stress.

In addition to GSH, several low molecular weight thiol-containing or non-thiol-containing compounds have displayed antioxidant function in vitro and in vivo, including ubiquinone, uric acid, and dihydrolipoic acid. They will be discussed in conjunction with acute and chronic exercise effects.

Antioxidant Protection During Acute Exercise

Adequate antioxidant protection is crucial for the cell to avoid oxidative damage caused by ROS. The following section will explore the response and regulation of enzymatic and non-enzymatic antioxidants to an acute bout of exercise.

Antioxidant Enzymes

Strenuous aerobic exercise is associated with increased ROS production in skeletal muscle, the liver (Davies et al., 1982; Jackson et al., 1985), and the heart (Kumar et al., 1992). SOD, CAT, and GPX provide the primary defense against ROS generated during exercise, and activities of these enzymes increase in response to exercise in both animal and human studies (Jenkins, 1983; Ji, 1995; Sen et al., 1994b). Because exercise duration is relatively brief (up to several hours), the increased catalytic activity is probably caused by modifying the existing enzyme molecules via allosteric and/or covalent mechanisms rather than by synthesizing new enzyme proteins (Ji, 1995). Because of the wide range of endogenous enzyme activity and levels of ROS production in the various tissues, antioxidant enzymes have demonstrated different exercise responses.

An acute bout of exercise increases SOD activity in a number of tissues, including the liver (Alessio and Goldfarb, 1988; Ji et al., 1988a, 1988b, 1990), skeletal muscle (Ji et al., 1990; Lawler et al., 1993; Quintanilha and Packer, 1983), the heart (Ji and Mitchell, 1994; Quintanilha and Packer, 1983), the lung (Quintanilha et al., 1982), red blood cells (Lukaski et al., 1990; Mena et al., 1991; Ohno et al., 1988), and platelets (Buczynski et al., 1991). With a few exceptions (Quintanilha and Packer, 1983), most studies also indicate that acute exercise increases CuZn-SOD rather than Mn-SOD activity. This activation of SOD was proposed to be caused by increased $O_2^{\bullet}$ production during exercise, based on the in vitro SOD kinetics that partial occupancy of the enzyme by $O_2^{\bullet}$ can increase its activity (Ji, 1993). Because we now know that CuZn-SOD has a quick turnover rate and a short $t^{1/}_{2}$, in the range of minutes (see previous section), de novo synthesis of new enzyme protein cannot be ruled out in explaining SOD responses to acute exercise lasting a few hours. Radák et al. (1995) showed that enzyme activities and immunoreactive enzyme contents of both CuZn- and Mn-SOD in rat soleus and tibialis muscles were significantly elevated after a single bout of exhaustive treadmill running lasting 60 to 70 min. Interestingly, CuZn-SOD activity and content gradually returned to the resting levels 1 to 3 d later, whereas Mn-SOD activity and protein content continued to increase during the postexercise period. This finding indicates that the stimulating effects of exercise on CuZn-SOD and Mn-SOD gene expression may differ in terms of threshold required and time course of induction.

GPX activity has demonstrated variable responses to an acute bout of exercise in the various types of skeletal muscle. Several studies have shown no change in this enzyme in skeletal muscle after acute exercise (Brady et al., 1979; Ji et al., 1990; Leeuwenburgh and Ji, 1995; Vihko et al., 1978), whereas others have reported significant elevation of GPX activity (Ji and Fu, 1992; Ji et al., 1992; Leeuwenburgh and Ji, 1996; Oh-ishi et al., 1996; Quintanilha, 1984). Furthermore, heart (Quintanilha, 1984) and platelet (Buczynski et al., 1991) GPX activities have been shown to increase after exercise, but liver GPX seems to be unaffected in all studies reported (Ji, 1995). Although no clear explanation can be provided, the discrepancies may be related to the exercise intensity (i.e., % $\dot{V}O_2max$), which varied among the aforementioned studies. Muscle fiber–specific responses of GPX have also been noticed. For example, Ji et al. (1992) found that GPX activity increased as a function of treadmill speed in DVL and SVL but not in the soleus. However, Radák et al. (1995) reported increased GPX activity 1 d after an acute bout of treadmill running to exhaustion in rat soleus but not tibialis muscle. The mechanism responsible for the increased GPX activity with acute exercise is unknown.

Most studies have revealed no significant alteration in CAT activity with acute exercise (Ji, 1995; Meydani and Evans, 1993). However, there are exceptions, as CAT activity was found to increase significantly after an acute bout of exercise to exhaustion or at high intensity in rats (Ji and Fu, 1992; Ji et al., 1992). Only DVL muscle showed this activation, whereas SVL, soleus, liver, and heart did not display any appreciable change with exercise. With a catalytic mechanism similar to SOD, CAT activity may be expected also to increase due to an increased H_2O_2 production during exercise. However, CAT is located primarily in the peroxisomes, whereas the main source of H_2O_2 during short-term acute exercises is the mitochondria (Ji and Leichtweis, 1997). Furthermore, mitochondrial and cytosolic GPX are probably more effective in competing with CAT for the H_2O_2 produced in these two cell compartments because of closer proximity to the source of ROS and lower K_m.

To examine the mechanism by which acute exercise affects gene regulation of antioxidant enzymes, some researchers have measured the relative mRNA abundance for the various enzymes in the skeletal muscles. Oh-ishi et al. (1997) reported a significant down-regulation of mRNA levels for both CuZn- and Mn-SOD isozymes in soleus muscle of untrained rats, but no exercise down-regulation was observed in the trained rats. We recently investigated the effects of a single bout of prolonged exercise on the mRNA abundance of muscle antioxidant enzymes in rats (Gore et al., 1997). mRNA abundance of CuZn-SOD, Mn-SOD, and CAT was not altered by exercise, but exercise decreased GPX mRNA levels by 21.6 and 60.8% ($p < .05$) in DVL and SVL, respectively. These data demonstrate that despite increased enzyme activity, an acute bout of exhaustive exercise may decrease the mRNA abundance of GPX, Mn-SOD, and CuZn-SOD.

Overall, antioxidant enzymes may be selectively activated during an acute bout of strenuous exercise. This activation may depend on the oxidative stress imposed on the specific tissues as well as the intrinsic antioxidant defense capacity. Skeletal muscle may be subjected to a greater level of oxidative stress during exercise than are the liver and heart due to increased oxygen consumption. It is estimated that muscle oxygen uptake can increase up to 100-fold in active contracting muscle cells compared to resting levels (Meydani and Evans, 1993). Therefore, the muscle needs greater antioxidant protection against potential oxidative damage that occurs during or after exercise. Understanding the mechanisms involved in the increased antioxidant enzyme activity during exercise remains a challenge. Although evidence is emerging that mammalian tissues can up-regulate gene expression of antioxidant enzymes in response to acute oxidative stress, the signal transduction pathway is still unknown. The finding that acute exercise decreases mRNA levels for several antioxidant enzymes adds to this paradox. Much greater progress has been made regarding gene regulation of SOD than GPX or CAT, due at least in part to the availability of antibodies to muscle SOD isozymes (Ohno et al., 1994). Until we obtain more specific knowledge about the pattern of gene expression of GPX and CAT, the explanation that ROS may cause posttranslational modulation of enzymes remains viable.

GSH Homeostasis During Exercise

Muscle GSH status is regulated both by GSH utilization, which is controlled by GPX and GR, and by GSH transport into the cell, controlled by the γ-glutamyl cycle enzymes. Adenosine triphosphate (ATP) and NADPH are required to ensure an adequate GSH level and redox status. Heavy aerobic exercise increases ROS production while decreasing intracellular ATP and NADPH levels due to competition from other metabolic demands, such as muscular contraction. This may result in a decreased capacity to regenerate GSH from GSSG and an accumulation of GSSG in the cell. Indeed, an acute bout of exhaustive exercise significantly increases GSSG content in rat skeletal muscle in a fiber-specific and dose-responsive manner (Ji and Fu, 1992; Ji et al., 1993; Lew et al., 1985). In these studies, accumulation of GSSG in exercising muscle was associated with activation of GPX and GR, which suggests an increased hydroperoxide production. Studies of the respiratory muscles involved in resistive breathing showed similar results (Anzueto et al., 1993). In contrast to heavy exercise, prolonged exercise at moderate intensity resulted in no accumulation of GSSG in skeletal muscle of mice (Leeuwenburgh and Ji, 1995) and rats (Leeuwenburgh and Ji, 1996). This may be explained by a continuous GSH output from the liver under the hormonal stimulation that ensured adequate GSH transport to skeletal muscle involved in exercise.

The ratio of GSH to GSSG, an indicator of intracellular redox status, was dramatically decreased in human skeletal muscle biopsy after marathon running (Corbucci et al., 1984). However, in rodent studies, most authors found only moderate reduction or no change in the GSH:GSSG ratio after acute exercise (Ji and Fu, 1992; Ji et al., 1993; Leeuwenburgh and Ji, 1995; Lew et al., 1985). The reported values of muscle GSH:GSSG ratio vary greatly from study to study (from 10 to several hundred), which probably reflects the different GSH assays used by the investigators.

The physiological role of GSH in preventing cellular oxidative damage is best illustrated when tissue GSH homeostasis is perturbed by physiological, nutritional, or pharmacological interventions and then subjected to acute exercise. Fasting profoundly affects liver GSH content and whole-body GSH homeostasis (Godin and Wohaieb, 1988; Tateishi et al., 1977). GSH levels in plasma, the lung, and skeletal muscle are particularly affected by fasting because of the low hepatic GSH reserve (Leeuwenburgh et al., 1997). Leeuwenburgh and Ji (1996) reported that rats fasted for 48 h could lose as much as 50% of liver total GSH content and showed a decreased hepatic GSH:GSSG ratio. Muscle GSH concentration was also reduced in the fasted rats but was not further affected by exercise, because hepatic GSH output supplied a stable blood GSH concentration.

Administration of L-buthionine SR-sulfoximine (BSO), an irreversible inhibitor of GCS, has been shown to decrease total GSH content in the liver, lung, blood, and plasma by ~50% and in skeletal muscle and heart by 80 to 90% (Leeuwenburgh and Ji, 1995; Meister, 1991). Exhaustive exercise decreased the GSH:GSSG ratio in skeletal muscle, especially in GSH-depleted animals (Sen et al., 1994b). Furthermore, running endurance time decreased 50% in the GSH-depleted rats. GSH depletion also significantly increased lipid peroxidation in the heart, skeletal muscles, and plasma of rats.

Although GSH homeostasis is essential for the body to cope with oxidative stress, GSH supplementation during exercise has only moderate promise because of strong feedback inhibition on GCS by GSH (Meister, 1991). Repeated GSH injection in human subjects raised plasma and kidney GSH significantly but did not increase GSH content in other tissues (Leeuwenburgh and Ji, 1998; Sen et al., 1994a). Endurance performance increased in acutely GSH-supplemented mice (Leeuwenburgh and Ji, 1998; Novelli et al., 1991). Reid et al. (1994) showed that administration of N-acetyl-L-cysteine (NAC), a cysteine analog, reduced low-frequency fatigue in electrically stimulated human leg muscles. These studies suggest that ROS production during aerobic exercise may be an important mechanism for muscle fatigue and that thiol supplementation may enhance muscle performance.

Vitamins and Low-Molecular- Weight Antioxidants

Antioxidant vitamins and other low-molecular-weight antioxidants play an important role in breaking free radical chain-reaction and keeping cellular homeostasis during acute exercise. Unlike antioxidant enzymes and GSM system, levels of these antioxidants are not as tightly regulated, and thus, could be heavily influenced by exercise.

Vitamin E

The importance of vitamin E during exercise is best illustrated in studies where animals are depleted of tissue vitamin E by receiving a vitamin E–deficient diet beginning at the early stage of life. Davies et al. (1982) found that vitamin

E deficiency exacerbated muscle and liver free radical production and enhanced lipid peroxidation and mitochondrial dysfunction in exhaustively exercised rats. Endurance performance has also been reported to decrease in rats fed a vitamin E–deficient diet (Davies et al., 1982; Gohil et al., 1986). Vitamin E deficiency has been shown to enhance lipid peroxidation, disturb GSH/GSSG redox status, and cause early fatigue in the diaphragm muscle during resistance breathing in rats (Anzueto et al., 1993). However, an acute bout of exercise does not seem to affect muscle vitamin E levels profoundly.

Vitamin C

The importance of vitamin C in protecting against exercise-induced oxidative stress is not well established, partly because most mammalian species synthesize vitamin C, making a deficiency study rather difficult. Vitamin C also performs numerous functions that are not related to those of an antioxidant (Bendich and Langseth, 1995). By reducing dietary vitamin C content to 10% of the normal values (0.2 g/kg), Packer et al. (1986) demonstrated that myocardial capacity to oxidize pyruvate, 2-oxoglutarate, and succinate was significantly reduced in guinea pigs (which cannot synthesize vitamin C). As a result, running time to exhaustion was significantly shortened in vitamin C–deficient animals.

It is well known that given at high doses vitamin C can behave as a prooxidant (Yu, 1994). This is because ascorbate reacts with transition metal ions to form ROS, including $\cdot OH$ (Halliwell and Gutteridge, 1989; Yu, 1994). Thus, it is interesting that in the study of Packer et al. (1986), a group of guinea pigs supplemented with twice the normal amount of dietary vitamin C also exhibited similar metabolic defects in the heart and early fatigue during prolonged exercise, possibly due to oxidative stress caused by excessive vitamin C. Because one of the primary antioxidant functions of vitamin C is to recycle vitamin E, Gohil et al. (1986) investigated the effect of dietary vitamin C supplementation on vitamin E–deficient rats during training. Vitamin C could not prevent a decrease of endurance time and mitochondrial dysfunction caused by vitamin E deficiency.

Ubiquinone (Q_{10})

As an electron carrier, ubiquinone is abundant in the mitochondrial inner membrane. Gohil et al. (1987) showed that training could significantly increase ubiquinone content in skeletal muscle and adipose tissues. Reduced ubiquinone acts as an antioxidant in vitro, and its role as an antioxidant in vivo has been proposed (Beyer, 1994b). Tissue slices from rats fed a high-ubiquinone diet demonstrated more resistance to hydroperoxide-induced lipid peroxidation than those from rats fed a control diet (Leibovitz et al., 1990). These antioxidant properties have prompted several studies that used dietary supplementation of

Q_{10} to evaluate its protective function during exercise. For example, Shimomura et al. (1991), using a rat model, reported that Q_{10} administration attenuated muscle creatine kinase and lactate dehydrogenase release caused by downhill running. However, these studies did not clearly establish the role of Q_{10} as an antioxidant in vivo. Furthermore, few data are available regarding the interaction of Q_{10} with other antioxidants during exercise.

Uric Acid

Uric acid is the end product of purine metabolism and appears in high concentrations in the circulation after heavy muscular contraction and in the effluent of ischemia-reperfused organs (Sjodin et al., 1990). This results because an insufficient intramuscular ATP supply causes excessive adenine nucleotide degradation and accumulation of hypoxanthine and xanthine (Hellsten, 1994; Hellsten-Westing et al., 1993; Norman et al., 1987; Sahlin et al., 1991). These purine metabolites are released from the muscle into the blood, and a portion of these compounds presumably is converted to uric acid by xanthine oxidase (XO) located in the endothelial cells of the blood vessels. Uric acid's function as a potential antioxidant has been emphasized (Yu, 1994). Besides being an excellent scavenger of •OH, uric acid may preserve plasma ascorbic acid under oxidative stress (Sevanian et al., 1985). Because an acute bout of exercise increases blood uric acid concentrations in human subjects (Radák et al., 1995; Sahlin et al., 1991), it is not unreasonable to speculate that the increased uric acid may protect against blood-borne sources of ROS, thus reducing oxidative stress to erythrocytes and other tissues. However, there are currently no data to substantiate the antioxidant function of uric acid in exercise.

α-Lipoic Acid

α-Lipoic acid is a well-known cofactor for the oxidative decarboxylation catalyzed by ketoacid dehydrogenases. Recently, a great deal of attention has been given to the antioxidant potential of its reduced form, dihydrolipoic acid (DHLA). Both α-lipoic acid and DHLA have exhibited specific scavenging capacity for a variety of free radicals, such as $O_2^{-•}$, •OH, $^1O_2$, peroxyl radical, and hypochlorous radical (Packer et al., 1995). α-Lipoic acid and DHLA are chelators of transition metal ions, thereby preventing damaging free radical chain reactions. DHLA is capable of regenerating other antioxidants such as vitamin E and vitamin C from their radical forms either directly or indirectly via the GSH-GSSG redox cycle. Thus, DHLA prevents vitamins E and C deficiencies possibly by increasing intracellular GSH levels (Packer et al., 1995). Perhaps the most intriguing and complex biological function of DHLA is its proposed effect on gene expression of antioxidant enzymes via the regulation of NF-κB. DHLA influences both the dissociation of the inhibitory subunit

IκB from NF-κB complex and the binding of the activated NF-κB (p50 and p65) to DNA. The overall effect could be either stimulatory or inhibitory depending on the redox state of the cell and the relative concentrations of α-lipoic acid and DHLA (Packer et al., 1995; Sen, 1995).

Antioxidant Response to Chronic Exercise

The benefit of exercise in preventing various diseases is well known. However, chronic exercise also represents a form of oxidative stress to the organisms and therefore can alter the balance between pro-oxidants and antioxidants. Thus, the following questions are relevant when we consider the effect of chronic exercise on cellular antioxidant systems:

1. Do organisms involved in long-term exercise show a deficit of their antioxidant reserve margin?

2. Can an antioxidant defense system be induced to meet the increased challenge?

3. Is it beneficial to supplement exogenous antioxidants during chronic exercise?

Chronic Exercise May Deplete Certain Antioxidants

Although an acute bout of exercise does not seem to significantly affect vitamin E content in tissues, vitamin E concentration has been shown to decrease in a number of tissues, such as skeletal muscle, liver, and heart, after endurance training in rats (Aikawa et al., 1984; Packer et al., 1989; Tiidus and Houston, 1993). More dramatic changes were observed when tissue vitamin E levels were expressed per unit of mitochondrial protein content (Gohil et al., 1987). The reduction of mitochondrial vitamin E after training probably reflects the increased free radical production at the electron transport chain on the mitochondrial inner membrane.

Endurance training significantly decreases GSH content in rat soleus muscle (Leeuwenburgh et al., 1997). Rigorous swimming training in rats caused a similar reduction of myocardial GSH content (Leichtweis et al., 1997). Other muscle tissues investigated, such as DVL and gastrocnemius muscle, showed increased GSH with training (Leeuwenburgh et al., 1997; Sen et al., 1992). As highly oxidative muscles, soleus and myocardium share many metabolic and biochemical characteristics, such as mitochondrial enzyme activities and GSH content. Despite a 4- to 5-fold high GPX activity, these two tissues have 60 to 70% and 32% lower GGT and GCS activity, respectively, than DVL. It is conceivable that the oxidation of GSH may far exceed the capacity of these tissues to import GSH from extracellular sources, resulting in a net deficit after training.

Endurance Training Induces Antioxidant Enzymes

Although an acute bout of exercise may activate certain antioxidant enzymes without de novo synthesis of new protein, the protective margin could be narrowed significantly in organisms undergoing intensive exercise training, particularly if training intensity is adjusted according to increased $\dot{V}O_2$max. The higher relative workload requires an increasing level of mitochondrial oxygen consumption, which could increase production of ROS. Heavy training may provoke tissue injury, which could trigger inflammatory response and secondary ROS production due to neutrophil activation (Ji and Leichtweis, 1997). Furthermore, training may deplete nonenzymatic antioxidant reserves, as discussed previously. Thus, as a long-term strategy, cells may activate de novo synthesis of antioxidant enzymes to cope with oxidative stress.

Many authors have reported that SOD activity in skeletal muscle increases significantly after training (Higuchi et al., 1985; Jenkins, 1983; Leeuwenburgh et al., 1994, 1997; Oh-ishi et al., 1997; Powers et al., 1994; Sen et al., 1992) (figure 2.1). However, many studies failed to detect a SOD training adaptation even though authors used similar animal training models (Alessio and Goldfarb, 1988; Ji, 1993; Laughlin et al., 1990). Furthermore, Tiidus et al. (1996) failed to find a training effect of SOD in human leg muscle after 8 wk of bicycle training. The discrepancies may be explained by the following factors:

1. Different SOD isozymes studied
2. Different SOD assays used
3. Different training intensity and frequency used, which imposed different oxidative stress to the muscle
4. Different muscle fiber types tested

Powers et al. (1994) investigated the influences of exercise intensity, duration, and muscle fiber type on training response of SOD in rats. Prominent training adaptation of SOD occurred in soleus muscle, whereas in red gastrocnemius muscle increased SOD activity was observed only at high training intensity. In soleus, SOD activity increased as a function of exercise duration rather than intensity.

Because SOD is present in both cytosol (CuZn-SOD) and mitochondria (Mn-SOD), researchers have attempted to identify which isozyme form is induced by training. Using cyanide as an inhibitor of CuZn-SOD, Higuchi et al. (1985) demonstrated that Mn-SOD is primarily responsible for the increased SOD activity with training. Ji et al. (1988b) found no training effect on Mn-SOD in rat hindlimb muscle, when enzyme activity was expressed per mitochondrial protein. Because mitochondrial protein content increases with training, total muscle Mn-SOD apparently was induced in the trained state. The

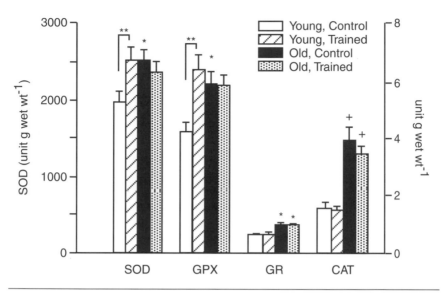

Figure 2.1   Changes of antioxidant enzyme activities in response to aging and endurance training in rat vastus lateralis muscle. Values are means ± *SEM* with the 4-11 rats group. SOD, superoxide dismutase (g ww); GPX, glutathione peroxidase; GR, glutathione reductase (μmol/min/g ww); CAT, catalase (K $\times$ 10^{-3}/g ww). *$p < .05$; +$p < .01$, old versus young rats. **$p < .05$, trained versus control rats.

Reprinted, by permission, from R. Chandwaney et al., 1998, "Oxidative stress and mitochondrial function in skeletal muscle: Effects of aging and exercise training," *Age* 21:109-117.

availability of monoclonal antibodies to SOD isozymes provides a powerful tool to investigate the gene regulation of SOD. Oh-ishi et al. (1997) studied the relationship among SOD isozyme activity, protein content, and mRNA abundance in rat soleus muscle after endurance training. Resting CuZn-SOD activity significantly increased with training, but the enzyme protein content was not altered. There was no significant change in CuZn-SOD mRNA level with training. In contrast, Mn-SOD showed both increased activity and protein content, but mRNA levels were not affected. These data suggest that training induction of both CuZn-SOD and Mn-SOD is caused by posttranscriptional mechanisms and that posttranslational modulation may play a role in CuZn-SOD gene expression. In general agreement with Oh-ishi et al. (1997), a study in our laboratory (Hollander et al., 2000) showed that Mn-SOD is inducible by endurance training, but the induction is muscle fiber specific. Significant increases in both Mn-SOD activity (+70%) and enzyme protein content (+26%) were found in rat DVL muscle after 10 wk of treadmill running. Only marginal increases in Mn-SOD protein, but not activity, were found in soleus muscle. CuZn-SOD activity and protein content showed no effect in either type of muscle. Resting mRNA levels for both Mn- and CuZn-SOD were unaltered with training. These two studies point to the importance of Mn-SOD training adaptation, which seems to support the notion that mitochondrial inner membrane is a major source of ROS production during endurance exercise (Ji and Leichtweis, 1997). SOD training adaptation is not limited to skeletal muscle, and several detailed reviews are available on this topic (Jenkins, 1993; Ji, 1995; Ohno et al., 1994).

Some authors have demonstrated that CAT activity increases in skeletal muscle after training (Hollander et al., 2000; Jenkins, 1993; Oh-ishi et al., 1997; Quintanilha, 1984). However, most studies have reported no change in muscle CAT with training (Jenkins, 1988; Ji, 1995; Meydani and Evans, 1993), and a few studies even reported a decrease (Laughlin et al., 1990; Leeuwenburgh et al., 1994). In contrast, a more consistent training adaptation has been reported on GPX (Ji et al., 1988a, 1988b; Laughlin et al., 1990; Lawler et al., 1993; Leeuwenburgh et al., 1994, 1997; Oh-ishi et al., 1997; Powers et al., 1994; Sen et al., 1992), with the exception of Tiidus et al. (1996), who failed to confirm GPX training adaptation in human subjects. GPX adaptation also demonstrates a muscle fiber–specific pattern, with type 2a muscle being the most responsive to training. Powers et al. (1994) showed a 45% increase in GPX activity in red gastrocnemius muscle (type 2a) after endurance training in rats, and the level of increment appeared to depend on running time rather than speed. Soleus and white gastrocnemius muscles revealed no training effect of GPX, regardless of training intensity and duration. Consistent with these findings, Leeuwenburgh et al. (1997) reported a 62% increase in GPX activity in DVL (type 2a) muscle in response to treadmill training for 2 h/d at moderate intensity (25 m/min, 10%). Soleus and myocardium showed no GPX training effect. Sen et al. (1992) found a small but significant training induction of GPX (16%) when they studied mixed vastus lateralis muscle in rats. Although GPX is expressed as a uniform enzyme in the various cellular compartments, there is evidence that mitochondrial GPX undergoes a greater training adaptation than the cytosolic GPX in rat skeletal muscle (Ji et al., 1988b). Because GPX has a wider range of substrate specificity with respect to peroxides and a lower K_m than CAT, increased GPX activity facilitates the removal of both H_2O_2 and lipid peroxides generated in the mitochondrial inner membrane (Nanji et al., 1995).

Why do different antioxidant enzymes display different characteristics of training adaptation? The answer may be multifaceted depending on the specific pattern of gene expression for each enzyme, the threshold required for induction, and their interactions. De novo synthesis of an enzyme is energy demanding and relatively slow and probably is reserved as the last means to cope with oxidative stress. SOD activities appear sufficiently high and relatively uniform across tissues and various muscle types (table 2.1), which suggests that the removal of superoxide anion may not be a rate-limiting step. In comparison, GPX destroys the end products of ROS generation pathway, that is, hydrogen peroxide and organic peroxide (including lipid and nucleotide peroxides), and its activity is relatively low. This may explain why GPX usually displays a greater training adaptation than SOD and CAT. After studying the kinetics of SOD, CAT, and GPX in vitro, Remacle et al. (1992) proposed that GPX is the most important antioxidant enzyme for cell survival because of its higher sensitivity to intracellular ROS levels and its greater adaptability to oxidative stress.

Researchers do not agree on the physiological significance of a training induction of antioxidant enzymes. It is conceivable that an up-regulation of antioxidant enzymes offers greater protection to tissues during exercise. In humans, a higher antioxidant enzyme activity was reported to correlate with $\dot{V}O_2$max, and trained athletes were shown to have greater SOD and CAT activities in skeletal muscle (Jenkins, 1993). To establish a clear benefit of training, future studies should investigate muscle functional improvement as a result of antioxidant enzyme adaptation.

Adaptation of Glutathione System to Training

Some body tissues are capable of increasing GSH content as an adaptation to exercise-induced oxidative stress. However, the effect of training on GSH content seems to vary greatly between animal species and tissues (Leeuwenburgh et al., 1997). High levels of endurance training have been shown to increase GSH content in the hindlimb muscles of dogs (Kretzschmar and Muller, 1993; Marin et al.,1990; Sen et al., 1992) and rats (Leeuwenburgh et al., 1997; Sen et al., 1992). Increased GSH content in the trained muscle groups may be explained by an enhanced ability to take up GSH from extramural sources, partly reflected by increased activities of GGT, GCS, and GS (Marin et al., 1990; Sen et al., 1992). A closer look reveals that training adaptation of GSH in muscle is muscle fiber specific. The reason for this muscle fiber type variation in adaptation may be related to the rate of GSH utilization versus the capacity of GSH uptake within each fiber type. Activities of the γ-glutamyl cycle enzymes seem to play an important role. For example, Leeuwenburgh et al. (1997) reported that DVL muscle, which has a high GGT activity, demonstrated a training adaptation of GSH content. In contrast, soleus and SVL, which have low GGT activity, showed no training effect (Leeuwenburgh et al., 1997). The authors noticed no significant difference in GCS activity between various types of skeletal muscle, which suggests that the translocation of amino acids, rather than intracellular assembly of GSH, is limiting.

Physically trained human subjects and animals, compared to their untrained counterparts, generally demonstrate a greater tolerance of exercise-induced disturbance of blood GSH (Ji et al., 1992; Lew et al., 1985; Robertson et al., 1991). Furthermore, erythrocyte GSH content has been shown to increase significantly after 20 wk of physical training in previously sedentary men (Evelo et al., 1992). Kretzschmar et al. (1991) found that both young and old trained individuals had higher resting plasma GSH concentrations than their sedentary counterparts and suggested that the observed training adaptation resulted from increased GSH export from both the liver and skeletal muscle. Blood GSH concentration has been reported to be higher in trained runners compared to sedentary subjects and appears to be elevated with increased running distance (Robertson et al., 1991). However, highly trained runners also demonstrated a significantly higher blood GSSG concentration at rest.

Is Antioxidant Supplementation Necessary?

Supplementation of GSH and GSH analogs has been used for therapeutic purposes such as drug-induced oxidative stress and radiation therapy (Meister, 1991). However, access of the target tissues to exogenously supplemented GSH is limited by GGT activity, which in most tissues, except for the kidney, is quite low (Meister and Anderson, 1983). Furthermore, GCS, the rate-limiting enzyme of GSH synthesis, is strongly feedback inhibited by GSH. To overcome these limitations, cysteine analogs such as NAC and L-2-oxothiazolidine-4-carboxylate (OTC) have been used to promote GSH synthesis. GSH monoester also has been used to transport GSH directly into the cell (Meister, 1991). Human subjects receiving 400 mg NAC/d for 2 d with an additional 800 mg before exercise showed an attenuated blood GSSG level during a maximal treadmill test (Sen et al., 1994b). Reid et al. (1994) showed that NAC administration improved muscle contractile functions and reduced low-frequency fatigue in humans. Oral supplementation of NAC and GSH prevented the increase in GSSG levels during exercise in rats (Sastre et al., 1992). NAC supplementation by intravenous bolus has been shown to attenuate the rate of diaphragmatic fatigue during repetitive isometric contraction (Shindoh et al., 1990). However, none of the mentioned studies demonstrated an increase in tissue or blood GSH with NAC administration.

Supplementation of free GSH has generated limited promise in raising tissue GSH content. Although repeated injection of GSH raised plasma and kidney GSH significantly, it did not increase GSH content in skeletal muscle, heart, liver, or lung in rats or mice (Sen et al., 1994a; Leeuwenburgh and Ji, 1998). Exercise endurance was reported to be unchanged with GSH supplementation in rats (Sen et al., 1994a). Novelli et al. (1991), however, showed that acute GSH injection doubled swimming endurance time in mice. Because total swimming time in that study was only 2 to 3 min, and tissue GSH levels were not measured, it is difficult to evaluate the contribution of GSH to endurance performance. Leeuwenburgh and Ji (1998) reported that intraperitoneal injection of GSH and GSH ethyl ester (6 mmol/kg) increased swimming endurance from 4 to 6 h, although no change in tissue GSH levels was observed. Sastre et al. (1992) reported that oral administration of GSH prevented an exercise-induced oxidation of GSH to GSSG. Thus, supplementing GSH during exercise appears to hold some promise, but the mechanisms of action and desirable protocols remain to be examined.

Dietary supplementation of vitamin E may increase tissue resistance to exercise-induced lipid peroxidation. Sumida et al. (1989) demonstrated a protective effect of vitamin E supplementation (300 mg/d) in reducing serum malondialdehyde (MDA) concentration and enzyme markers of tissue damage during exercise. Kumar et al. (1992) showed that dietary supplementation of vitamin E for 60 d abolished exercise-induced free radical production and lipid peroxidation in rat myocardium. Goldfarb et al. (1994) reported that rats supplemented with 250 IU vitamin E/kg diet for 5 wk had lower thiobarbituric acid reactive substance (TBARS) and lipid peroxide levels in plasma and leg muscles than controls after 1 h of treadmill exercise. Kanter et al. (1993) showed that daily supplementation of a vitamin mixture containing 600 mg α-tocopherol for 6 wk significantly decreased human subjects' levels of serum MDA and expired pentane both at rest and after 30 min of treadmill exercise at 60 and 90% $\dot{V}O_2$max. Greater protection by vitamin E may occur several days after acute heavy exercise, because urinary lipid peroxide levels were lower when supplemented and placebo groups of human subjects were compared (Meydani and Evans, 1993). These findings indicate that physically active individuals should consider increasing daily dietary vitamin E intake. Despite the aforementioned beneficial effects, no study has demonstrated improved physical performance as a result of vitamin E supplementation (Kanter, 1995).

The effect of dietary supplementation of vitamin C has not been well studied in human subjects involved in physical exercise. Although some authors have claimed that large doses of vitamin C intake reduced fatigue and muscle damage, no specific oxidative stress markers were measured; therefore, it is difficult to determine whether the observed benefits were related to the antioxidant functions of vitamin C (Kanter, 1995).

Age-Related Changes
in Antioxidant Defense

It has been proposed that aging is caused by the cumulative action of deleterious oxygen free radicals throughout the lifespan (Harman, 1956). It is well known that aging influences both free radical production and cellular antioxidant defense systems in various organs and tissues (Ames et al., 1993; Meydani and Evans, 1993; Nohl, 1993; Yu ,1994). However, the response of antioxidant enzymes to aging seems to be highly tissue specific, and skeletal muscle especially displays some unique changes during aging.

Adaptation of Antioxidant Enzymes with Aging in Skeletal Muscle

Although most postmitotic tissues show a decline of cellular antioxidant defenses (Matsuo, 1993), senescent skeletal muscle appears to increase antioxidant enzyme activities (Ji et al., 1990; Lammi-Keefe et al., 1984; Lawler et al., 1993; Leeuwenburgh et al., 1994; Luhtala et al., 1994; Vertechy et al., 1989). Lammi-Keefe et al. (1984) were the first to report increased SOD activity with aging in several skeletal muscles. Ji et al. (1990) showed that activities of all major antioxidant enzymes, such as SOD, CAT, and GPX, were significantly higher in DVL muscle of old versus young rats. In addition, the activity of GST, GR, and G6PDH also increased with age. These changes occurred despite a general decline of muscle mitochondrial oxidative capacity. Leeuwenburgh et al. (1994) reported similar increases in the activities of all the antioxidant enzymes with aging in both DVL and soleus muscles in rats. In addition, γ-glutamyl transpeptidase activity was significantly elevated in DVL, indicating that aged muscle has greater potential of GSH uptake. Age adaptation of antioxidant enzymes appears to be muscle fiber specific, with the most prominent increases found in type 1 muscles such as soleus, followed by type 2a muscles such as DVL; type 2b muscles showed little effect (Leeuwenburgh et al., 1994; Oh-ishi et al., 1996). Different enzymes also showed variable responses. For example, Lawler et al. (1993) reported a significant increase in GPX but not SOD activity in the gastrocnemius and soleus muscles in old versus young rats.

The up-regulation of muscle antioxidant enzymes has been proposed to reflect an age-associated increase in free radical production (Ji, 1993); however, the precise mechanism seldom has been examined. One possibility is that mitochondria from aged muscles produce greater levels of ROS (Nohl and Hegner, 1978), which may induce antioxidant enzyme gene expression. This scenario is consistent with the finding that mitochondrial fractions of antioxidant enzyme activity increased more in senescent skeletal muscle (Ji et al., 1990; Luhtala et al., 1994) and myocardium (Ji et al., 1993), compared to tissues from younger animals. Aged muscle may be more susceptible to injury, which can also trigger an acute phase response, causing further ROS production (Meydani and Evans, 1993). Thus, aged muscles may be in a chronic inflammatory state wherein steady-state ROS production is elevated (Cannon and Blumberg, 1994).

Luhtala et al. (1994) discovered that elevation of muscle antioxidant enzymes during aging was markedly affected by caloric restriction in Fischer 344 rats. The progressive increases in CAT and GPX activities from 11 to 34 mo of age were prevented by a 30% reduction of food intake, and an increase in Mn-SOD was also prevented. Because caloric restriction is known to attenuate free radical generation, ROS may indeed induce muscle antioxidant enzymes at old age. However, these age-related increases in antioxidant enzyme activities did not seem to prevent lipid oxidation in senescent muscles (Ji et al., 1990; Leeuwenburgh et al., 1994; Starnes et al., 1989).

Gene regulation of antioxidant enzymes in the aging muscle has been studied sparsely. Oh-ishi et al. (1996) reported that 24-mo-old rats had higher activities of CuZn-SOD, Mn-SOD (after normalized with CS activity), GPX, and CAT in soleus muscle compared with 4-mo-old rats. CuZn-SOD protein content was also elevated in the aged muscle; however, no significant change was noticed in the relative abundance of mRNA for both SOD isozymes. Studies in our laboratory revealed no significant age difference in immunoreactive protein content for Mn-SOD, and an increase in CuZn SOD protein in gastrocnemius and soleus muscles whereas mRNA levels for Mn-SOD were significantly decreased in these muscles despite prominent increases in their activities with old age (Hollander et al., 1997). These findings suggest that regulation for at least CuZn-SOD in aging muscle may involve posttranscriptional mechanisms. Further, posttranslational modification of Mn SOD cannot be ruled out, because increased enzyme activity was observed without cocommitant increase in enzyme protein (Hollander et al., 2000).

Glutathione and Vitamins

Aging is associated with a decline of cellular thiol reserve in most tissues (Matsuo, 1993). However, data from our laboratory suggest that skeletal muscle and heart may be spared this effect. Leeuwenburgh et al. (1994) showed that aging did not significantly alter GSH content or GSH:GSSG ratio in rat DVL muscle, whereas in soleus there was a 37% increase in GSH content in old rats along with a higher GSH:GSSG ratio. Fiebig et al. (1996) showed a significant increase in total glutathione content (GSH + GSSG) in the hearts of 27- versus 5-mo-old rats. The elevated myocardial GSH content was accompanied with a 2 fold increase in GGT activity, suggesting a greater potential of the γ-glutamyl cycle. In contrast, Stio et al. (1994) and Vega et al. (1992) showed a decrease in GSH content in the myocardium. A possible reason for the discrepancy may be related to whether GSH levels are expressed per tissue weight or protein level. It is well known that aging generally reduces tissue protein content and increases connective tissues.

Aging has been shown to increase vitamin E concentration in a number of tissues such as liver, lung, and brain (Matsuo, 1993). The limited data on skeletal muscle reveal no age-related change in α-tocopherol levels (Starnes et al., 1989). Interestingly, vitamin E concentration seems to increase in the heart in senescence (Weglicki et al., 1969). Vitamin C levels generally decline at an older age (Matsuo, 1993).

Influence of Exercise on the Antioxidant System in Aged Tissues

Although an acute bout of exercise has been shown to activate various antioxidant enzymes in several tissues of young animals (see previous discussion), there are few data regarding exercise response of antioxidant enzymes at old age. Ji et al. (1990, 1993) found no significant alteration in most antioxidant enzymes in the DVL muscle and heart of 26- and 31-mo-old rats after 1 h treadmill running. It is possible that aged animals cannot reach a high workload that elicits sufficient oxidative stress found in the young muscles. Furthermore, aged muscles already have higher antioxidant enzyme activities and GSH levels, which may also influence their response to acute exercise (Ji et al., 1990; Luhtala et al., 1994).

Senescent skeletal muscles lose considerable regenerative ability because of a decreased rate of protein turnover, satellite cell population, and proliferative capacity (Carlson, 1995). Increased protein breakdown, partly explained by oxidative damage and consequently selective degradation, coupled with decreased protein synthesis progressively decreases oxidative enzyme levels and energy production in aging skeletal muscle (Hansford, 1983; Meydani and Evans, 1993). The age-associated deterioration of muscle protein content and mitochondrial oxidative capacity can be effectively restored by endurance training (Fitts et al., 1984). The question arises as to whether training can also improve antioxidant defense capacity in aged skeletal muscle. Ji et al. (1993) showed that, along with increasing the activities of muscle citrate synthase, malate dehydrogenase, and lactate dehydrogenase, training increased GPX activity 60% in 27.5-mo-old Fischer 344 rats. Hammeren et al. (1993) also reported a significant increase in GPX activity with training in several skeletal muscles of old Fischer 344 rats. However, Leeuwenburgh et al. (1994) could not find antioxidant enzyme adaptation in either DVL or soleus muscle from old rats. The failure to induce antioxidant enzyme levels in aged muscle obviously cannot be accounted for by a reduced cell proliferative capacity, because training successfully increased mitochondrial oxidative enzyme activity and protein content (Fitts et al., 1984; Hammeren et al., 1993; Ji et al., 1993). A possible explanation is that antioxidant enzyme activities in the senescent muscle are already high and therefore the training threshold is raised, requiring greater intensity to provoke a training effect (figure 2.1).

There is a general concern that aged animals have a deficiency of antioxidant nutrients (Katz and Robinson, 1986). Exercise training potentially may exacerbate this nutritional deficiency due to increased consumption or decreased dietary intake, or both (Packer 1986). However, a clear consensus in this important area of gerontology is lacking. Kretzschmar et al. (1991) reported that trained men at age 36 to 57 had higher levels of GSH in the plasma than their sedentary counterparts; the authors proposed that increased muscle GSH output helped maintain plasma GSH homeostasis with training at old age. Leeuwenburgh et al. (1994) showed no differences in muscle GSH content between trained and untrained old rats, although aging increased GSH levels in the muscle. Starnes et al. (1989) compared muscle α-tocopherol levels between trained and untrained 24-mo-old rats and found a significant decrease in the trained muscle. Old trained rats also displayed a higher level of muscle TBARS when challenged by ascorbic acid and ferrous ion. These findings are consistent with the earlier data concerning training effect on muscle vitamin E levels in young rats (Aikawa et al., 1984; Gohil et al., 1987). Together, these data indicate that aged skeletal muscle does not necessarily lose adaptability to training. Aged muscle is still capable of up-regulating antioxidant enzyme synthesis and maintaining GSH status. However, aged muscle may be susceptible to deficits of antioxidant vitamins that the body cannot synthesize. It is conceivable that physically active elderly individuals may benefit from dietary antioxidant supplementation.

Training Reduces Muscle Oxidative Stress in Old Age

When an individual exercises at a given workload, oxygen consumption is no different between trained and untrained muscle. Therefore, endurance-trained aged muscle has the advantage, compared to untrained muscle, of distributing the flux of oxidative phosphorylation (electron flux) among increased mitochondrial electron transport chain (Davies et al., 1982). Theoretically, this should decrease ROS production. Furthermore, trained muscles have higher levels of antioxidant enzyme activity and GSH content, providing a greater ROS removal. Thus, training at old age should reduce muscle oxidative injury during acute physical exertion. Indeed, DVL and soleus muscles from aged rats displayed significantly lower MDA levels than their sedentary counterparts after endurance training (Leeuwenburgh et al., 1994). The same training effect was found in the aged myocardium after endurance training (Fiebig et al., 1996). However, these data should be interpreted with caution, because muscle MDA levels could also be influenced by the rate of efflux from the muscle (Halliwell and Gutteridge 1989). The low MDA content may result from enhanced output due to circulatory training adaptation in the old animals.

The effect of endurance training on mitochondrial respiratory function and resistance to ROS has been studied in aged skeletal muscle (Chandwaney et al., 1998). Although training elicited no apparent improvement on state 3 respiration rate and respiratory control index (RCI), the magnitude of RCI inhibition by equal doses of $O_2^{\bullet}$ and H_2O_2 was significantly smaller in trained versus control rats. If the results from this in vitro study could be extrapolated to an in vivo condition, aged skeletal muscle undergoing exercise training may be more capable of maintaining adequate mitochondrial function when exposed to acute oxidative stress.

Conclusion

ROS production increases during heavy exercise and in senescence. The antioxidant defense system is vitally important in protecting against oxidative stress in skeletal muscle and other aerobic tissues. Perhaps the distinction between inducible and noninducible antioxidants should be developed. Inducible antioxidants, including antioxidant enzymes and the GSH system, have demonstrated prominent adaptive responses to chronic exercise and aging in at least skeletal muscle, provided that nutritional status is adequate. Noninducible antioxidants, however, heavily depend on dietary intake and are thus susceptible to deficiency. Understanding the unique characteristics and regulatory mechanisms of various antioxidants will help us develop strategies to enhance cellular antioxidant capacity through physiological and nutritional means. However, there appears to be no single strategy that can improve every antioxidant system.

References

Aebi, H. 1984. Catalase. In: *Methods in Enzymology*, L. Packer (ed.), pp. 121-125. Academic Press, Orlando, FL.

Aikawa, K.M., A.T. Quintanilha, B.O. deLumen, G.A. Brooks, and L. Packer. 1984. Exercise endurance training alters vitamin E tissue levels and red blood cell hemolysis in rodents. *Bioscience Report* 4:253-257.

Alessio, H.M., and A.H. Goldfarb. 1988. Lipid peroxidation and scavenger enzymes during exercise: Adaptive response to training. *Journal of Applied Physiology* 64:1333-1336.

Ames, B.N., M.K. Shigenaga, and T.M. Hagen. 1993. Oxidant, antioxidants, and degenerative diseases of aging. Proceedings of the National Academy of Sciences of the USA 90:7915-7922.

Anzueto, A., F.H. Andrade, L.C. Maxwell, S.M. Levine, R.A. Lawrence, W.J. Gibbons, and S.G. Jenkinson. 1993. Diaphragmetic function after resistive breathing in vitamin E-deficient rats. *Journal of Applied Physiology* 74:267-271.

Asuncion, J.G., A. Millan, R. Pla, L. Bruseghini, A. Esteras, F.V. Pallardo, J. Sastre, and J. Vina. 1996. Mitochondrial glutathione oxidation correlates with age-associated oxidative damage to mitochondrial DNA. *FASEB Journal* 10:333-338.

Bendich, A., and L. Langseth. 1995. The health effects of vitamin C supplementation: A review. *Journal of the American College of Nutrition* 14:124-136.

Beyer, R.E. 1994a. The role of ascorbate in antioxidant protection of biomembranes: Interaction with vitamin E and coenzyme Q. *Journal of Bioenergetics and Biomembranes* 26:349-358.

Beyer, R.E. 1994b. The relative essentiality of the antioxidant function of coenzyme Q— The interactive role of DT-diaphorase. *Molecular Aspects of Medicine* 15:S117-S129.

Brady, P.S., L.J. Brady, and D.E. Ullrey. 1979. Selenium, vitamin E and the response to swimming stress in rats. *Journal of Nutrition* 109:1103-1109.

Buczynski, A., J. Kedziora, W. Tkaczewski, and B. Wachowicz. 1991. Effect of submaximal physical exercise on antioxidative protection of human blood platelets. *International Journal of Sports Medicine* 12:52-54.

Burton, G.W., and K.V. Ingold. 1984. Beta carotene: An unusual type of lipid antioxidant. *Science* 224:569-573.

Cannon, J.G., and J.B. Blumberg. 1994. Acute phase immune responses in exercise. In: *Exercise and Oxygen Toxicity*; C.K. Sen (ed.), pp. 447-479. Elsevier Science, New York.

Carlson, B.M. 1995. Factors influencing the repair and adaptation of muscles in aged individuals: Satellite cells and innervation. *Journal of Gerontology* 50:96-100.

Chance, B., C.H. Sies, and A. Boveris. 1979. Hydroperoxide metabolism in mammalian organs. *Physiological Review* 59:527-605.

Chandwaney, R., S. Leichtweis, C. Leeuwenburgh, and L.L. Ji. 1998. Oxidative stress and mitochondrial function in skeletal muscle: Effects of aging and exercise training. *Age* 21:109-117.

Corbucci, G.G., G. Montanari, M.B. Cooper, D.A. Jones, and R.H.T. Edwards. 1984. The effect of exertion on mitochondrial oxidative capacity and on some antioxidant mechanisms in muscle from marathon runners (Abstract). *International Journal of Sports Medicine* 5:135S.

Das, K.C., Y. Lewis-Molock, and C.W. White. 1995. Thiol modulation of TNF and IL-1 induced Mn SOD gene expression and activation of NF-κB. *Molecular and Cellular Biochemistry* 148:45-57.

Davies, K.J.A., A.T. Quantanilla, G.A. Brooks, and L. Packer. 1982. Free radicals and tissue damage produced by exercise. *Biochemical and Biophysical Research Communications* 107:1198-1205.

Deneke, S.M., and B.L. Fanburg. 1989. Regulation of cellular glutathione. *American Journal of Physiology* 257:L163-L173.

Dillard, C.J., R.F. Litov, W.M. Savin, F.F. Mumelin, and A.L. Tappel. 1978. Effect of exercise, vitamin E, and ozone on pulmonary function and lipid peroxidation. *Journal of Applied Physiology* 45:927-932.

Diplock, A.T. 1991. Antioxidant nutrients and disease prevention: An overview. *American Journal of Clinical Nutrition* 53:189S-193S.

Evelo, C.T.A., N.G. Palmen, Y. Artur, and G.M.E. Janssen. 1992. Changes in blood glutathione concentrations, and in erythrocyte glutathione reductase and glutathione *S-*

transferase activity after running training and after participation in contests. *European Journal of Applied Physiology* 64:354-358.

Fiebig, R., C. Leeuwenburgh, M. Gore, and L.L. Ji. 1996. The interactive effects of aging and training on myocardial antioxidant enzymes and oxidative stress. *Age* 19:83-89.

Fitts, R.H., J.P. Troup, and F.A. Witzmann. 1984. The effect of aging and exercise on skeletal muscle function. *Mechanisms of Ageing and Development* 27:161-172.

Flohe, L. 1982. Glutathione peroxidase brought into focus. In: *Free Radicals in Biology and Medicine*, W. Pryor (ed.), pp. 223-253. Academic Press, New York.

Fridovich, I. 1985. Quantitation of superoxide dismutase. In: *Handbook of Methods for Oxygen Free Radical Research*, R.A. Greenwald (ed.), pp. 211-215. CRC Press, Boca Raton, FL.

Fridovich, I. 1995. Superoxide radical and superoxide dismutases. *Annual Review of Biochemistry* 64:97-112.

Godin, D.V., and S.A. Wohaieb. 1988. Nutritional deficiency, starvation, and tissue antioxidant status. *Free Radical Biology and Medicine* 5:165-176.

Gohil, K., L. Packer, B. deLumen, G.A. Brooks, and S.E. Terblanche. 1986. Vitamin E deficiency and vitamin C supplementation: Exercise and mitochondrial oxidation. *Journal of Applied Physiology* 60:1986-1991.

Gohil, K., L. Rothfuss, J. Lang, and L. Packer. 1987. Effect of exercise training on tissue vitamin E and ubiquinone content. *Journal of Applied Physiology* 63:1638-1641.

Goldfarb, A.H., M.K. McIntosh, B.T. Boyer, and J. Fatouros. 1994. Vitamin E effects on indexes of lipid peroxidation in muscle from DHEA-treated and exercised rats. *Journal of Applied Physiology* 76:1630-1635.

Gore, M., R. Fiebig, J. Hollander, and L.L. Ji. 1997. Acute exercise alters mRNA abundance of antioxidant enzyme and nuclear factor B activation in skeletal muscle, heart and liver. *Med. Sci. Sports Exer.* 29:5229.

Gorecki, M., Y. Beck, J.R. Hartman, M. Fischer, L. Weiss, Z. Tochner, S. Slavin, and A. Nimrod. 1991. Recombinant human superoxide dismutases: Production and potential therapeutical uses. *Free Radical Research Communication* 12-13:401-410.

Griffiths, O.W., and A. Meister. 1979. Glutathione: Interorgan translocation, turnover, and metabolism. *Proceeding of the National Academy of Sciences of the USA* 76:5606-5610.

Habig, W.H., M.J. Pabst, and W.B. Jakoby. 1984. Glutathione *S*-transferases. *Journal of Biological Chemistry* 249:7130-7139.

Halliwell, B., and J.M.C. Gutteridge. 1989. *Free Radicals in Biology and Medicine*. Clarendon Press, Oxford.

Hammeren, J., S. Powers, J. Lawler, D. Criswell, D. Lowenthal, and M. Pollock. 1993. Exercise training-induced alterations in skeletal muscle oxidative and antioxidant enzyme activity in senescent rats. *International Journal of Sports Medicine* 13:412-416.

Hansford, R.G. 1983. Bioenergetics in aging. *Biochimica et Biophysica Acta* 726:41-80.

Harman, D. 1956. Aging: A theory based on free radical and radiation chemistry. *Journal of Gerontology* 11:298-300.

Hellsten, Y. 1994. Xanthine dehydrogenase and purine metabolism in man: With special reference to exercise. *Acta Physiologica Scandinavica* 621:1-73.

Hellsten-Westing, Y., P.D. Balsom, B. Norman, and B. Sjodin. 1993. The effect of high-intensity training on purine metabolism in man. *Acta Physiologica Scandinavica* 149:405-412.

Higuchi, M., L.J. Cartier, M. Chen, and J.O. Holloszy. 1985. Superoxide dismutase and catalase in skeletal muscle: Adaptive response to exercise. *Journal of Gerontology* 40:281-286.

Hollander, J., M. Gore, R. Fiebig, J. Bejma, and L.L. Ji. 1997. Superoxide dismutase gene expression in skeletal muscle: fiber-specific adaptation to training. *Am. J. Physiol.* 277:R856-R862, 1999.

Hollander, J., J. Bejma, T. Ookawara, H. Ohno, and L.L. Ji. 2000. Superoxide dismutase gene expression: fiber-specific effect of age. *Mech. Age. Devel.*

Holmes, R.S., and J.A. Duley. 1975. Biochemical and genetic studies of peroxisomal multiple enzyme systems: Alpha-hydroxyacid oxidase and catalase. In: *Isozymes: I. Molecular Structure*, D.L. Markert (ed.), pp. 191-211. Academic Press, New York.

Holmes, R.S., and C.J. Master. 1970. Epigenetic interconversions of the multiple forms of mouse liver catalase. *FEBS Letters* 11:45-48.

Jackson, M.J., D.A. Jones, and R.H.T. Edwards. 1983. Vitamin E and skeletal muscle. In: *Biology of Vitamin E*, R. Porter and J. Whelan (eds.), pp. 224-239. Pitman Books, London.

Jackson, M.L., R.H.T. Edwards, and M.C.R. Symons. 1985. Electron spin resonance studies of intact mammalian skeletal muscle. *Biochimica et Biophysica Acta* 847:185-190.

Jenkins, R.R. 1983. The role of superoxide dismutase and catalase in muscle fatigue. In: *Biochemistry of Exercise*; H.G. Knuttgen, J.A. Vogel, and J. Poortmans (eds.), pp. 467-471. Human Kinetics, Champaign, IL.

Jenkins, R.R. 1988. Free radical chemistry: Relationship to exercise. *Sports Medicine* 5:156-170.

Jenkins, R.R. 1993. Exercise, oxidative stress and antioxidants: A review. *International Journal of Sport Nutrition* 3:356-375.

Ji, L.L. 1993. Antioxidant enzyme response to exercise and aging. *Medicine and Science in Sports and Exercise* 25:225-231.

Ji, L.L. 1995. Exercise and oxidative stress: Role of the cellular antioxidant systems. In: *Exercise and Sport Sciences Reviews*, J.O. Holloszy (ed.), pp. 135-166. Williams & Wilkins, Baltimore.

Ji, L.L. 1998. Antioxidant enzyme response to exercise and training in skeletal muscle. In: *Oxidative Stress in Skeletal Muscle*, A.Z. Reznick (ed.), pp. 105-127. Birkhauser Verlag, Basel, Switzerland.

Ji, L.L., D. Dillon, and E. Wu. 1990. Alteration of antioxidant enzymes with aging in rat skeletal muscle and liver. *American Journal of Physiology* 258:R918-R923.

Ji, L.L., and R.G. Fu. 1992. Responses of glutathione system and antioxidant enzymes to exhaustive exercise and hydroperoxide. *Journal of Applied Physiology* 72:549-554.

Ji, L.L., R.G. Fu, and E.W. Mitchell. 1992. Glutathione and antioxidant enzymes in skeletal muscle: Effects of fiber type and exercise intensity. *Journal of Applied Physiology* 73:1854-1859.

Ji, L.L., A. Katz, R.G. Fu, M. Parchert, and M. Spencer. 1993. Alteration of blood glutathione status during exercise: The effect of carbohydrate supplementation. *Journal of Applied Physiology* 74:788-792.

Ji, L.L., and S. Leichtweis. 1997. Exercise and oxidative stress. *Age* 20:91-106.

Ji, L.L., and E.W. Mitchell. 1994. The effect of adrimycin on heart mitochondrial function in rested and exercised rats. *Biochemical Pharmacology* 47:877-885.

Ji, L.L., F.W. Stratman, and H.A. Lardy. 1988a. Antioxidant enzyme systems in rat liver and skeletal muscle. *Archives of Biochemistry and Biophysics* 263:150-160.

Ji, L.L., F.W. Stratman, and H.A. Lardy. 1988b. Enzymatic downregulation with exercise in rat skeletal muscle. *Archives of Biochemistry and Biophysics* 263:137-149.

Kanter, M.M. 1995. Free radicals and exercise: Effects of nutritional antioxidant supplementation. In: *Exercise and Sport Sciences Reviews*, J.O. Holloszy (ed.), pp. 375-398. Williams & Wilkins, Baltimore.

Kanter, M.M., L.A. Nolte, and J.O. Holloszy. 1993. Effect of an antioxidant vitamin mixture on lipid peroxidation at rest and postexercise. *Journal of Applied Physiology* 74:965-969.

Katz, M.L., and W.G. Robinson. 1986. Nutritional influences on autooxidation, lipofuscin accumulation, and aging. In: *Free Radicals, Aging, and Degenerative Diseases*, J.E. Johnson, R. Walford, D. Harman, and J. Miguel (eds.), pp. 221-262. Alan R. Liss, New York.

Kretzschmar, M., and D. Muller. 1993. Aging, training and exercise: A review of effects of plasma glutathione and lipid peroxidation. *Sports Medicine* 15:196-209.

Kretzschmar, M., U. Pfeifer, G. Machnik, and W. Klinger. 1991. Influence of age, training and acute physical exercise on plasma glutathione and lipid peroxidation in man. *International Journal of Sports Medicine* 12:218-222.

Kretzschmar, M., U. Pfeifer, G. Machnik, and W. Klinger. 1992. Glutathione homeostasis and turnover in the totally hepactomized rat: Evidence for a high glutathione export capacity of extrahepatic tissues. *Experimental Toxicology and Pathology* 44:273-281.

Kumar, C.T., V.K. Reddy, M. Plasad, K. Thyagaraju, and P. Reddanna. 1992. Dietary supplementation of vitamin E protects heart tissue from exercise-induced oxidant stress. *Molecular and Cellular Biochemistry* 111:109-115.

Lammi-Keefe, C.J., P.B. Swan, and P.V.J. Hegarty. 1984. Copper-zinc and manganese superoxide dismutase activities in cardiac and skeletal muscles during aging in male rats. *Gerontology* 30:153-158.

Laughlin, M.H., T. Simpson, W.L. Sexton, O.R. Brown, J.K. Smith, and R.J. Korthuis. 1990. Skeletal muscle oxidative capacity, antioxidant enzymes, and exercise training. *Journal of Applied Physiology* 68:2337-2343.

Lawler, J.M., S.K. Powers, T. Visser, H. Van Dijk, M.J. Korthuis, and L.L. Ji. 1993. Acute exercise and skeletal muscle antioxidant and metabolic enzymes: Effect of fiber type and age. *American Journal of Physiology* 265:R1344-R1350.

Leeuwenburgh, C., R. Fiebig, R. Chandwaney, and L.L. Ji. 1994. Aging and exercise training in skeletal muscle: Response of glutathione and antioxidant enzyme systems. *American Journal of Physiology* 267:R439-R445.

Leeuwenburgh, C., J. Hollander, S. Leichtweis, R. Fiebig, M. Gore, and L.L. Ji. 1997. Adaptations of glutathione antioxidant system to endurance training are tissue and muscle fiber specific. *American Journal of Physiology* 272:R363-R369.

Leeuwenburgh, C., and L.L. Ji. 1995. Glutathione depletion in rested and exercised mice: Biochemical consequence and adaptation. *Archives of Biochemistry and Biophysics* 316:941-949.

Leeuwenburgh, C., and L.L. Ji. 1996. Alteration of glutathione and antioxidant status with exercise in unfed and refed rats. *Journal of Nutrition* 126:1833-1843.

Leeuwenburgh, C., and L.L. Ji. 1998. Glutathione and glutathione ethyl ester supplementation of mice alter glutathione homeostasis during exercise. *Journal of Nutrition* 128:2420-2426.

Leibovitz, B., M.L. Hu, and A.L. Tappel. 1990. Dietary supplements of vitamin E, beta-carotene, coenzyme Q10 and selenium protect tissues against lipid peroxidation in rat tissue slices. *Journal of Nutrition* 120:97-104.

Leichtweis, S.B., C. Leeuwenburgh, D. Parmelee, R. Fiebig, and L.L. Ji. 1997. Rigorous swim training impairs mitochondrial function in post-ischaemic rat heart. *Acta Physiologica Scandinavica* 160:139-148.

Lew, H., S. Pyke, and A. Quintanilha. 1985. Change in the glutathione status of plasma, liver and muscle following exhaustive exercise in rats. *FEBS Letters* 185:262-266.

Lu, S.C., J.L. Gee, J. Kulenkamp, and N. Kaplowitz. 1992. Insulin and glucocorticoid dependence of hepatic γ-glytamylcysteine synthetase and glutathione synthesis in the rat. *Journal of Clinical Investigation* 90:260-269.

Lu, S.C., J. Kulenkamp, C. Garcia-Ruiz, and N. Kaplowitz. 1991. Hormone-mediated downregulation of hepatic glutathione synthesis in the rat. *Journal of Clinical Investigation* 88:260-269.

Luhtala, T., E.B. Roecher, T. Pugh, R.J. Feuers, and R. Weindruch. 1994. Dietary restriction attenuates age-related increases in rat skeletal muscle antioxidant enzyme activities. *Journal of Gerontology* 49:B321-B328.

Lukaski, H., B.S. Hoverson, S.K. Gallagher, and W.W. Bolonchuck. 1990. Physical training and copper, iron, and zinc status of swimmers. *American Journal of Clinical Nutrition* 51:1093-1099.

Machlin, L.J., and A. Bendich. 1987. Free radical tissue damage: Protective role of antioxidant nutrients. *FASEB Journal* 1:441-445.

Marin, E., O. Hanninen, D. Muller, and W. Klinger. 1990. Influence of acute physical exercise on glutathione and lipid peroxide in blood of rat and man. *Acta Physiologica Hungarica* 76:71-76.

Mascio, P.D., M.E. Murphy, and H. Sies. 1991. Antioxidant defense systems: The role of carotenoids, tocopherols, and thiols. *American Journal of Clinical Nutrition* 53:194S-200S.

Matsuo, M. 1993. Age-related alterations in antioxidant defense. In: *Free Radicals in Aging*, B.P. Yu (ed.), pp. 143-181. CRC Press, Boca Raton, FL.

McCord, J.M., and I. Fridovich. 1969. Superoxide dismutase. *Journal of Biological Chemistry* 244:6049-6055.

Meister, A. 1991. Glutathione deficiency produced by inhibition of its synthesis, and its reversal: Applications in research and therapy. *Pharmacology and Therapeutics* 51:155-194.

Meister, A., and M.E. Anderson. 1983. Glutathione. *Annual Review of Biochemistry* 52:711-760.

Mena, P., M. Maynar, J.M. Gutierrez, J. Maynar, J. Timon, and J.E. Campillo. 1991. Erythrocyte free radical scavenger enzymes in bicycle professional racers: Adaptation to training. *International Journal of Sports Medicine* 12:563-566.

Meydani, M., and W.J. Evans. 1993. Free radicals, exercise, and aging. In: *Free Radicals in Aging*, B.P. Yu, (ed.), pp. 183-204. CRC Press, Boca Raton, FL.

Moscow, J.A., C.S. Morrow, R. He, G.T. Mullenbach, and K.H. Cowan. 1992. Structure and function of the 5′-flanking sequence of the human cytosolic selenium-dependent glutathione peroxidase gene (hgpx 1). *Journal of Biological Chemistry* 267:5949-5958.

Nanji, A.A., B. Griniuviene, S.M.H. Sadrzadeh, S. Levitsky, and J.D. McCully. 1995. Effect of type of dietary fat and ethanol on antioxidant enzyme mRNA induction in rat liver. *Journal of Lipid Research* 36:736-744.

Niki, E., N. Noguchi, H. Tsuchihashi, and N. Gotoh. 1995. Interaction among vitamin C, vitamin E, and beta-carotene. *American Journal of Clinical Nutrition* 62:1322S-1326S.

Nohl, H. 1993. Involvement of free radicals in aging: A consequence or cause of senescence. *British Medical Bulletin* 49:653-667.

Nohl, H., and D. Hegner. 1978. Response of mitochondrial superoxide dismutase, catalase, and glutathione peroxidase activities to aging. *Mechanisms of Ageing and Development* 11:145-151.

Norman, B., A. Sovelli, L. Kaijser, and E. Jansson. 1987. ATP breakdown products in human muscle during prolonged exercise to exhaustion. *Clinical Physiology* 7:503-510.

Novelli, G.P., S. Falsini, and G. Braccioti. 1991. Exogenous glutathione increases endurance to muscle effort in mice. *Pharmacology Research* 23:149-155.

Oh-ishi, S., T. Kizaki, J. Nagasawa, T. Izawa, T. Komabayashi, N. Nagata, K. Suzuki, N. Taniguchi, and H. Ohno. 1997. Effects of endurance training on superoxide dismutase activity, content, and mRNA expression in rat muscle. *Clinical and Experimental Pharmacology and Physiology* 24:326-332.

Oh-ishi, S., T. Kizaki, H. Yamashita, N. Nagata, K. Suzuki, N. Taniguchi, and H. Ohno. 1996. Alteration of superoxide dismutase iso-enzyme activity, content, and mRNA expression with aging in rat skeletal muscle. *Mechanisms of Ageing and Development* 84:65-76.

Ohno, H., K. Suzuki, J. Fujii, H. Yamashita, T. Kizaki, S. Oh-ishi, and N. Taniguchi. 1994. Superoxide dismutases in exercise and disease. In: *Exercise and Oxygen Toxicity*, C.K. Sen, L. Packer, and O. Hanninen (eds.), pp. 127-161. Elsevier Science, New York.

Ohno, H., T. Yahata, Y. Sato, K. Yamamura, and N. Taniguchi. 1988. Physical training and fasting erythrocyte activities of free radical scavenging enzyme systems in sedentary men. *European Journal of Applied Physiology* 57:173-176.

Packer, L. 1986. Oxygen radicals and antioxidants in endurance exercise. In: *Biochemical Aspects of Physical Exercise*, G. Benzi, L. Packer, and N. Siliprandi (eds.), pp. 73-92. Elsevier Science, New York.

Packer, L. 1991. Protective role of vitamin E in biological systems. *American Journal of Clinical Nutrition* 53:1050S-1055S.

Packer, L., A.L. Almada, L.M. Rothfuss, and D.S. Wilson. 1989. Modulation of tissue vitamin E levels by physical exercise. *Annals of the New York Academy of Sciences* 570:311-321.

Packer, L., K. Gohil, B. DeLumen, and S.E. Terblanche. 1986. A comparative study on the effects of ascorbic acid deficiency and supplementation on endurance and mitochondrial oxidative capacities in various tissues of the guinea pig. *Comparative Biochemistry and Physiology* 83B:235-240.

Packer, L., E.H. Witt, and H.J. Tritschler. 1995. α-Lipoic acid as a biological antioxidant. *Free Radical Biology and Medicine* 19:227-250.

Powers, S.K., D. Criswell, J. Lawler, L.L. Ji, D. Martin, R. Herb, and G. Dudley. 1994. Influence of exercise intensity and duration on antioxidant enzyme activity in skeletal muscle differing in fiber type. *American Journal of Physiology* 266:R375-R380.

Quintanilha, A.T. 1984. The effect of physical exercise and/or vitamin E on tissue oxidative metabolism. *Biochemical Society Transactions* 12:403-404.

Quintanilha, A.T., and L. Packer. 1983. Vitamin E, physical exercise and tissue oxidative damage. *Ciba Foundation Symposium* 101:56-69.

Quintanilha, A.T., L. Packer, J.M.S. Davies, T. Racanelli, and K.J.A. Davies. 1982. Membrane effects of vitamin E deficiency: Bioenergetics and surface charge density studies of skeletal muscle and liver mitochondria. *Annals of the New York Academy of Sciences* 399:32-47.

Radák, Z., K. Asano, M. Inoue, T. Kizaki, S. Oh-ishi, K. Suzuki, N. Taniguchi, and H. Ohno. 1995. Superoxide dismutase derivative reduces oxidative damage in skeletal muscle of rats during exhaustive exercise. *Journal of Applied Physiology* 79:129-135.

Reed, D. 1986. Regulation of reductive processes by glutathione. *Biochemical Pharmacology* 35:7-13.

Reid, M.B., D.S. Stokic, S.M. Koch, F.A. Khawli, and A.A. Lois. 1994. N-acetylcysteine inhibits muscle fatigue in humans. *Journal of Clinical Investigation* 94:2468-2474.

Remacle, J., D. Lambert, M. Raes, E. Pigeolet, C. Michiels, and O. Toussaint. 1992. Importance of various antioxidant enzymes for cell stability. *Biochemical Journal* 286:41-46.

Robertson, J.D., R.J. Maughan, G.G. Duthie, and P.C. Morrice. 1991. Increased blood antioxidant systems of runners in response to training. *Clinical Science* 80:611-618.

Sahlin, K., K. Ekberg, and S. Cizinsky. 1991. Changes in plasma hypoxanthine and free radical markers during exercise in man. *Acta Physiologica Scandinavica* 142:273-281.

Sastre, J., M. Asensi, E. Gasco, F.V. Pallardo, J. Ferrero, T. Furukawa, and J. Vina. 1992. Exhaustive physical exercise causes oxidation of glutathione status in blood: Prevention by antioxidant administration. *American Journal of Physiology* 263:R992-R995.

Sen, C.K. 1995. Oxidants and antioxidants in exercise. *Journal of Applied Physiology* 79:675-686.

Sen, C.K., M. Atalay, and O. Hanninen. 1994a. Exercise-induced oxidative stress: Glutathione supplementation and deficiency. *Journal of Applied Physiology* 77:2177-2187.

Sen, C.K., E. Marin, M. Kretzschmar, and O. Hanninen. 1992. Skeletal muscle and liver glutathione homeostasis in response to training, exercise and immobilization. *Journal of Applied Physiology* 73:1265-1272.

Sen, C.K., T. Rankinen, S. Vaisanen, and R. Rauramaa. 1994b. Oxidative stress following human exercise: Effect of N-acetylcysteine supplementation. *Journal of Applied Physiology* 76:2570-2577.

Sevanian, A., K.J.A. Davies, and P. Hochstein. 1985. Conservation of vitamin C by uric acid in the blood. *Free Radical Biology and Medicine* 1:117-124.

Shimomura, Y., M. Suzuki, S. Sugiyama, Y. Hanaki, and T. Ozawa. 1991. Protective effect of coenzyme Q10 on exercise-induced muscular injury. *Biochemical and Biophysical Research Communications* 176:349-355.

Shindoh, C., A. DiMarco, A. Thomas, P. Manubay, and G. Supinski. 1990. Effect of N-acetylcysteine on diaphragm fatigue. *Journal of Applied Physiology* 68:2107-2113.

Sies, H. 1985. *Oxidative Stress*. Academic Press, London.

Sies, H., and P. Graf. 1985. Hepatic thiol and glutathione efflux under the influence of vasopressin, phenylephrine and adrenaline. *Biochemical Journal* 226:545-549.

Sjodin, B., H. Westing, and S. Apple. 1990. Biochemical mechanisms for oxygen free radical formation during exercise. *Sports Medicine* 10:236-254.

Sohal, R.S., and R. Weindruch. 1996. Oxidative stress, caloric restriction, and aging. *Science* 273:59-63.

Starnes, J.W., G. Cantu, R.P. Farrar, and J.P. Kehrer. 1989. Skeletal muscle lipid peroxidation in exercise and food-restricted rats during aging. *Journal of Applied Physiology* 67:69-75.

Stio, M., T. Iantomasi, F. Favilli, P. Marraccini, B. Lunghi, M.T. Vincenzini, and C. Treves. 1994. Glutathione metabolism in heart and liver of the aging rat. *Biochemical and Cellular Biology* 72:58-61.

Sumida, S., K. Tanaka, H. Kitao, and F. Nakadomo. 1989. Exercise-induced lipid peroxidation and leakage of enzymes before and after vitamin E supplementation. *International Journal of Biochemistry* 21:835-838.

Tateishi, N., T. Higashi, A. Naruse, K. Nakashima, and Y. Sakamoto. 1977. Rat liver glutathione: Possible role as a reservoir of cysteine. *Journal of Nutrition* 107:51-60.

Tiidus, P.M., and M.E. Houston. 1993. Vitamin E status does not affect the responses to exercise training and acute exercise in female rats. *Journal of Nutrition* 123:834-840.

Tiidus, P.M., J. Pushkarenko, and M.E. Houston. 1996. Lack of antioxidant adaptation to short-term aerobic training in human muscle. *American Journal of Physiology* 271:R832-R836.

Vega, J.A., C. Cavallotti, W.L. Collier, G. DeVincentis, I. Rossodivita, and F. Amenta. 1992. Changes in glutathione content and localization in rat heart as a function of age. *Mechanisms of Ageing and Development.* 64:37-38.

Vertechy, M., M.B. Cooper, O. Chirardi, and M.T. Ramacci. 1989. Antioxidant enzyme activities in heart and skeletal muscle of rats of different ages. *Experimental Gerontology* 24:211-218.

Vihko, V., A. Salminen, and J. Rantamaki. 1978. Oxidative lysosomal capacity in skeletal muscle of mice after endurance training. *Acta Physiologica Scandinavica* 104:74-79.

Vina, J., J. Sastre, M. Asensi, and L. Packer. 1995. Assay of blood glutathione oxidation during physical exercise. *Methods in Enzymology* 251:237-243.

Visner, G.A., W.C. Dougall, J.M. Wilson, I.M. Burr, and H.S. Nick. 1990. Regulation of manganese superoxide dismutase by lipopolysaccharide, interleukin-1, and tumor necrosis factor. *Journal of Biological Chemistry* 265:2856-2864.

Weglicki, W.B., Z. Luna, and P.P. Nair. 1969. Sex and tissue specific differences in concentrations of α-tocopherol in mature and senescent rats. *Nature* 221:185-186.

Whitsett, J.A., J.C. Clark, J.R. Wispe, and G.S. Pryhuber. 1992. Effects of TNF-alpha and phorbol ester on human surfactant protein and Mn SOD gene transcription in vitro. *American Journal of Physiology* 262:688-693.

Yu, B.P. 1994. Cellular defenses against damage from reactive oxygen species. *Physiological Reviews* 74:139-162.

Zhang, N. 1996. Characterization of the $5'$-flanking region of the human Mn SOD gene. *Biochemical and Biophysical Research Communications* 220:171-180.

Chapter 3

The Role of Antioxidant Nutrition in Exercise and Aging

Eli Carmeli
Physical Therapy Program, Sackler Faculty of Medicine, Tel-Aviv University, Ramat Aviv, Israel

Gila Lavian
Laboratory of Clinical Neurophysiology, Rambam Medical Center and Technion, Faculty of Medicine, Haifa, Israel

Abraham Z. Reznick*
Department of Anatomy and Cell Biology, The Bruce Rappaport Faculty of Medicine, Technion, Haifa, Israel

As early as the mid-18th century, Lavoisier observed that a candle cannot burn and an aerobic organism cannot live without oxygen. Indeed, oxidation was once considered a "burning" process. In chemical terms, oxidation is the reaction of an element with oxygen. Modern chemistry defines oxidation as the removal of electrons from an element or a substrate, whereas gain of electrons by a substrate is called reduction. Reduction always accompanies oxidation. The reaction between two elements or molecules in which one donates and the other accepts electrons is called an oxidation-reduction reaction, with the two reactants forming an oxidation-reduction couple.

The same principles apply to biochemical systems. The magnitude of the reactants' tendency to donate or accept electrons may be evaluated by measuring the redox potential (E_0^+) of the oxidation-reduction system as compared to a standard hydrogen electrode potential. Table 3.1 displays redox potential values of some biological oxidation systems (Mayes, 1996).

*Acknowledgment: The work and report described in this chapter was supported by the KROL Foundation, Lakewood, NJ, USA.

Table 3.1   Displays Redox Potentials Values of Some Biological Oxidation Systems

System	E_0 volts
H^+/H_2	−0.42
$NAD^+/NADH$	−0.32
Lipoate; ox/red	−0.29
Acetoacetate/3-hydroxybutyrate	−0.27
Pyruvate/lactate	−0.19
Oxaloacetate/malate	−0.17
Fumarate/succinate	+0.03
Cytochrome b; Fe^{3+}/Fe^{2+}	+0.08
Ubiquinone; ox/red	+0.10
Cytochrome c_1; Fe^{3+}/Fe^{2+}	+0.22
Cytochrome a; Fe^{3+}/Fe^{2+}	+0.29
Oxygen/water	+0.82

Oxidative Stress in Physiological Conditions and Its Consequences

Biological oxidation systems include a series of reactions initiated by the removal of hydrogen from the substrate, followed by its multiple transfer via several redox couples, and its final "attachment" to molecular oxygen. If the specific process is respiration, the electrons, carried by hydrogen atoms, will be transferred through the cytochrome system to be attached to oxygen. Electrons pass from one redox couple to another, leaving the reduced member of one couple and passing to the oxidized member of another one characterized by a higher redox potential. Oxygen, which has the highest redox potential, is the final hydrogen acceptor, which also concludes the respiratory chain. Respiration may be defined as the process by which energy, in the form of adenosine triphosphate (ATP) molecules, is produced from a substrate in the form of ATP molecules, by aerobic organisms using oxygen.

It is hard to believe, therefore, that oxygen, the essential element for life, may harm the organism. Indeed, the phenomenon has been called the "oxygen paradox." The atomic properties of oxygen may help explain the paradox. Molecular oxygen that consists of two atoms may behave as a free radical.

Free radicals are defined as any chemical species that has one or more unpaired electrons. According to this definition, the hydrogen element with one unpaired electron spinning around its single proton nucleus and most of the transient metal ions are free radicals (Halliwell, 1998). Adding a single electron to a molecule of oxygen creates a reduction process that occurs easily in the tissues, and a series of reactive structures such as O_2^-, H_2O_2, and HO• may lead to the generation of additional reactive oxygen species (ROS). The respiratory chain itself is, indeed, a source of ROS: About 2 to 4% of the reacting oxygen may be oxidized to superoxide radical O_2^-.

All these reactive elements, but especially the hydroxyl radical HO•, may produce oxidative damage to other molecular components of the cell, including peroxidation of membrane phospholipids, modification of nuclear deoxyribonucleic acid (DNA), or alteration of proteins causing enzymatic changes and proteolysis. The challenge presented by the reactive species, ROS, to the different cellular components is called oxidative stress (Halliwell and Gutteridge, 1990; Sohal and Weindrich, 1996; Stadtman, 1992).

However, the enzymatic processes of the respiratory chain are not the only sources of free radical production. Additional sources of free radicals in vivo include ionizing radiation (including x-radiation, ultraviolet [UV] light, ultrasound, and microwave; Sies, 1993), phagocytosis, prostaglandin synthesis, and nonenzymatic reactions of oxygen with organic compounds (Halliwell and Gutteridge, 1992).

In spite of its extremely short (10^{-9} s) half-life (Sies, 1993), the hydroxyl radical, HO•, seems to be the most aggressive and damaging of all other species. Possible sources of this metabolite are radiolysis of water as well as Fe^{2+}- or Cu^{2+}-catalyzed cleavage of hydrogen peroxide (H_2O_2). The Fe^{2+}-dependent decomposition of H_2O_2 is called the Fenton reaction (Berlett and Stadtman, 1997; Halliwell and Gutteridge, 1990, 1992).

The anion free radical, O_2^-, is formed in almost all aerobic cells. It may be produced accidentally, for example, by leakage of electrons from electron transport chains in mitochondria, chloroplasts, or endoplasmic reticulum, or it may be produced purposefully by cells such as phagocytes, lymphocytes, endothelial cells, and fibroblasts (Halliwell and Gutteridge, 1990). This radical may react as both oxidant and reductant. It may be removed in the presence of a specific enzyme, superoxide dismutase (SOD), which results in the formation of hydrogen peroxide, another highly reactive ROS, and molecular oxygen:

3.1 $$2 \cdot O_2^- + 2H^+ \rightleftharpoons H_2O_2 + O_2$$

Hydrogen peroxide, by definition, is not a radical because it has no unpaired electrons.

However, pure H_2O_2 easily crosses biological membranes and may give rise to other ROS. Hydrogen peroxide is mainly produced via the reaction shown in equation 3.1 in cells such as bacteria, phagocytic cells, spermatozoa, and others. Cellular organelles such as mitochondria and chloroplasts or microsomal cell fraction preparation are sources of H_2O_2 as well (Halliwell and Gutteridge, 1990). H_2O_2 in vivo may also result from the activity of oxidative enzymes such as amino acid oxidase or xanthine oxidase.

Hydrogen peroxide may be enzymatically removed by the action of catalase or peroxidases such as glutathione peroxidase (GPX), thiol-specific peroxidases, and other peroxidases. Catalase converts peroxide to oxygen and water, as shown in equation 3.2.

3.2 $2H_2O_2 \Leftrightarrow 2H_2O + O_2$

However, in the presence of iron, H_2O_2 may generate hydroxyl radicals even at physiological levels.

The hydroxyl radical reacts with almost all biological molecules; however, one of its favorite targets is the DNA molecule, where it causes multiple modifications involving all four purine and pyrimidine bases. DNA changes are produced as well by other oxygen-derived species, such as singlet $^1O_2$ and peroxyl radicals; however, these metabolites exclusively attack guanidine (Wagner et al., 1992). Such changes in the DNA bases may cause mutagenesis and carcinogenesis. The excretion of modified DNA bases in human urine is consistent with the view that DNA bases are altered in vivo. The extent of DNA damage per cell caused by ROS is estimated to be 10,000 modified bases per day (Ames et al., 1991). The organism is normally provided with recognition and repair systems for altered DNA.

An additional remarkable effect of HO•, the hydroxyl radical, is the initiation of a peroxidative chain reaction of polyunsaturated fatty acids (PUFA), which are components of the cell membranes. The result of these reactions, catalyzed by iron, is the formation of lipid hydroxyl, alkoxyl, and peroxyl radicals. If uncontrolled, these reactions may spread, causing denaturation of membrane lipids, which may lead to damage of phospholipid dependent enzymes, loss of membrane selective permeability, and lysis of the membrane (Williams, 1995).

Cellular proteins are another target for free radical attacks. In the past, less importance has been accorded to the age-related increase in protein damage. This may be explained by the fact that a constant protein turnover takes place in vivo, including normal as well as abnormally formed or damaged proteins; the turnover of the abnormally formed or damaged proteins is even more rapid than that of normal proteins (Stadtman, 1992).

Protein modifications are mainly initiated by the hydroxyl radical; however, the availability of O_2 itself or other oxygen free radicals, such as $\bullet O_2^-$ or its protonated form $HO\bullet_2^-$, will determine the course of the reactions (Berlett and Stadtman, 1997).

Proteins may be affected during oxidation in several ways:

a. Oxidation of the protein backbone, leading to protein fragmentation

b. Formation of protein-protein cross linkages

c. Oxidation of amino acid side chains

d. Generation of carbonyl derivatives

Oxidation of the Protein Backbone

The polypeptide backbone attack by the hydroxyl radical starts by abstraction of a hydrogen atom of an amino acid residue to form a carbon-centered radical $(RC\bullet)$. In the presence of O_2, the reaction chain is further propagated through the formation of an alkylperoxyl radical, an alkylperoxide, and an alkoxyl radical $(RO\bullet)$, which will be converted to a hydroxyl protein derivative.

Peroxyl radicals, $ROO\bullet$, are relatively stable, with a half-life of 7 s (Sies, 1993). However, these molecules may diffuse, carrying the radical oxidant to act as oxidant and oxidize other molecules to other target sites. Alkoxyl radicals, as well, may start new peroxidation cycles by abstracting further hydrogen from amino acid residues. They may also lead to the cleavage of peptide bonds, which causes protein fragmentation.

The production of the hydroxyl radical itself from H_2O_2, as mentioned previously, as well as the reactions responsible for the formation of the hydroxyl protein via the intermediary radicals, also may be catalyzed by the metal ions Fe^{2+} or Cu^{2+}.

Formation of Protein-Protein Cross Linkages

The intermediate radicals formed by oxidation of the protein backbone may undergo side reactions with other amino acid residues to generate new carbon-centered radicals, which in the absence of O_2 may react one with another to form protein-protein cross-linked derivatives $(R^1CC\ R^2)$.

Oxidation of Amino Acid Side Chains

All amino acid residues of proteins are susceptible to oxidation by $HO\bullet$. However, the two sulfur-containing amino acid residues, cysteine and methionine, are particularly sensitive to oxidation. Oxidative conditions, even if they are mild, will lead to the production of disulfides and methionine sulfoxide (MeSOX), respectively, from these amino acid residues.

The redox status of the sulfhydryl groups may affect the structure and/or biological activity of several enzymes including glucose 6-phosphate dehydrogenase (G-6PD), phosphoglycerate kinase, glyceraldehyde-3-phosphate dehydrogenase (Brodie and Reed, 1990; Fernando et al., 1992), lipooxygenase (Levine et al., 1996), pyruvate kinase and other enzymes (Axelsson and Mannervik, 1983; Hyslop et al., 1988), and receptor (Bhatnagar et al., 1990) or hormone activity, for example, human parathyroid hormone (Nabuchi et al., 1995). Assigning such important roles to the sulfhydryl groups implies a special "attitude" of the organism concerning its redox status. Indeed, the oxidation of cysteine and methionine residues is the only modification of proteins that can be repaired, most biological systems containing disulfide reductase and MeSOX reductase for this purpose. Moreover, it has been proposed that the cyclic oxidation-reduction of methionine may protect against irreversible modification (Berlett and Stadtman, 1997; Levine et al., 1996).

Cysteine and methionine are also characterized by a particular vulnerability to oxidation by peroxynitrite, another free radical species, formed via the reaction of O_2^- with nitric oxide (NO). Equation 3.3 illustrates one source of peroxynitrite:

3.3 $\quad \cdot O_2^- + NO\cdot \Rightarrow ONOO^-$

NO• is a normal product of arginine metabolism, which has an important role in a variety of crucial functions in the organism such as synaptic transmission involved in the process of memory, smooth muscle relaxation, and macrophage killing activity (Halliwell and Gutteridge, 1995; Kandel and Hawkins, 1992).

Aromatic amino acid residues are also favorite targets for ROS species as well as for peroxynitrite. The list includes phenylalanine, tyrosine, and tryptophan, which undergo oxidation and form hydroxyderivatives, and histidine, which when oxidized converts asparagine to aspartic acid residue.

An extremely important reaction is the nitration of tyrosine residues. This reaction is a prerequisite for the cyclic interconversion between the phosphorylated and the unphosphorylated forms of tyrosine (Hunter, 1995). This conversion is crucial for signal transduction in the cell.

Generation of Protein Carbonyl Derivatives

An additional oxidative challenge may be presented to proteins by a series of substances containing carbonyl groups. These substances themselves result from oxidative processes. The list of carbonyl-containing agents includes reactive aldehydes (RA), products of lipid peroxidation such as 4-hydroxy-2-nonenal and malondialdehyde (MDA), or substances that result from the reduction of sugars or their oxidation products with lysine residues. An additional source for carbonyl derivatives of proteins is metal catalyzed oxidation of side chain amine groups via a Fenton-like reaction. The formation of carbonyl-containing proteins has been used to detect and evaluate the extent of ROS-mediated protein oxidation (Berlett and Stadtman, 1997).

The preceding data may lead to the following conclusions:

a. The production of reactive oxidant species may occur under normal aerobic conditions, even following ordinary metabolism.

b. Considering the extensive damage that may be produced, unlimited production of oxidative species is incompatible with life.

c. The very existence of life implies that organisms have efficient defense systems against oxidative stress. Indeed, a diversity of strategies enable organisms to face oxidative challenge.

Following are several examples of such strategies, some are more active whereas some more passive, whose aim is to reduce or eliminate oxidative damage:

a. Avoidance of factors that cause the formation of free radicals is best exemplified in the behavior of plankton, which leaves the ocean surface and descends to its depth to avoid high levels of solar radiation (Sies, 1993).

b. Another procedure to prevent oxidative attack involves DNA, which is isolated from reactive oxidative agents by being "packed" into a chromatin wrapping. Moreover, less important alternative targets are exposed to encounter the attack launched by oxidative species.

c. An additional example of an active procedure may be found in the microorganism *E. coli*. These bacteria secrete catalase that defends them from the H_2O_2 attack of phagocytic cells (Haas and Goebel, 1992). Only colonies are protected, but not individual cells. This may imply that the need for protection, better achieved by grouping of cells, could have played a role in the transition from unicellular to multicellular organisms during evolution (Ma and Eaton, 1992).

d. Additional defense strategies may be seen in vital enzymes such as cytochrome oxidase or ribonucleotide reductase. Despite being endowed with the property to reduce oxygen, the cytochrome oxidase will not release superoxide or other radicals, even though it contains Fe^{2+} and Cu^{2+}. The ribonucleotide reductase contains, as well, a component able to form radicals: the tyrosyl group. However, similar to DNA, the tyrosyl group is also contained in an appropriate wrapping.

e.   An example of a more active defense procedure is the action of protein metal chelators, such as ferritin, transferrin, ceruloplasmin, and metallothionin, useful in particular for trapping Fe^{2+} and Cu^{2+}, which initiate lipid peroxidation and DNA fragmentation.

f.   An extremely interesting defense mechanism performed by labile tissue cells involves the intestinal mucosa. These cells, exposed to a multitude of damaging agents and foreign substances, the xenobiotica, accumulate harmful products, preventing them from spreading to other cells. The frequent elimination of these whole cells from the organism, due to the high turnover of the tissue, removes the damaging agents.

These antioxidative strategies will, indeed, lower oxidative challenge, but their efficiency is far from being 100%. Therefore, the organism must use its different repair systems to reduce damage.

The arsenal of the organism charged with fighting oxidative stress is composed of specialized substances generally called antioxidants. An antioxidant may be defined as "any substance that, when present at low concentrations compared to those of an oxidizable substrate, significantly delays or inhibits oxidation of that substrate" (Halliwell and Gutteridge, 1999, p. 106). A wider definition is "any compound that protects biological systems against the potential harmful effect of processes or reactions that can cause excessive oxidation" (Krinsky, 1992, p. 248). Antioxidants may be enzymatic or nonenzymatic compounds. The enzymes are SOD, catalase (CAT), and GPX. The nonenzymatic antioxidants may be lipid- or water-soluble compounds. Examples of lipid-soluble antioxidants are α-tocopherol (vitamin E), α-tocotrienol, β-carotene, ubiquinol, bilirubin, and 2-hydroxy estrone or estradiol. Ascorbic acid (vitamin C), uric acid, glutathione (GSH), cysteine, and creatinine function as water-soluble anti-oxidants (Krinsky, 1992). As explained in the following sections, a whole series of compounds of different origins may be available for antioxidative defense.

It is tempting to categorize oxidants as the "bad guys" and antioxidants as the "good guys." However, such a categorization is false and oversimplified, because neither element is all good or all bad.

The early stages of evolution may help explain the present status of various organisms. At the beginning of life on earth, the atmosphere contained very little oxygen. The first beings were, indeed, anaerobic microorganisms. The later development of photosynthesizing organisms enriched the atmosphere with O_2. According to evolutionary principles, this change probably killed most anaerobic species, sparing only those that could defend themselves against or adapt to the new environmental challenges. The means for survival included the development of appropriate tools against oxidative stress. Here, most likely, are the origins of the antioxidant strategies in general and specific compounds in particular.

Moreover, aerobic organisms turned the "newly arrived" O_2 to their profit, by using it in several vital metabolic processes. Energy production, where oxygen serves as the final electron acceptor in the respiratory chain, is the most impressive example (Halliwell, 1998). These evolutionary facts lead to two assumptions:

   a. Oxidants are not all bad; on the contrary, they may trigger or even be involved actively in certain vital functions.
   b. Antioxidants are beneficial as long as they do not exceed a certain critical level, beyond which they may be involved in pathophysiological processes.

Following are a few examples of biological processes that may describe "the schizophrenic nature" (McCord, 1995) of the radical metabolites and the dual behavior of antioxidants. The ambivalence of the superoxide radical may be seen in its paradoxical role of initiating but also interrupting lipid peroxidation. As we have mentioned, the oxidation of membrane PUFA has a deleterious effect on cell functions. The oxidation process occurs in PUFA in a similar way as it occurs in proteins. The oxidant species abstract an allylic hydrogen from a polyunsaturated side chain and create a lipid radical. In the presence of O_2, a lipid dioxyl radical, LOO, is produced. Abstraction of an additional hydrogen from another PUFA side chain results in a lipid hydroxyperoxide, LOOH, and a new lipid radical, L•. The lipid radical will attack again and perpetuate the chain. Ferrous ion has an important role in this propagation process. It can reduce the lipid hydroperoxide molecule (LOOH) and produce the alkoxyl radical, LO•, which will itself initiate an additional chain reaction. Any oxidative species can start the first reaction; however, excessive superoxide radical may supply ferrous iron, by mobilizing it from reservoirs such as ferritin or by reducing the ferric ion (Fe^{3+}), and indirectly contribute to the propagation of lipid peroxidation (McCord, 1995).

3.4 $O_2^- + \text{Ferritin-Fe}^{3+} \Rightarrow O_2 + \text{Ferritin} + Fe^{2+}$

3.5 $O_2^- + Fe^{3+} \Rightarrow O_2 + Fe^{2+}$

Antioxidant agents may interrupt the chain reaction by quenching the intermediary products. Glutathione peroxidase, for example, may prevent the production of alkoxyl via the reaction of lipid hydroperoxide (LOOH) with Fe^{2+}, by reducing it to alcohol, whereas vitamins E and C may scavenge the lipid dioxyl radical (LOO•). The paradox is that the same superoxide radical that indirectly propagates lipid peroxydation reactions may also interrupt the chain of lipid peroxidations. This will be the result of "pairing" its free electron with the free electron of the radical LO• or LOO•, thus eliminating the aggressive behavior of the radical.

3.3 $LO• + O_2^- + H^+ \Rightarrow LOH + O_2$

3.3 $LOO• + O_2^- + H^+ \Rightarrow LOOH + O_2$

Hence, an adequate supply of this relatively unreactive radical, O_2^-, will ensure the scavenging of lipid radicals and prevent propagation of the chain reaction. As shown in equation 3.1, SOD is the enzyme responsible for removing $•O_2^-$. An excess of SOD will decrease the concentration of the superoxide radical and thus perpetuate lipid peroxidation. Hence, too much of the "good" antioxidant is not desirable. On the other hand, if SOD levels are too low, an excess of $•O_2^-$ will supply Fe^{2+} ions that will propagate the oxidative reactions (Aust et al., 1985; Halliwell and Gutteridge, 1985). Experimental data confirm the hypothesis of an optimal SOD level (Nelson et al., 1994).

One of the well-studied dual effects of the superoxide radical involves killing bacteria but also producing inflammatory reactions. This apparent paradox may be explained as follows: The superoxide radical, as mentioned previously, is the only ROS produced "by design" (McCord, 1995). The cells responsible for its production are the phagocytic cells, which are endowed with bactericidal properties. A genetic disorder called chronic granulomatosis disease (CGD), characterized by a reduced $•O_2^-$ formation by neutrophils, is manifested by multiple recurrent local infections, leading to septicemia and early death (Hohn and Lehrer, 1975).

Hence, the principal role of the neutrophil is, indeed, killing by using arms such as $•O_2^-$, H_2O_2, and hypochlorous acid (HClO). Considering the life-threatening condition of unopposed bacterial development in the organism, the killing role of the neutrophils must be a massive and not a selective role. Therefore, unfortunately, host cells will also succumb to their injuries, resulting in the classic inflammatory reaction characterized by swelling, redness, heat, and pain (Markert et al., 1985). Autoimmune diseases are, in fact, caused by an exaggerated reaction of the organism's defense system toward an imaginary pathogenic agent, sometimes an altered protein of the organism itself, that the defense system considers a threat. In this case, superoxide will be overproduced.

One of the features of Down's syndrome is an elevated (by 50%) level of SOD, which decreases superoxide levels. Individuals with Down's syndrome are, indeed, more sensitive to bacterial infections (Anneren and Bjorksten, 1984). We may again conclude that an optimal concentration of superoxide, permitted by an optimal level of SOD, characterizes the normal state of the organism.

Another example of a dual role for oxidants and antioxidants is the apparently paradoxical role of the superoxide radical in cell growth and proliferation, apoptosis, and malignant transformation. Cell division seems to be triggered by mild oxidative stress. This may be one factor that induces repair of tissues and scar formation at wounds, where phagocytes are activated using their oxidative ammunitions (Sies et al., 1992). The proliferation of the cells ceases when the oxidative activity of the phagocytes stops. Teleologically speaking, no more repair is needed. However, some genetically modified cells may be exposed to a permanent oxidative stimulation, as a result, for example, of reduced formation of the antioxidant SOD. The cells will continuously proliferate; that is, they will undergo malignant transformation (Oberley et al., 1991). On the contrary, overproduction of the enzyme, in this case manganese-containing SOD, by genetic manipulations confers to mice a particular resistance to radiation-induced neoplastic transformation (St. Clair et al., 1992).

Apoptosis, the process of programmed cell death, which may be triggered also by malignancy, will interrupt the uncontrolled proliferation of cells by forcing them to commit suicide (Buttke et al., 1994). Hence, oxidative stress could lead to positive processes, such as cell growth and apoptosis, but also to unwanted and dangerous ones, such as malignancy.

All these examples point to the conclusion that quenching of free radicals is surely important, but a total quenching will also quench life. The important issue is not the absolute levels of oxidants or antioxidants but maintaining an adequate balance between them. Any deviation from such a balance will lead to pathologies.

An obvious graphic presentation of the ideal balance is a "golden triangle," which has oxidants, antioxidants, or biomolecules at each apex (figure 3.1). In a normal situation, a balanced equilibrium exists among the three elements of the triangle. However, any disturbance of this equilibrium, caused either by a decrease in antioxidants or by an excess of oxidants, will result in pathogenic processes. Any one of these conditions will cause oxidative stress to the biomolecules that will be manifested as cell and tissue damage (Reznick et al., 1998). More detailed description of the major antioxidants, their origins, and their means of action will shed more light on physiological situations in which this balance plays a crucial role.

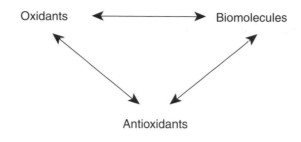

Figure 3.1 The interaction of oxidants, antioxidants, and biomolecules.

The Antioxidant Defense Systems

Concomitantly with the evolution of oxygen as a vital agent of life, reactive oxygen species turned out incompatible with life. This detrimental effect of oxygen was encountered with nature developing a myriad of anti-oxidant defense systems. These were selected by their unique biological and chemical properties to face the oxidative challenge in all levels and forms of life.

The Major Natural Antioxidants

The major natural antioxidants are depicted in figure 3.2. They can neutralize and scavenge most of the reactive oxygen species, stop lipid peroxidation, and protect proteins and nucleic acids from oxidation and damage. In general, one can divide these antioxidants into two main categories:

1. those found in the intracellular systems, and

2. those present mainly in the extracellular fluids.

The intracellular antioxidants can be further divided into two different types:

a. lipid-soluble, membrane-located antioxidants, and

b. intercellular antioxidants that are present in the cytosol and are in principle water soluble.

Antioxidants found in membranes are vitamins E and A; ubiquinol (Q_{10}); and several carotenoids, the major one β-carotene. These antioxidants are considered lipophilic in nature.

However, because the quantity of vitamin E per unit amount of lipid molecules in membranes is the highest of all membrane antioxidants, vitamin E is considered the most important lipid-soluble antioxidant (figure 3.2).

Intracellular systems membranes

Vitamin E

Ubiquinol

Carotenoids

Cytosol

Vitamin C

Glutathione

Superoxide dismutase

Catalase

Glutathione peroxidase

Glutathione-transferase

Ferritin

Extracellular fluids

Vitamin C

Uric acid

Bilirubin

Caeruloplasmin

Transferrin

Lactoferrin

Albumin

Haptoglobin

Superoxide dismutase

Lipoproteins

α, γ-tocopherols, tocotrienols

Ubiquinol-10, carotenoids

(β-carotene, lycopene, phytofluene)

Figure 3.2   The major natural antioxidants.

The cytosolic antioxidants are numerous and may be classified into three categories:

1. Small biomolecules such as gluthathione
2. Antioxidant enzymes such as catalase, superoxide dismutase, and gluthathione peroxidase
3. Iron-binding proteins such as ferritin

Because oxidative damage takes place initially and primarily in mitochondrial membranes and also in other cellular membranes, the lipid-soluble antioxidants are considered the first line of defense, whereas the aqueous-soluble antioxidants act as a second line of defense.

The extracellular fluid antioxidants are found in most body fluids such as plasma, saliva, lavage fluids, and synovial fluids. The extracellular fluid antioxidants can be found in purely soluble conditions or incorporated into lipoproteins found in plasma. The latter possess lipid-soluble properties, as shown in figure 3.2.

The Antioxidant Network Concept: Main Biological Reductants and Their Redox Potentials

Table 3.2 lists the important reductants found in biological systems and their redox potentials. The table shows that the more negative the redox

---

Table 3.2 Main Biological Reductants and Their Redox Potentials

Reductants	E_0^+ (mv)
Nicotinamide adenine dinucleotide	−320
Dihydrolipoic acid	−290
Glutathione	−240
Ascorbate	+80
Ubiquinol	+100
Trolox (soluble form of vitamin E)	+370

---

potential (E_0^+) of an antioxidant, the more potent a reductant it is. Thus, once an antioxidant acts as a reductant, it becomes oxidized. For example, as shown in figure 3.3, when vitamin E reduces lipid peroxyl radical (ROO•), vitamin E becomes a radical (vitamin E radical). A more potent reductant such as ascorbate (vitamin C) can then reduce the vitamin E radical back to the reduced form of vitamin E in the form of tocopherol or tocotrienol. This has been termed the "recycling" of vitamin E, and the interaction of vitamins with each other in a cascade-like manner has been called the "network of vitamins" action (Haramaki et al., 1998). Indeed, this recycling of vitamins has been observed in several biological systems: in red blood cell membranes (Reinhold et al., 1990), in low density lipoprotein (LDL) (Kagan et al., 1992), in plasma exposed to cigarette smoke (Reznick et al., 1997), and in ischemia-reperfusion of the heart (Haramaki et al., 1998). In the heart model of ischemia and reperfusion, Haramaki et al., 1998 showed that heart tissue hydrophilic antioxidants such as ascorbate and glutathione are the first to be oxidized. In the reperfusion period, there was a marked increase of dehydroascorbate and glutathione disulfide, the oxidized form of ascorbate and glutathione. On the other hand, the lipophilic antioxidants such as ubiquinol and particularly vitamin E were oxidized only under harsh oxidative conditions, when H_2O_2 was added to the buffer solution during reperfusion. The results of the Haramaki et al. study indicated that only after all other antioxidants are depleted is vitamin E used. Thus, vitamin E may serve as the ultimate antioxidant, protecting the integrity of cellular membranes (figure 3.3).

Similar results were obtained when levels of vitamins C and E were monitored in plasma exposed to cigarette smoke (Reznick et al., 1997). GSH and dihydrolipoic acid (DHLA) could quite efficiently protect vitamins C and E from oxidation by cigarette smoke. However, only after complete disappear-

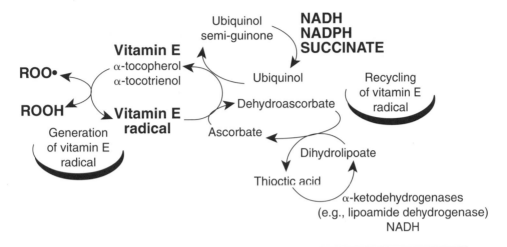

Figure 3.3 Interaction of vitamin C and ubiquinol with the vitamin E cycle.

ance of vitamin C was vitamin E depleted (Reznick et al., 1997). This graded use of antioxidants, found in the aforementioned studies, supports the redox-based antioxidant network concept found in many biological systems. The obvious implications of this concept for antioxidant nutrition is that a combined supplementation of several antioxidant vitamins is probably more efficient than supplementation of a single antioxidant vitamin.

Exogenous and Endogenous Antioxidants— The Role of Nutrition

Strong reducing capacity is the most important feature of an antioxidant needed to scavenge the unpaired electron of a free radical. Most of the antioxidant vitamins and other antioxidant compounds are based on a certain chemical structure: a derivative of an aromatic ring of polyphenols, phenols, quinols, and catecols (Karlsson, 1997). All these compounds have hydroxyl groups bound to a benzene ring-based structure. These active hydroxyl groups feature in the quenching of lipid radicals. What is unique about this benzene ring of various phenols and quinols is that these structures are synthesized exclusively in plants and microorganisms. Essentially, animals do not possess the necessary enzymes to synthesize these structures. However, animals may acquire these from essential amino acids such as phenylalanine. Thus, many antioxidants such as vitamins A and E, β-carotene, and ubiquinol are based on these phenol-like structures. This discussion shows that the above compounds can be supplied to the animal body mainly through nutrition. On the other hand, figure 3.2 shows that the list of antioxidants consists of a variety of other biomolecules such as proteins, enzymes, and small molecules, the majority of which are synthesized endogenously in the animal body. These are indeed the endogenous antioxidants that presumably handle all the oxidative challenges imposed on our biological systems. Substantial evidence shows that under several physiological and pathological conditions, among them strenuous exercise, high-altitude training, immobilization, and a variety of diseases, endogenous antioxidants do not completely prevent oxidative damage. That these situations can oxidatively damage proteins, DNA, and lipids may indicate that the balance between oxidants and antioxidants is shifted toward oxidants and that the endogenous antioxidants are not capable of neutralizing the oxidants' effects. In such cases, the role of dietary antioxidants becomes very important, because they could supply the extra antioxidant capacity needed to overcome oxidative attack (Halliwell, 1998).

Table 3.3 summarizes the main compounds and chemicals that are considered exogenous dietary antioxidants. The table shows that most dietary anti-

Table 3.3 Major Dietary Antioxidants (AO)

Dietary antioxidant	Food source	Main features
Vitamin E	Palm oil and sunflower seeds	–Main lipophilic AO –May protect against cardiovascular disease –Deficiency may cause neurodegeneration
Ubiquinol	Soybean oil	–Important lipophilic AO of LDL –May protect from atherosclerosis by attenuating LDL oxidation
Vitamin C	Citrus fruits	–Major hydrophilic AO –The main plasma AO
β-carotene and other carotenoids	Carrots, parsley, and other vegetables	–Carotenoids are good singlet-oxygen quenchers. –Precursors of vitamin A
Sulfur amino acids and allylic sulfides	Onion and garlic	–Precursors of glutathione
Flavanoids and other plants; polyphenols	Green tea, red wine, vegetables, and fruits	–Inhibit lipid peroxidation –May explain the French paradox of protection against atherosclerosis
Lycopene	Tomatoes	–General antioxidant properties

LDL, low density lipoprotein.

oxidants are derived from fruits and vegetables and may act by a variety of mechanisms. In the last few years, researchers have attempted to prove the significance and vitality of these dietary supplements in boosting endogenous antioxidant systems. This will be discussed in the following sections.

Exercise, Oxidative Stress, and Antioxidant Supplementation

Several studies have discussed the effect of physical exercise on production of oxygen radicals (Jackson, 1994). Physical exercise, which is associated with a remarkably enhanced rate of oxygen utilization, increases accumulation of free radicals as a response to the need for oxygen. Whether all forms of exercise increase oxygen radicals is yet to be proved. However, it is now well accepted that certain exercise regimens may play different roles in oxygen free radical production.

Free radical activity may be detected in various tissues affected by exercise. Skeletal muscle generates free radical activity in a variety of ways, and the level and accumulation of oxygen radicals are influenced by the mode, intensity, and duration of contraction. Skeletal muscle may contract in three biomechanical ways: When the muscle is shortened, the contraction is defined as dynamic concentric. When the muscle is lengthened, the contraction is defined as dynamic eccentric. When muscle length remains fixed, the contraction is considered static isometric. Few studies have indicated that repetitive eccentric contractions initiate more cellular damage than others (Armstrong et al., 1991; Newham et al., 1986). In addition, different energy requirements, oxygen consumption, and mechanical loads on soft tissue involved in different types of exercise affect the generation of free radicals. Also, occasional exercise, which involves extensive use of large-muscle groups, causes more tissue damage than routine exercise. The sources of free radicals are primary and secondary. The major primary sources are mitochondria, reaction of xanthine oxidase, prostanoid metabolism, and nicotinamide adenine dinucleotide phosphate (reduced form) (NADPH) oxidase activity. The mitochondria are the prime source of free radicals. As mentioned before, 2 to 4% of the total oxygen consumption of the mitochondria undergo one electron reduction with the production of the superoxide radical. One electron reduction of superoxide produces hydrogen peroxide, which is also produced by isolated mitochondria. Therefore, aerobic exercise may increase the production of superoxide radicals as well as H_2O_2. Ischemia may lead to conversion of xanthine oxidase from xanthine dehydrogenase and also in a breakdown of ATP with the formation of adenosine monophosphate (AMP) via the adenylate kinase reaction.

This enzyme is localized in the capillary epithelium of human muscle, providing a potential source for superoxide production adjacent to skeletal muscle tissue. Exercising muscles also show a failure of calcium homeostasis. Calcium-activated proteases are activated in muscle after exercise and have been shown to trigger the transition from xanthine dehydrogenase to xanthine oxidase in immobilized muscle (Kondo et al., 1993). Prostaglandins are released from muscle after exercises (McArdle and Jackson, 1994); however, their physiological role is yet to be investigated. The enzyme NADPH oxidase occurs in neutrophils and is an important source ROS in neutrophils. It is also found in muscle, but its role is not yet clear (Duncan, 1991).

Secondary sources of radicals are phagocytic processes of white cells, following muscle damage, which disrupt iron-containing proteins (i.e., myoglobin) released to the bloodstream after endurance running. Exercise may induce a mechanism that provides a source of low molecular mass iron such as ferum. Indeed, training increases the fragility of erythrocyte and induces hemolysis, gastrointestinal bleeding, and changes of serum ferritin, thus providing the source for redox active iron.

It is well accepted that aerobic metabolism generates reactive oxygen species such as superoxide anion or hydrogen peroxide, which are capable of causing oxidative stress. Because exercise elevates oxygen consumption, these reactive oxygen categories are produced during physical activity at much higher rates than at rest. Under certain circumstances, too much aerobic metabolism might cause more harm than good.

Exercise and Lipid Peroxidation

Some studies have reported an increase of lipid peroxidation in exercised muscles. However, the influence of exercise on the process of oxidative stress is controversial among researchers. Several studies reported no change in lipid peroxidation after aerobic training, yet others demonstrated changes after anaerobic, isometric, or submaximal exercise (Alessio, 1994; Ortenblad, 1997). Differences in exercise conditions (intensity, duration) and methodologies (oxidative measurements and type of samples) could be responsible for these differences in results. In summary, results concerning lipid peroxidation due to exercise are inconsistent, and different methods for determining oxidative stress may be responsible for the inconsistent data. However, most investigators agree that higher intensity and longer duration exercise may cause more lipid peroxidation.

Exercise and Oxidation of Glutathione, Proteins, and DNA

Elevations of protein oxidation, oxidation of DNA, and glutathione oxidation due to exercise have been reported. Witt et al. (1992) reported that endurance training doubled the protein carbonyl concentration of lower-limb muscles.

Reznick et al. (1992a) found a significant increase of protein carbonyl concentration after a single bout of exhaustive endurance exercise. Inoue et al. (1993) and Radák and Goto (1998) showed a marked increase of muscle oxidative DNA damage after aerobic exercise. Oxidation of glutathione during and after exercise has been reported by several authors (Hellsten et al., 1996; Ji et al., 1992; Sastre et al., 1992; Sen, 1995).

Elevation of muscle enzyme activities also has been reported as a good index of muscle damage related to oxidative stress. Kanter et al. (1988) and Atalay et al. (1996a) reported an increase of creatine kinase and lactate dehydrogenase after endurance exercise. In addition, inherent protective systems are supposed to detoxify some of the free radicals. These are physiological scavenger enzymes, which are located within cells and protect against reactive oxidative stress. However, these endogenous antioxidant enzymes may not be sufficient to combat the oxidative challenge imposed by certain types of exercise. Ji (1998) reported changes of the enzyme activities due to exercise. Therefore, some athletes and health-conscious exercisers may occasionally adopt a diet concentrated with nutritional antioxidants. How long these changes in nutritional intake are useful remains unclear. Strenuous physical exercise most probably induces oxidative stress. Short and prolonged exercise can under certain conditions overwhelm antioxidant defenses, which include vitamins E and C and thiol antioxidants, that are linked in an antioxidant network of reductant and antioxidant enzymes. Many other studies suggest that antioxidant supplementation can improve performance, and others suggest that bolstering antioxidant defenses may ameliorate exercise-induced damage in the long term rather than the short term.

Fair and Berra (1995) presented evidence that supported the use of antioxidant supplementation in treatment for coronary heart disease. Giuliani and Cestaro (1997), Fielding and Meydani (1997), and Benderitter et al. (1996) discussed the role of vitamins as part of the antioxidant system and their relationship with exercise. Alessio et al. (1997) suggested that exercise-induced oxidative stress was lower when trained subjects received daily vitamin C for 2 wk, compared to nontrained subjects. Kim et al. (1996a) found that an exercise-induced increase of antioxidant enzymes modulated the extent of free radical damage in the heart. Atalay et al. (1996b) demonstrated for the first time that supplementation of the enzyme-reduced glutathione can induce neutrophil mobilization and decrease exercise-induced leukocyte margination. Therefore, exogenous and endogenous reduced glutathione can regulate exercise-induced stimulation of the neutrophil oxidative burst.

Venditti et al. (1996) concluded that different gender susceptibility to the effect of physical exercise might be related to different capacities to oppose oxidative stress effectively; female subjects showed more decrease of antioxidant capacities after prolonged exercise than did male subjects. Clarkson (1995) recommended that endurance athletes consume antioxidants and minerals through food sources rather than as supplements. Oostenbrug et al. (1997) suggested that high-intensity jump training is associated with elevated activities of SOD and two muscle enzymes: glutathione peroxidase and glutathione reductase.

Effects of Vitamin E and Other Antioxidants

Table 3.4 summarizes research findings on the effects of vitamin E and other antioxidants on exercise-induced oxidative stress. According to Kanter (1988), antioxidant supplementation, particularly with vitamins C and E and especially

Table 3.4   Recent Studies Showing Effects of Vitamin E and Other Antioxidants on Exercise-Associated Oxidative Stress

Findings	Study
Vitamin E supplementation favorably affects markers of lipid peroxidation following exercise.	Kanter (1994)
Vitamin E protects against acute effects of ozone on lung function in heavily exercising amateur cyclists.	Grievink et al. (1998)
β-Carotene supplementation to humans attenuates oxidative stress due to exercise.	Sumida et al. (1997)
Vitamin C reduces oxidative parameters of exercising humans.	Alessio et al. (1997)
Administration of vitamins C, E, and Q to long-distance runners increases the antioxidant potential of low density lipoprotein.	Vasankari et al. (1997)
Vitamin E has an effect on exercise, red blood cells, and lipid peroxidation.	Oostenbrug et al. (1997)
Vitamin E decreases lipid peroxidation and protein oxidation due to exercise-induced oxidative stress.	Sen et al. (1997)
Vitamin C reduces the incidence of post-race upper-respiratory tract infection.	Peters-Futre (1997)
Vitamin E totally abrogates the inhibitory effects mediated by thermal stress.	Franci et al. (1996)
Supplementation with ubiquinol-10 causes cellular damage during intense exercise.	Malm et al. (1996)
Derivatives of superoxide dismutase prevent oxidative damage induced by exercise.	Radák et al. (1996)
Vitamin E protects against oxidative stress induced by strenuous exercise.	Rokitzki et al. (1994)

Findings	Study
Endogenous glutathione plays a role in circumventing exercise-induced oxidative stress.	Sen et al. (1994)
Vitamins C, E, and β-carotene reduce the level of expired pentane and serum malondialdehyde.	Kanter et al. (1993)
Vitamin C, N-acetyl-L-cysteine, and glutathione prevent oxidation of blood glutathione after exercise.	Sastre et al. (1992)
Vitamin E inhibits protein oxidation in resting and and exercised skeletal muscles.	Reznick et al. (1992a), Reznick and Packer (1994)
Lipid peroxides in urine after exercise are lower after vitamin C and E supplementation.	Witt et al. (1992)
Vitamin E does not attenuate exercise-induced muscle damage.	Warren et al. (1992)
Provision of selenium reduces lipid peroxides.	Dragan et al. (1990)

to elderly active people, favorably affects markers of lipid peroxidation following exercise. Grievink et al. (1998) found that administration of antioxidant vitamins such as C, E, and β-carotene protects against the acute effects of ozone on lung function in heavily exercising amateur cyclists. Baseline urinary excretion of 8-hydroxy-deoxyguanosine before exercise tends to be lower after β-carotene supplementation. Hence, a single bout of incremental exercise to exhaustion does not induce oxidative DNA damage. Nevertheless, elevated β-carotene plasma levels due to high β-carotene diet were reduced considerably after a single bout to exhaustion (Sumida et al., 1997).

Supplementation of 1 g/d of vitamin C for 2 wk decreased exercise-induced oxidative stress after 30 min of submaximal exercise (Alessio et al., 1997). Daily supplementation of vitamins C and E and ubiquinone for 4 wk increased serum and LDL antioxidant potential after endurance exercise; the major increase of the antioxidant potential was mainly due to elevation of serum vitamin E (Vasankari et al., 1997). Three wk of fish oil supplementation, with and without vitamin E, decreased the rate of LDL oxidation but did not improve endurance performance (Oostenbrug et al., 1997). Eight wk of vitamin E and fish oil supplementation, to resting and exercised rats, markedly decreased lipid peroxidation and protein oxidation (Sen et al., 1997). Supplementation of vitamin C after moderate submaximal exercise reduced the incidence of post-race upper-respiratory tract infection (Peters-Futre, 1997).

Supplementation of vitamin E, but not C, restored immune-suppressed responses after intense exercise. Especially α-tocopherol abrogated the inhibitory effects mediated by thermal stress (Franci et al., 1996). Supplementation of ubiquinol-10 caused cellular damage during intense exercise, which suggests that antioxidants also may have some adverse effects (Malm et al., 1996). Supplementation of SOD derivative was found to protect against strenuous exercise-induced oxidative stress in the liver and kidney (Radák et al., 1996). Five mo of vitamin E administration protected against oxidative stress induced by aerobic exercise training in 30 top-class cyclists (Rokitzki et al., 1994). Repeated administration of glutathione affected exhaustive exercise performance in rats (Sen et al., 1994). Six wk administration of vitamins C and E and β-carotene significantly lowered resting and 30 min treadmill running-induced levels of expired pentane and serum MDA (Kanter, 1994). Administration of vitamin C and glutathione prevented oxidation of blood glutathione, lactate, and pyruvate levels after exhaustive concentric physical exercise in trained rats and humans (Sastre et al., 1992). Supplementation of vitamin E protected against resting and postexercise-induced protein oxidation in skeletal muscle (Reznick and Packer, 1994; Reznick et al., 1992a). Supplementation of vitamin E and coenzyme Q_{10} decreased muscle protein carbonyl formation caused by oxidative stress following endurance training (Witt et al., 1992). Supplementation of 40 IU vitamin E for 5 wk did not decrease exercise-induced oxidative stress (Warren et al., 1992). Fourteen d of selenium treatment induced significant changes of lipid peroxides in 33 top long-distance swimmers (Dragan et al., 1990). A general conclusion that we can draw from these findings is that the majority of studies have found beneficial effects of antioxidant nutrients, especially vitamin E, on exercise-induced oxidative stress.

Aging, Oxidative Stress, and Provision of Antioxidants

Over the last several years, there has been an enormous increase of information concerning free radical damage accumulation with advanced age. In many studies, oxidative components of DNA, lipids, and proteins have been shown to accumulate with age. Harman referred to the interaction between free radicals and aging by the following statement: "The aging process may be simply the sum of random changes produced by free radical reactions" (Harman, 1992, p. 1).

Many studies indicate that age-related diseases as well as the aging process itself may result from genetic instability, whereby cells change their proper state of differentiation (Cutler, 1991; Halliwell and Gutteridge, 1995). There is also considerable evidence that formation of reactive oxygen species may be a primary cause of aging. The major factor associated with age-related diseases is the increase of oxidative DNA damage. This DNA damage may induce expression of certain genes that regulate cell proliferation or could block the expression of other genes and thus permit damage with increasing age. As previously mentioned, free radicals or their precursors could be generated by endogenous processes as well as by numerous exogenous agents. These free radical–associated processes may lead to permanent damage, defined as DNA sequence changes or mutations. The accumulation of DNA damage may accelerate aging, lower specific metabolic rate, alter dietary caloric intake, and decrease life span (Simic, 1992). Many cell types from amoebae to some mammalian cells do not show a time-related involution, whereas other cells (i.e., neurons) change considerably with age. Miquel et al. (1998) wrote that aging is the nonprogrammed but unavoidable "side effect" of oxyradical damage to the membrane and to the genome of the mitochondria of irreversibly differentiated cells. Damage to mitochondrial DNA blocks the rejuvenation of the mitochondrial population and leads to bioenergetic decline and cellular death.

Other studies indicate a positive correlation between tissue concentration of antioxidants and life span (Block et al., 1992; Cutler, 1991). Furthermore, it has been found that some extrinsic factors of increased free radical–related oxidative stress reduce life span (Radák and Goto, 1998), whereas extrinsic factors, such as caloric and protein restriction, increase life span (Youngman, 1993; Yu et al., 1998). Antioxidant compounds are closely linked in interlocking cycles of regeneration. Oxidants attack the body tissues, and antioxidants should protect the tissues, although they sometimes fail to do so. Hence, free radicals may overload the tissues and accelerate aging. To successfully combat oxidative stress, one needs to combine preventive and ameliorative approaches.

The elderly population is at risk for antioxidant deficiency as a result of decreased intake, inefficient food absorption, diminished retention or storage capacity, and increased elimination. Therefore, the elderly may receive the most benefit from antioxidant prevention regimen. Lipid peroxidation is known to increase with aging. The evidence cited most often is the accumulation of "age pigments" or lipofuscins. Furthermore, animals fed vitamin E–deficient diets have shown an increased formation of lipofuscins.

Some studies suggest that maintaining an appropriate antioxidant/oxidant balance is important in maintaining health in aging. It is possible that the continuous presence of small amounts of free radicals in aged individuals is an important factor in aging (Quiroga et al., 1992). Aerobic cells contain various amounts of the three main antioxidant enzymes: SOD, CAT, and GPX. With age, the levels of these antioxidant enzymes do not necessarily change, so it is not possible to explain the aging process only by decreased activity of these three enzymes (Remacle et al., 1992). However, cells are very susceptible to free radical attacks with age. The free radicals have some influence not only on the cell mitotic capacity but on the rate of aging itself. High oxygen concentrations speed the aging process of cells or even cause the cell to undergo cellular toxicity, which decreases global cell population and reduces its mitotic capacity.

Some authors claim that aging may be driven by passive rather than active biochemical mechanisms. In other words, aging may be a by-product and not necessarily a genetically driven process. Cutler (1991) wrote that "detecting the presence of cellular oxidative stress represents an ideal candidate to test the passive aging process differentiated state" (p. 1). Apoptosis is believed to be a mechanism for cells to destroy themselves, and consequentially it may be important in the development of some age-related diseases such as Alzheimer's and Parkinson's disease (Ceballos et al., 1990; Zemlan et al., 1989). In addition, the passive aging process could be caused by a series of side effects of basic metabolic processes. However, the active mechanisms, which are also called health maintenance processes, are postulated to be regulated by a small number (<100) of longevity determinant genes (Orr and Sohal, 1994).

Possible involvement of active oxygen species has been shown in the generation of altered proteins that accumulate in tissues of aged animals (Goto et al., 1995). Many altered proteins are known to increase with age (Simic, 1992), and oxidative modifications conceivably could alter proteins with age. However, we have no evidence to explain the increase in half-life time of proteins with age, which has important consequences on aging by extending the life of molecules in cells or tissues. During aging a number of enzymes accumulate as catalytically inactive or less active forms. The age-related changes in catalytic activity are due in part to reactions of the proteins with "active" oxygen species such as singlet oxygen, or with oxygen free radicals that are produced during exposure to metal ion–catalyzed oxidation system. The levels of oxidized proteins increase progressively with age and might represent 30 to 50% of the total cellular proteins. This increase in the oxidized protein level is caused by a decrease in the neutral protease activity that degrades oxidized protein (Stadtman et al., 1992).

There is also clear evidence of a decline in respiratory chain function with age. This effect has been documented in several tissues and appears predominantly to affect complex I (subunits derived from nuclear and mitochondrial genomes) such as in Parkinson's disease. Mutations of mitochondrial DNA accumulate as well with age, and there is evidence of oxidative stress and damage in senescent tissues. However, the connection among respiratory chain decline, oxidative damage, and mitochondrial DNA is unclear, and it is premature to accept without reservation the mitochondrial aging theory.

Many different diseases related to aging appear to be attenuated by dietary antioxidant levels, indicating the biological significance of antioxidants in protecting against age-related diseases. Halliwell (1989) demonstrated age-related oxidative stress in the central nervous system (CNS). Low levels of endogenous antioxidants and high levels of oxidizable substrate (polyunsaturated fatty acids) make the aged brain a vulnerable target for oxidative stress that leads to molecular damage. Free radicals play an important role in senile neurodegenerative disease, such as Parkinson's disease (Ceballos et al., 1990), Alzheimer's disease (Evans et al., 1992), and amyotrophic lateral sclerosis (Bowling et al., 1993). The CNS is particularly susceptible to oxidant damage due to a high rate of oxidative metabolic activity, when ischemic tissue is reoxygenated. Oxidative stress has a dual action on calcium uptake and discharge of insulin. Depending on the concentration and duration of action, oxidative stressors depolarize insulin-producing cells, stimulate calcium uptake with insulin secretion, and inhibit calcium uptake and insulin release. Diabetes mellitus type I is probably associated with free radicals and itself may generate additional free radicals. Several studies have shown a release of oxygenated free radicals in diabetic retinas and have linked this action to a marked increase of lipoperoxidation (Doly et al., 1992).

Because of the ubiquitous nature of aging and reduced function of essentially all physiological processes, it appears that the only long-term solution to problems associated with human aging is to uniformly decrease the aging rate of the entire body. Some evidence suggests that aging involves genetic instability and that longer lived species have a more stable genetic apparatus and superior protection against reactive oxygen species that may affect our genetic machinery. Other studies may attribute aging to random passive processes that involve free radical–associated damage, and thus any supplementation of dietary antioxidants could possibly influence aging on both the genetic and random nongenetic levels. This is illustrated in figure 3.4.

Causes of aging

Extrinsic	Intrinsic
–Environmental stresses	–Programmed
Pollutants	Genetically associated processes
Nutrition	–Random
Exercise	Passive random biochemical processes
Psychological stresses	

Figure 3.4 Possible causes of aging.

Table 3.5   Recent Findings That Show Positive Effects of Vitamin E Supplementation on Aging

Findings	Study
Reverses age-related deficit in long-term potentiation in dentate gyrus in the brain.	Murray and Lynch (1998)
Effects cytokine production and T-cell biological function in the aged.	Meydani et al. (1998)
Suppresses low density lipoprotein oxidation.	Yu et al. (1998)
Ameliorates enhanced renal lipid peroxidation and accumulation of F2-isoprostanes in aging kidneys.	Reckelhoff et al. (1998)
Improves endothelium-dependent vasodilation in aortic strips of young and aging rats exposed to oxidative stress.	Guarnieri et al. (1996)
Prevents oxidative modification of brain and lymphocytes band 3 protein during aging.	Poulin et al. (1996)
In conjunction with PBN, improves cognitive performance of aged brain.	Socci et al. (1995)
Protects aged fibroblasts.	Farriol et al. (1994)
Scavenger of free radicals.	Sies et al. (1992)
Regulates vascular smooth muscle cell proliferation.	Boscoboinik et al. (1991)
Vitamin E is required in relation to dietary fish oil supplementation and oxidative stress.	Meydani (1992)
May reduce the oxidative damage in exercised muscles of aged animals.	Reznick et al. (1992)

PBN, phenyl-tert-butyl-nitrone.

In figure 3.4, the causes of aging are divided into two main categories: extrinsic and intrinsic. The extrinsic causes can affect aging directly and indirectly by influencing intrinsic causes. The intrinsic causes can be further divided into two main types, programmed causes and random causes, which affect aging at all biological levels (i.e., intra- and extracellular levels and tissue and organ levels). Free radical reactions are involved in all of these categories and thus affect aging extrinsically and intrinsically, in active and passive ways.

The Effect of Vitamin E on Aging

Table 3.5 summarizes recent findings concerning the effect of vitamin E on aging. Murray and Lynch (1998) reported that aged rats fed a diet supplemented with α-tocopherol exhibited an ability to sustain long-term potentiation in the hippocampus. Samples prepared from aging brain tissues exhibited decreased interleukin-1 beta, decreased lipid peroxidation, and increased membrane arachidonic acid concentration. These findings are consistent with the hypothesis that some age-related changes in the hippocampus might be caused by oxidative stress. Meydani et al. (1998) discussed the possibility that oxidant/antioxidant balance is an important determinant of immune cell function, including the state of membrane lipids, cellular proteins, and nucleic acids, as well as the control of signal transduction and gene expression in immune cells. Meydani et al. found that an age-related increase in free radicals is associated with dysregulation of immune responses such as cytokine production and T-cell–mediated function. Clear evidence of vitamin E's protective effect has been seen in its suppression of LDL oxidation both in vitro and in vivo. An additional beneficial effect is that vitamin E improves glucose transport and insulin sensitivity (Yu et al., 1998). Renal aging is associated with an increase in oxidant-sensitive heme oxygenase, advanced glycosylation end products (AGEs), and their receptors (RAGE). AGE-RAGE interaction has been shown to induce oxidative stress. Reckelhoff et al. (1998) reported that a diet high in vitamin E significantly increased glomerular filtration rate, suppressed F2 isoprostanes, and attenuated expression of heme oxygenase and RAGE. This diet also tended to decrease glomerular sclerosis. Hence, administration of an antioxidant such as vitamin E could attenuate the decline in renal function (Reckelhoff et al., 1998).

Guarnieri et al. (1996) confirmed that oxidative stress impairs endothelium-mediated vasodilation, although the mechanism by which α-tocopherol pretreatment protects the vessel against this damage is unexplained. Poulin et al. (1996) reported that vitamin E treatment could prevent the observed age-related decline in anion transport by lymphocytes and the generation of aged band 3 leading to senescent cell antigen formation. Socci et al. (1995) provided evidence that chronic antioxidant treatment could improve cognitive function during aging, thus supporting the free radical hypothesis of aging as related to brain function. Farriol et al. (1994) indicated that a 50 μmol/L dose of water-soluble vitamin E produced some cytoprotective effect on aged fibroblasts. Sies et al. (1992) described vitamin E as an efficient scavenger of peroxyl radicals in biological membranes such as in light density lipoproteins (LDLS) LDLS. Boscoboinik et al. (1991) reported that a physiological concentration of vitamin E specifically inhibited aorta smooth muscle cell proliferation and protein kinase C activity; thus, it might attenuate the onset mechanism of atherosclerosis. Reznick et al. (1992) reported that administration of vitamin E may alter the threshold of age-related response to exercise, delaying this threshold to an older age. The studies described in table 3.5 indicate that in some cases, vitamin E supplementation may have some beneficial effects on age-associated phenomena.

Effects of Other Antioxidants

Table 3.6 summarizes findings concerning effects of other antioxidants on aging systems. Many studies indicate that insufficient antioxidant intake may shorten the life span (Armeni et al., 1997; Byun et al., 1995; Feuers et al., 1993; Kim et al., 1996b; Sohal et al., 1994; Sohal and Weindruch, 1996; Xia

---

Table 3.6   Summary of Other Antioxidant Effects on Aging Systems

Findings	Study
Antioxidant supplementation influences aging parameters associated with oxidative damage.	Sohal et al. (1994) Sohal and Weindruch (1996) Feuers et al. (1993) Armeni et al. (1997) Yu (1996) Kim et al. (1996b) Byun et al. (1995) Xia et al. (1995)
Cancer treatment with thiol-containing antioxidants may improve life quality.	Hack et al. (1998)
Correlation of the rate of free radical formation and aging	Perez-Campo et al. (1998)
Garlic protects from oxidant injury.	Yamasaki and Lau (1997) Ide et al. (1997)
Pycnogenol protects vascular endothelial cells from oxidative injury.	Rong et al. (1995)
Antioxidants protect against oxidative stress.	Kanter (1994)
Antioxidants are associated with eye disease.	Seddon et al. (1994) Christen (1994) West et al. (1994) Taylor (1992) Rose et al. (1998) Varma et al. (1995)
The role of antioxidants in preventive geriatrics.	Wards (1994) Sram et al. (1993) Rikans and Hornbrook (1997) Short et al. (1997)
Some positive effect of vitamin A on measles infection.	Bendich (1993)

Findings	Study
Effect of vitamin E and selenium on the growing of peripheral neurons.	Koistinaho et al. (1990)
Effect of antioxidants in preventing atherosclerosis.	Miquel et al. (1998)
Effect of antioxidants on chronic diseases.	Coudray et al. (1997)
N-acetylcysteine slows aging process.	Brack et al. (1997)
Antioxidants improve the immune system.	Spencer et al. (1997)
Ubiquinol-10 (Coenzyme Q) affects low density lipoprotein oxidation.	Aejmelaeus et al. (1997)
Beta-catechin increases lifespan.	Kumari et al. (1997) Yoneda et al. (1995)
Antioxidants supplementation improves memory.	Perrig et al. (1997) de Angelis and Furlan (1995)
Antioxidants may slow down osteoarthritis.	McAlindon et al. (1996)
Antixoidants may increase life span.	Kitani et al. (1996)
Antioxidants may increase glucose transport.	Moustafa et al. (1995)
Rooibos tea has positive effect on the brain.	Inanami et al. (1995)
Melatonin may protect the brain from oxidative stress.	Reiter (1995)

et al., 1995; Yu, 1996). Lack of antioxidants may shift the redox state, increasing oxidative status and contributing to the loss of body cell mass in aging and other pathologies. Indeed, treatment of cancer patients with thiol-containing antioxidants could improve their quality of life (Hack et al., 1998).

In most animal groups, the endogenous levels of enzymatic and nonenzymatic antioxidants in tissues negatively correlate with maximum life span. The most longevous animals studied display relatively lower levels of antioxidants, which may indicate that these animals are exposed to lower levels of oxidative challenges in aging. Indeed, some studies suggest that the longer the life span of an organism, the lower the rate of oxygen radical production, particularly in the mitochondria. So there are two main features of long-living species: a high rate and capacity of DNA repair and a low rate of free radical production (Perez-Campo et al., 1998). Yamasaki and Lau (1997) and Ide et al. (1997) found that garlic extract and S-allyl cysteine can protect vascular endothelial cells from oxidant injury and inhibit oxidative modification of LDL. Pycnogenol was found to protect vascular endothelial cells from oxidant injury and thus may prevent disorders associated with oxidative damage leading to atherosclerosis (Rong et al., 1995).

Antioxidant supplementation may detoxify the peroxides produced during exercise and diminish muscle damage and soreness (Kanter, 1994). Because of its high concentration in the eye, vitamin C is the major substance in ocular protection (Christen, 1994; Rose et al., 1998; Seddon et al., 1994; Taylor, 1992; Varma et al., 1995; West et al., 1994). Indeed, some studies have indicated that increased consumption of foods rich in certain carotenoids, in particular dark green, leafy vegetables, may decrease the risk of developing advanced or oxidative age-related macular degeneration. Hence, antioxidant supplementation may protect against the cumulative effects of oxidative stress of cataract and macular degeneration. Several investigators indicated a significant inverse relation between the levels of antioxidant variables and the rate of biological aging. Animals that had high antioxidant variable scores had low rates of biological aging scores and vice versa. Thus, antioxidant activity influences the vulnerability to disease (Rikans and Hornbrook, 1997; Short et al., 1997; Sram et al., 1993; Wards, 1994). Bendich (1993) found that vitamin A supplementation decreased morbidity and mortality associated with measles infection in children. An adequate intake of antioxidants is particularly recommended in postmenopausal women because of their greater sensitivity to age-related oxygen stress. Dietary antioxidant supplementation (Miquel et al., 1998) could lower the high concentration of lipid peroxides in the serum of people with advanced age. Coudray et al. (1997) reported that cognitive dysfunction as a result of cerebrovascular accident caused by lipidemia can be diminished by administration of vitamin E and increased plasma selenium. Another study (Brack et al., 1997) on dietary intake of N-acetylcysteine showed that it affected longevity, a phenomenon that may involve differential expression of specific mRNA genes, as suggested by RNA fingerprinting experiments (Brack et al., 1997). Spencer et al. (1997) showed that therapeutic treatment with dietary antioxidants or with agents capable of activating the peroxisome proliferator–activated receptor alpha (PPAR) corrected abnormal nuclear NF-κB activity and reduced lipid peroxide levels. Therefore, effective therapeutic regimens against aging and inflammatory diseases might include administration of antioxidants that activate PPAR. Aejmelaeus et al. (1997) found that Q_{10} supplementation doubled the number of ubiquinol-10–containing LDL and might inhibit LDL oxidation.

Yoneda et al. (1995) and Kumari et al. (1997) emphasized the importance of a daily supplement of combined natural antioxidants for a long and disease-free life. de Angelis and Furlan (1995) and Perrig et al. (1997) reported that high serum levels of vitamin C and β-carotene were associated with better memory performance and might prevent amnesia. These results emphasize the important role that antioxidants play in brain aging, which may have implications for preventing progressive cognitive impairments. High intake of vitamins C and E and β-carotene may also reduce the risk of cartilage loss and disease progression in people with osteoarthritis (McAlindon et al., 1996). In addition, the role of SOD in protecting catecholaminergic neurons was found to be crucial for life-span extension (Kitani et al., 1996). Moustafa et al. (1995) investigated the protective capacity of vitamin C and butylated hydroxytoluene (BHT) against age-associated free radical damage. This protective role is probably mediated via an effect on lipid matrix of the cell membrane as well as on the intracellular cytoplasmic constituents. Age-related accumulation of lipid peroxides in the brain is closely related to the morphological changes revealed by magnetic resonance imaging. Thus, chronic Rooibos tea administration can prevent age-related accumulation of lipid peroxides in several regions of rat brain (Inanami et al., 1995). Reiter (1995) found that melatonin is very effective in scavenging the highly toxic hydroxyl radical and also stimulated an important antioxidant enzyme of the brain, glutathione peroxidase.

A general conclusion is that in some age-related diseases and in certain age-related processes, administration of antioxidants could be beneficial.

Studies That Show No Beneficial Effect of Antioxidants on Aging

Table 3.7 summarizes findings that antioxidants have no effect on aging. Teikari et al. (1998) found that supplementation of vitamin E, β-carotene, BHT, and vitamin C had no effect on the levels of protein oxidation as measured by protein-bound o-tyrosine and 3-nitrotyrosine in young and old female rats. Leeuwenburgh et al. (1998) showed that even very large variations in the content

Table 3.7   Recent Studies That Show No Effect of Antioxidants on Aging

Findings	Study
No effect on aged maculopathy.	Teikari et al. (1998)
No effect on aromatic-amino acids	Leeuwenburgh et al. (1998)
No effect on liver DNA oxidative damage.	Cadenas et al. (1997)
Vitamin E failed to reduce several types of cancer.	Thurman and Mooradian (1997)
No proof that vitamins C and E taken as supplements provide better protection than a diet rich in fruits and vegetables	Wards (1998)
No effect on reducing the levels of glycosylated hemoglobin in people with and without diabetes.	Shoff et al. (1993)
Antioxidant diet had no effect on runners' longevity.	Holloszy (1998)

of the antioxidant vitamins C and E in the diet and in liver had no influence on the steady-state level of oxidative damage to guanine in the liver DNA of normal unstressed guinea pigs. The efficacy and safety of vitamin supplementation for prevention of several types of cancer were questioned by Cadenas et al. (1997). Thurman and Mooradian (1997) found that fruits and vegetables protect against aging as well as do antioxidant vitamin supplements. Wards (1998) and Holloszy (1998) found that even large doses of vitamins C and E and β-carotene could not reduce protein glycosylation in diabetic people and had no effect on the longevity of runners. Shoff et al. (1993) demonstrated that antioxidants did not extend the lives of wheel-running male rats that had oxidative tissue damage. The discrepancies and negative findings about the effect of antioxidant supplementation on aging could be attributed to many reasons, such as the research models, the design of experiments, or certain conditions in which oxidative challenges can overwhelm antioxidant defense systems.

Dietary Restriction, Life Span, and Oxidation Stress

Yu (1996) provided an excellent review about modulation of aging and longevity by dietary restriction. The basic idea of this monograph was that dietary restriction (DR) is the single most accepted modality that can intervene with aging processes. This point of view has now been accepted by most gerontologists as well as by mainstream nutritionists. Furthermore, several lines of research concerning the cellular and molecular mechanisms of DR action have emphasized its usefulness in modulating oxidative stress in aging animals. Thus, by attenuating oxidative damage and bolstering the antioxidant defense systems, DR can retard aging and extend the life span (Yu, 1996).

In the 1930s, McCay performed the first experiments on DR with rats, finding that DR can extend mean life span of animals by 30 to 50% compared to an ad libitum feeding schedule (McCay, 1936). Very comprehensive studies performed in the last few years on the effect of DR on life span of primates such as monkeys have also shown that DR extends the mean life span (Lane et al., 1992). There may be two explanations for the positive effects of DR on aging: DR may retard basic biological processes of aging or it may suppress age-related pathologies. Accordingly, DR was found to affect age-related changes of several basic biological systems: the immune system, protein turnover, bone loss, and neural degeneration. DR also affects several pathologies, such as glomerulonephritis, cataract, and cardiomyopathy (Yu, 1996).

The first study that showed that DR can modulate oxidative damage parameters was conducted by Chipalkatti et al. in 1983. This study revealed that

DR reduced the age-related accumulation of brain lipofuscins and formation of MDA as an indicator of oxidative stress by approximately 50%. In other research work on aged animals, DR was found to reduce oxidative damage to DNA (Youngman, 1993), to lipid peroxidation (Koizumi et al., 1987), and to protein oxidation (Sohal et al., 1994). On the cellular and tissue levels, recent studies by Yu et al. (1998) showed that DR can significantly modulate the level of tissue iron and thus suppress age-related iron accumulation. In addition, previous studies showed that DR can alter membrane lipid composition by increasing the level of saturated lipids and reducing the level of polyunsaturated lipids in membrane of aging cells (Laganiere and Yu, 1989a). Another interesting aspect of DR modulation of oxidation status of aging tissues is the finding that DR can alter the activity of antioxidant enzymes and can boost other antioxidant defense systems. Koizumi et al. (1987) showed that in addition to suppressing lipid peroxidation, DR increased the level of catalase activity of aging mice. The effect of DR on SOD and GPX activities was less pronounced in the preceding studies. A similar study by Laganiere and Yu (1989b) showed that DR had a wide-ranging effect on several hepatic antioxidant systems in aging Fischer rats. Thus, DR affects age-related changes in GSH, GSH reductase, GPX, CAT, and ascorbic acid.

In summary, DR has the following possible effects on oxidative status of aging animals and on aging processes:

Decreased oxidative insults

Increased antioxidant systems

Decreased oxidative damage

Slowed aging processes

Increased life span

According to this simplified scheme, DR has a double effect on oxidation of biological systems: It can attenuate oxidative insults and boost the antioxidant defense systems. Thus, by decreasing oxidative damage of aging tissues, DR slows aging processes and therefore increases the life span of DR-treated animals.

Conclusion

Oxidative stress-related events are very important in many biological processes such as protein turnover, cellular signal transudation, and immune responses. On the other hand, they may also be harmful, producing detrimental free radicals. Nature and evolution have endowed organisms with a myriad of antioxidant defense systems to combat excessive oxidative stresses. These compounds, acting in concert as reductants, are capable of neutralizing most of the oxygen free radicals and thus prevent or minimize oxidative damages to biological systems.

Antioxidants are generally classified as endogenous antioxidants, those produced internally by animals and humans, and as exogenous antioxidants, those which the body is not able to produce and which must be provided from external sources. In a number of pathologies as well as in exercise and aging, studies have shown that endogenous antioxidants are not sufficient to protect against excessive oxidative stress. A variety of exogenous antioxidants such as vitamins C and E and β-carotene have proven beneficial in attenuating oxidative stress-associated changes. However, some other studies have questioned the role of antioxidant nutrition, because antioxidant supplementation showed meager benefits and sometimes was quite ineffective. Therefore, more work is needed to explain the discrepancies found in various studies.

References

Aejmelaeus, R., K.T. Metsa, P. Laippala, T. Solakivi, and H. Alho. 1997. Ubiquinol-10 and total peroxyl radical trapping capacity of LDL lipoproteins during aging. *Molecular Aspects of Medicine* 18:Sl 13-120.

Alessio, H.M. 1994. Lipid peroxidation processes in healthy and diseased models. In: *Exercise and Oxygen Toxicity*, C.K. Sen, L. Packer, and O. Hanninen (eds.), pp. 269-295. Elsevier Science, Amsterdam.

Alessio, H.M., A.H. Goldfarb, and G. Cao. 1997. Exercise-induced oxidative stress before and after vitamin C supplementation. *International Journal of Sport Nutrition* 7:1-9.

Ames, B.A., M.K. Shingenaga, E.M. Park. 1991. Oxidation damage and repair. *Chemical, biological and medical aspects*, K.J.A. Davis Ed. Pergamon, Elmsford, NY.

Anneren, G., and B. Bjorksten. 1984. Low superoxide levels in blood phagocytic cells in Down's syndrome. *Acta Pediatrica Scandinavica* 73:345-348.

Armeni, T., M. Tomasetti, S.B. Svegliati, F. Saccucci, M. Marra, C. Pieri, G.P. Littarru, G. Principato, and M. Battino. 1997. Dietary restriction affects antioxidant levels in rat liver mitochondria during aging. *Molecular Aspects of Medicine* 18:S247-250.

Armstrong, R.B., G.L.Warren, and J.A. Warren. 1991. Mechanisms of exercise-induced muscle fiber injury. *Sports Medicine* 12:184-207.

Atalay, M., P. Marnila, E.M. Lilius, O. Hanninen, and C.K. Sen. 1996a. Glutathione-dependent modulation of exhausting exercise-induced changes in neutrophil function of rats. *European Journal of Applied Physiology and Occupational Physiology* 74:342-347.

Atalay, M., T. Seene, O. Hanninen, and C.K. Sen. 1996b. Skeletal muscle and heart antioxidant defences in response to sprint training. *Acta Physiologica Scandinavica* 158:129-134.

Aust, D.D., L.A. Morehouse, and C.E. Thomas. 1985. Role of metals in oxygen radical reactions. *Free Radical Biology and Medicine* 1:3-25.

Axelsson, K., and B. Mannervik. 1983. An essential role of cytosolic thioltransferase in protection of pyruvate kinase from rabbit liver against oxidative inactivation. *FEBS Letters* 152:114-118.

Benderitter, M., F. Hadj-Saad, M. Lhuissier, V. Maupoil, J.C. Guilland, and L. Rochette. 1996. Effects of exhaustive exercise and vitamin B6 deficiency on free radicals oxidative process in male-trained rats. *Free Radical Biology and Medicine* 21:541-549.

Bendich, A. 1993. Physiological role of antioxidants in the immune system. *Journal of Dairy Sciences* 76:2789-2794.

Berlett, B.S., and E.R. Stadtman. 1997. Protein oxidation in aging, disease and oxidative stress. *Journal of Biological Chemistry* 272 (33): 20313-20316.

Bhatnagar, A., S.K. Srivastava, and G. Szabo. 1990. Oxidative stress alters specific membrane currents in isolated cardiac myocytes. *Circulation Research* 67:535-549.

Block, G., B. Patterson, and A. Subar. 1992. Fruit, vegetable, and cancer prevention: A review of the epidemiological evidence. *Nutrition and Cancer* 18:1-29.

Boscoboinik, D., A. Szewczyk, and A. Azzi. 1991. Alpha-tocopherol regulates vascular smooth cell proliferation and protein kinase C activity. *Archives of Biochemistry and Biophysics* 286:264-269.

Bowling, A.C., J.B. Schulz, R.H.J. Brown, and M.F. Beal. 1993. Superoxide dismutase activity, oxidative damage and mitochondrial energy metabolism in familial and sporadic amyotrophic lateral sclerosis. *Journal of Neurochemistry* 61:2322-2325.

Brack, C., T.E. Bechter, and M. Labuhn. 1997. N-acetylcysteine slows down aging and increases the life span of *Drosophila melanogaster. Cellular and Molecular Life Sciences* 53:960-966.

Brodie, A.E., and D.J. Reed. 1990. Cellular recovery of glyceraldehyde-3-phospate dehydrogenase activity and thiol status after exposure to hydroperoxides. *Archives of Biochemistry and Biophysics* 276:212-218.

Buttke, T.M., and P.A. Sandstrom. 1994. Oxidative stress as a mediator of apoptosis. *Immunology Today* 15:1-4.

Byun, D.S., J.T. Venkatraman, B.P. Yu, and G. Fernandes. 1995. Modulation of antioxidant activities and immune response by food restriction in aging Fischer 344 rats. *Aging* 7:40-48.

Cadenas, E. 1997. Basic mechanism of antioxidant. *Biofactors* 6:391-397.

Ceballos, I., M. Lafon, F.J. Agid, E. Nicole, P.M. Sinet, and Y. Agid. 1990. Superoxide dismutase and Parkinson's disease. *Lancet* 335:1035-1036.

Chipalkatti, S., A.K. De, and A.S. Aiyar. 1983. Effect of diet restriction on some biochemical parameters related to aging in mice. *Journal of Nutrition.* 113: 955-960.

Christen, W.J. Jr. 1994. Antioxidants and eye disease. *American Journal of Medicine* 97:14S-17S.

Clarkson, P.M. 1995. Micronutrients and exercise: Antioxidants and minerals. *Journal of Sports Sciences* S11-24.

Coudray, C., A.M. Roussel, F. Mainard, J. Arnaud, and A. Favier. 1997. Lipid peroxidation level and antioxidant micronutrient in a pre-aging population. *Journal of the American College of Nutrition* 16:584-591.

Cutler, R.G. 1991. Human longevity and aging: Possible role of reactive oxygen species. *Annals of the New York Academy of Sciences* 621:1-28.

de Angelis, L., and C. Furlan. 1995. The effect of ascorbic acid and oxiracetam on scopolamine-induced amnesia in a habituation test in aged mice. *Neurobiology of Learning and Memory* 64:119-124.

Doly, M.M., T.L. Droy, and P. Barquet. 1992. Oxidative stress in diabetic retina. In: *Free Radicals and Aging*, I. Emerit and B. Chance (eds.), pp. 299-307. Birkhauser Verlag, Basel, Switzerland.

Dragan, I., V. Dinu, M. Mohora, E. Cristea, E. Ploesteanu, and V. Stroescu. 1990. Studies regarding the antioxidant effects of selenium on top swimmers. *Revue Roumaine de Physiologie* 27(1):15-20.

Duncan, C.J. 1991. In: *Calcium, Oxygen Radicals and Cellular Damage*, C.J. Duncan (ed.), pp. 97-113. Cambridge University Press, Cambridge.

Evans, P.H., Y. Eiji, J. Klinowski, and E. Peterhans. 1992. Oxidative damage in Alzheimer's dementia, and the potential etiopathogenic role of aluminosilicates, microglia and micronutrient interactions. In: *Free Radicals and Aging*, I. Emerit, and B. Chance, (eds.), pp. 178-189. Birkhauser Verlag, Basel, Switzerland.

Fair, J.M., and K. Berra. 1995. Life style changes and coronary heart disease: The influence of nonpharmacologic interventions. *Journal of Cardiovascular Nursing* 9:12-24.

Farriol, M., M. Mourelle, and S. Schwartz. 1994. Effect of vitamin C and E analog on aged fibroblasts. *Revista Espanola de Fisiologia* 50:253-257.

Fernando, M.R., H. Nanri, S. Yoshitake, K.N. Kuno, and S. Minakami. 1992. Thioredoxin regenerates protein inactivated by oxidative stress in endothelial cells. *European Journal of Biochemistry* 209:917-922.

Feuers, R.J., R. Weindruch, and R.W. Hart. 1993. Caloric restriction, aging, and antioxidant enzymes. *Mutation Research* 295:191-200.

Fielding, R.A., and M. Meydani. 1997. Exercise, free radical generation, and aging. *Aging Milano* 9:12-18.

Franci, O., F. Ranfi, C. Scaccini, A. Amici, N. Merendion, G. Tommasi, and E. Piccolella. 1996. Differential effect of alpha-tocopherol and ascorbate on oxidative injury induced in immune cells by thermal stress. *Journal of Biological Regulators and Homeostatic Agents* 10:54-59.

Giuliani, A., and B. Cestaro. 1997. Exercise, free radical generation and vitamins. *European Journal of Cancer* 1:S55-S67.

Golan, R., and A.Z. Reznick. 1994. Aging and exercise-induced oxidative stress. In: *Exercise and Oxygen Toxicity*, C.K. Sen., L. Packer, and O. Hanninen (eds.), pp. 235-248. Elsevier Science, Amsterdam.

Goto, S., M. Yoshikawa, K. Yamada, and Y. Ushio. 1995. Survival of neurons containing the enzyme nicotinamide adenine dinucleotide phosphate diaphorase in static slice culture of adult rat striatum. *Neuroscience Letters*. 195:129-132.

Grievink, L., S.M. Jansen, V.P. Van't, and B. Brunekreef. 1998. Acute effects of ozone on pulmonary function of cyclists receiving antioxidant supplements. *Occupational and Environmental Medicine* 55:13-17.

Guarnieri, C., E. Giordano, C. Muscari, L. Grossi, and C.M. Caldarera. 1996. Alphatocopherol pretreatment improves endothelium-dependent vasodilatation in aortic strips of young and old aging rats exposed to oxidative stress. *Molecular and Cellular Biochemistry* 157:223-228.

Haas, A., and W. Goebel. 1992. Microbial strategies to prevent oxygen dependent killing by phagocytes. *Free Radical Research Communications* 16:137-157.

Hack, V., R. Breitkreutz, R. Kinscheerf, H. Rohrer, P. Bartsch, F. Taut, A. Benner, and W. Droge. 1998. The redox state as a correlate of senescence and wasting and as a target for therapeutic intervention. *Blood* 92:59-67.

Halliwell, B. 1989. Oxidants and the central nervous system: Some fundamental questions. *Acta Neurologica Scandinavica* 126:23-33.

Halliwell, B. 1998. Free radicals and oxidative damage in biology and medicine. In: *Oxidative Stress in Skeletal Muscle*, A.Z. Reznick (ed.), pp. 1-27. Birkhauser Verlag, Basel, Switzerland.

Halliwell, B., and J.M.C. Gutteridge. 1990. Role of free radicals and catalytic metal ions in human disease: An overview. *Methods in Enzymology* 186:1-85.

Halliwell, B., and J.M.C. Gutteridge. 1992. Biologically relevant metal ion-dependent hydroxyl radical generation. *FEBS Letters* 307:108-112.

Halliwell, B., and J.M.C. Gutteridge. 1995. The importance of free radicals and catalytic metal in human diseases. *Molecular Aspects of Medicine* 8:89-92.

Halliwell, B., and Gutteridge, J.M.C. 1999. Free radicals in biology and medicine. *Oxford Science Pub.* p. 106.

Haramaki, N., D.B. Stewart, S. Aggarwal, H. Ikeda, A.Z. Reznick, and L. Packer. 1998. Networking antioxidants in the isolated rat heart are selectively depleted by ischemia-reperfusion. *Free Radical Biology and Medicine* 25:329-339.

Harman, D. 1992. Free radical theory of aging: History. In: *Free radicals and aging*. Emerit Ingrid and Chance Britton (eds.), pp. 1-10. Birkhauser Verlag, Basel, Switzerland.

Hellsten, Y., F.S. Apple, and B. Sjodin. 1996. Effect of sprint cycle training on activities of antioxidant enzymes in human skeletal muscle. *Journal of Applied Physiology* 81:1484-1487.

Hohn, D.C., and R.I. Lehrer. 1975. NADPH oxidase deficiency in x-linked chronic granulomatouse disease. *Journal of Clinical Investigation* 55:707-713.

Holloszy, J.O. 1998. Longevity of exercising male rats: Effect of an antioxidant supplemented diet. *Mechanisms of Ageing and Development* 100:211-219.

Hunter, T. 1995. Protein kinases and phosphatases. *Cell* 80:225-236.

Hyslop, P.A., D.W. Hinshaw, W.A. Halsey, I.U. Schraufstatter, R.D. Sauerheber, R.G. Spragg, J.H. Jackson, and C.G. Cochrane. 1988. Mechanisms of oxidant-mediated cell injury: The glycolytic and mitochondrial pathway of ADP phosphorylation are major intracellular targets inactivated by hydrogen peroxide. *Journal of Biological Chemistry* 263:1665-1675.

Ide, N., A.B. Nelson, and B.H. Lau. 1997. Aged garlic extract and its constituents inhibit Cu (2+)-induced oxidative modification of LDL. *Planta Medica* 63:263-264.

Inanami, O., T. Asanuma, N. Inukai, T. Jin, S. Shimokawa, N. Kasai, M. Nakano, F. Sato, and M. Kuwabara. 1995. The suppression of age-related accumulation of lipid peroxides in rat brain by administration of Rooibos tea. *Neuroscience Letters* 196:85-88.

Inoue, T., Z. Mu, K. Sumikawa, K. Adachi, and T. Okochi. 1993. Effect of physical exercise on the content of 8-hydroxydeoxyguanosine in nuclear DNA prepared from human lymphocytes. *Japanese Journal of Cancer Research* 84:720-725.

Jackson, M.J. 1994. Exercise and oxygen radical production by muscle. In: *Exercise and Oxygen Toxicity*, C.K. Sen, L. Packer, and O. Hanninen (eds.), pp. 49-57. Elsevier Science, Amsterdam.

Ji, L.L. 1998. Antioxidant enzyme response to exercise and training in the skeletal muscle. In: *Oxidative Stress in Skeletal Muscle*, A.Z. Reznick (ed.), pp. 103-125. Birkhauser Verlag, Basel, Switzerland.

Ji, L.L., R. Fu, and E.M. Mitchell. 1992. Glutathione and antioxidant enzymes in skeletal muscle. *Journal of Applied Physiology* 73:1854-1859.

Kagan, V.E., E.A. Serbinova, T. Forte, G. Scita, and L. Packer. 1992. Recycling of vitamin E in human low density lipoproteins. *Journal of Lipid Research* 33:385-397.

Kandel, E.R., and R.D. Hawkins. 1992. The biological basis of learning and individuality. *Scientific American* 276 (3):78-86.

Kanter, M.M. 1994. Free radicals, exercise, and antioxidant supplementation. *International Journal of Sport Nutrition* 4:205-220.

Kanter, M.M., G.R. Lesmes, L.A. Kaminsky, J.L. Ham-Saeger, and N.D. Nequin. 1988. Serum creatine kinase and lactate dehydrogenase changes following an eighty kilometer race. Relationship to lipid peroxidation. *European Journal of Applied Physiology and Occupational Physiology.* 57:60-63.

Kanter, M.M., L.A. Nolte, and J.O. Holloszy. 1993. Effects of an antioxidant vitamin mixture on lipid peroxidation at rest and post exercise. *Journal of Applied Physiology* 74:965-969.

Karlsson, J. 1997. *Antioxidants and Exercise.* Human Kinetics, Champaign, IL.

Kim, J.D., R.J. McCarter, and B.P. Yu. 1996a. Influence of age, exercise, and dietary restriction on oxidative stress in rats. *Aging* 8:123-129.

Kim, J.D., B.P. Yu, R.J. McCarter, S.Y. Lee, and J.T. Herlily. 1996b. Exercise and diet modulate cardiac lipid peroxidation and antioxidant defenses. *Free Radical Biology and Medicine* 20:83-88.

Kitani, K., K. Miyasaka, S. Kanai, M.C. Carrillo, and G.O. Ivy. 1996. Upregulation of antioxidant enzyme activities by deprenyl. Implication for life span extension. *Annals of the New York Academy of Sciences* 786:391-409.

Koistinaho, J., H. Alho, and A. Hervonen. 1990. Effect of vitamin E and selenium supplement on the aging peripheral neurons of the male Sprague-Dawley rat. *Mechanisms of Ageing and Development* 51:63-72.

Koizumi, A., R. Weindruch, and R.L. Walford. 1987. Influence of dietary restriction and age on liver enzyme activities and lipid peroxidation in mice. *Journal of Nutrition* 117:361-367.

Kondo, H., J. Kodama, T. Kishibe, and Y. Itokawa. 1993. Oxidative stress during recovery from muscle atrophy. *FEBS Letters* 326:189-191.

Krinsky, N.I. 1992. Mechanisms of action of biological antioxidants. *Proceedings of the Society for Experimental Biology and Medicine* 200:248-254.

Kumari, M.V., T. Yoneda, and M. Hiramatsu. 1997. Effect of beta catechin on the life span of senescence accelerated mice. *Biochemistry and Molecular Biology International* 41:1005-1011.

Laganiere, S., and B.P. Yu. 1989a. Effect of chronic food restriction in aging rats: Part I. *Mechanisms of Ageing and Development* 48:207-219.

Laganiere, S., and B.P. Yu. 1989b. Effect of chronic food restriction in aging rats: Part II. *Mechanisms of Ageing and Development* 48:221-230.

Lane, M.A., D.K. Ingram, Q.G. Cutler, J.J. Knapka, D.E. Barnard, and G.S. Roth. 1992. Dietary restriction in non human primates: Program report on the NIH study. *Annals of the New York Academy of Sciences* 673:36-45.

Leeuwenburgh, C., P. Hansen, A. Shaish, J.O. Holloszy, and J.W. Heinecke. 1998. Markers of protein oxidation by hydroxyl radical and reactive nitrogen species in tissues of aging rats. *American Journal of Physiology* 274(2 Pt 2):R453-R461.

Levine, R.L., L. Mosoni, B.S. Berlett, and E.R. Stadtman. 1996. Methionine residues as endogenous antioxidants in proteins. *Proceedings of the National Academy of Sciences of the USA* 93:15036-15040.

Ma, M., and J.W. Eaton. 1992. Multicellular oxidant defense in unicellular organisms. *Proceedings of the National Academy of Sciences of the USA* 89:7924-7928.

Malm, C., M. Svensson, B. Sjoberg, B. Ekblom, and B. Sjodin. 1996. Supplementation with ubiquinone-10 causes cellular damage during intense exercise. *Acta Physiologica Scandinavica* 157:511-512.

Markert, M., G.A., Glass and B.M. Babior. 1985. Respiratory burst oxidase from human neutrophils: Purification and some properties. *Proceedings of the National Academy of Sciences of the USA* 82:3144-3148.

Mayes, P.A. 1996. Biological oxidation. In Harper's biochemistry, Stamford C.T. Edt. Appleton and Lange., 116-122.

McAlindon, T.E., P. Jacques, Y. Zhang, M.T. Hannan, P. Aliabadi, B. Weissman, D. Rush, D. Levy, and D.T. Felson. 1996. Do antioxidant micronutrients protect against the development and progression of knee osteoarthritis? *Arthritis and Rheumatism* 39:648-656.

McArdle, A., and M.J. Jackson. 1994. Intracellular mechanisms involved in damage to skeletal muscle. *Basic and Applied Myology* 4:43-50.

McCay, C.M., M.F. Crowell, and L.A. Maynard. 1936. The effect of retarded growth upon the length of life span and upon the ultimate body size. *Journal of Nutrition* 10:63-79.

McCord, J.M. 1995. Superoxide radical: Controversies, contradictions and paradoxes. *Proceedings of the Society of Experimental Biology and Medicine* 209:112-117.

Meydani, M. 1992. Vitamin E requirement in relation to dietary oil and oxidative stress in elderly. *EXS* 62:411-418.

Meydani, S.N., M.S. Santos, D. Wu, and M.G. Hayek. 1998. Antioxidant modulation of cytokines and their biologic function in the aged. *Zeitschrift fur Emahrungswissenschaft* 37 (Suppl. 1):35-42.

Miquel, J., A.B. Ramirez, A. Soler, A. Diez, M.A.C. Gutierrez, J.D. Alperi, E.Q. Ripll, A. Bernd, and E. Q. Almagro. 1998. Increase with age of serum lipid peroxides. Implication for the prevention of atherosclerosis. *Mechanisms of Ageing and Development* 100:17-24.

Moustafa, S.A., J.E. Webster, and F.E. Mattar. 1995. Effects of aging and antioxidants on glucose transport in rat adipocytes. *Gerontology* 41(6): 301-307.

Murray, C.A., and M.A. Lynch. 1998. Dietary supplementation with vitamin E reverses the age-related deficit in long term potentiation in dentate gyrus. *Journal of Biological Chemistry* 273:12161-12168.

Murray, R.K., D.K. Granner, P.A. Mayes, and V.W. Rodwell. 1996. Biological oxidation. In: *Harper's Biochemistry, 24th (ed)*, section II, pp. 116-122. Appleton and Lange. Stamford, Connecticut.

Nabuchi, Y., E. Fujiwara, K. Ueno, H. Kuboniva, Y. Asoh, and H. Ushio. 1995. *Pharmacol. Res.*, 12, 2049-52.

Nelson, S.K., S.K. Bose, and J.M. McCord. 1994. The toxicity of high dose superoxide dismutase can both initiate and terminate lipid peroxidation in the superfused heart. *Free Radical Biology and Medicine* 16:195-200.

Newham, D.J., D.A. Jones, and R.H.T. Edwards. 1986. Plasma creatine kinase changes after eccentric and concentric contractions. *Muscle and Nerve* 9:59-63.

Oberley, T.D., R.G. Allen, J.L. Schultz, and L.J. Lauchner. 1991. Antioxidant enzymes and steroid-induced proliferation of kidney tubular calls. *Free Radical Biology and Medicine* 10:79-83.

Oostenbrug, G.S., R.P. Mensink, M.R. Hardeman, T.D. Vries, F. Brouns, and G. Homstra. 1997. Exercise performance, red blood cell deformability, and lipid peroxidation: Effect of fish oil and vitamin E. *Journal of Applied Physiology* 83:746-752.

Orr, W.C., and R.S. Sohal. 1994. Extension of life-span by over expression of superoxide dismutase and catalase in *Drosophila melanogaster. Science* 263:1128-1130.

Ortenblad, N., K. Madsen, and M. S. Djurhuus. 1997. Antioxidant status and lipid peroxidation after short-term maximal exercise in trained and untrained humans. *American Journal of Physiology* 272:RI258-RI263.

Perez-Campo, R., M. Lopez-Torres, S. Cadenas, C. Rojas, and G. Barja. 1998. The rate of free radical production as a determinant of the rate of aging. *Journal of Comparative Physiology and Biochemistry* 168:149-158.

Perrig, W.J., P. Perrig, and H.B. Stahelin. 1997. The relation between antioxidants and memory performance in the old and very old. *Journal of the American Geriatrics Society* 45:718-724.

Peters-Futre, E.M. 1997. Vitamin C, neutrophil function, and upper respiratory tract infection risk in distance runners: The missing link. *Exercise and Immunological Reviews* 3:32-52.

Poulin, J.E., C. Cover, M.R. Gustafson, and M.B. Kay. 1996. Vitamin E prevents oxidative modification of brain and lymphocyte band 3 proteins during aging. *Proceedings of the National Academy of Sciences of the USA* 93:5600-6503.

Quiroga, B.D., M.T. Lopez, and R.P. Campo. 1992. Relationship between antioxidants, lipid peroxidation and aging. In: *Free radicals and aging*, I. Emerit and B. Chance (eds.), pp. 109-123. Birkhauser Verlag, Basel, Switzerland.

Radák, Z., K. Asano, M. Inoue, T. Kizaki, S.O. Ishi, K. Suzuki, N. Taniguchi, and H. Ohno. 1996. Superoxide dismutase derivative prevents oxidative damage in liver and kidney of rats induced by exhausting exercise. *European Journal of Applied Physiology and Occupational Physiology* 72:189-194.

Radák, Z., and S. Goto. 1998. The effects of exercise, ageing and caloric restriction on protein oxidation and DNA damage in skeletal muscle. In: *Oxidative Stress in Skeletal Muscle*, A.Z. Reznick (ed.), pp. 87-102. Birkhauser Verlag, Basel, Switzerland.

Reckelhoff, J.F., V. Kanji, L.C. Racusen, A.M. Schmidt, S.D. Yan, J.M. Roberts, and A.K. Salahudeen. 1998. Vitamin E ameliorates enhanced renal lipid peroxidation and accumulation of F2-isoprostanes in aging kidneys. *American Journal of Physiology* 274:R767-R774.

Reinhold, H.S., W. Calvo, J.W. Hopewell, and A.P. van der Berg. 1990. Development of blood vessel-related radiation damage in the fimbria of the central nervous system. *Int. J. Radiat. Oncol. Biol. Phys.* 18 (1): 37-42.

Reiter, R.J. 1995. Oxidative processes and antioxidative defense mechanisms in the aging brain. *FASEB Journal* 9:526-533.

Remacle, J., C. Michiels, and M. Raes. 1992. The importance of antioxidant enzymes in cellular aging and degeneration. In: *Free radicals and aging*, I. Emerit and B. Chance (eds.), pp. 99-108. Birkhauser Verlag, Basel, Switzerland.

Reznick, A.Z., D. Han, and L. Packer. 1997. Cigarette smoke induced oxidation of human plasma proteins, lipid, and antioxidants: Selective protection by biothiols dihydrolipoic acid and glutathione. *Redox Report* 3:169-174.

Reznick, A.Z., and L. Packer. 1994. Oxidative damage to proteins spectrophotometric method for carbonyl assay. *Methods in Enzymology* 233:357-363.

Reznick, A.Z., L. Packer, and C.K. Sen. 1998. Strategies to assess oxidative stress. In: *Oxidative Stress in Skeletal Muscle*, A.Z. Reznick (ed.), pp. 43-58. Birkhauser Verlag, Basel, Switzerland.

Reznick, A.Z., E.H. Witt, M. Matsumoto, and L. Packer. 1992a. Vitamin E inhibits protein oxidation in skeletal muscles of resting and exercised. *Biochem. Biophys. Res. Comm.* 189, 801-806.

Reznick, A.Z., E.H. Witt, M. Silbermann, and L. Packer. 1992b. The threshold of age in exercise and antioxidants action. *EXS* 62:423-427.

Rikans, L.E., and K.R. Hornbrook. 1997. Lipid peroxidation, antioxidant protection and aging. *Biochimica et Biophysica Acta* 1362:116-127.

Rokitzki, L., E. Logemann, G. Huber, E. Keck, and J. Keul. 1994. Alpha-tocoperol supplementation in racing cyclists during extreme endurance training. *International Journal of Sport Nutrition* 4:253-264.

Rong, Y., L. Li, V. Shah, and B.H. Lau. 1995. Pycnogenol protects vascular endothelial cells from t-butyl hydroperoxide induced oxidant injury. *Biotechnology and Therapy* 5:117-126.

Rose, R.C., S.P. Richer, and A.M. Bode. 1998. Ocular oxidants and antioxidant protection. *Proceedings of the Society for Experimental Biology and Medicine* 217:397-407.

Sastre, J., M. Asensi, E. Gasco, F.V. Pallardo, J.A. Ferrero, T. Furukawa, and J. Vina. 1992. Exhaustive physical exercise causes oxidation of glutathione status in blood: Prevention by antioxidant administration. *American Journal of Physiology* 263:R992-R995.

Seddon, J.M., U.A. Ajani, R.D. Sperduto, R. Hiller, N. Blair, T.C. Burton, M.D. Farber, E.S. Gragoudas, and J. Haller. 1994. Dietary carotenoids, vitamins A, C, E, and advanced age-related macular degeneration. *Journal of the American Medical Association* 272:1413-1420.

Sen, C.K. 1995. Oxidants and antioxidants in exercise. *Journal of Applied Physiology* 79:675-686.

Sen, C.K., M. Atalay, J. Agren, D.E. Laaksonen, S. Roy, and O. Hanninen. 1997. Fish oil and vitamin E supplementation in oxidative stress at rest and after physical exercise. *Journal of Applied Physiology* 83:189-195.

Sen, C.K., M. Atalay, and O. Hanninen. 1994. Exercise-induced oxidative stress: Glutathione supplementation and deficiency. *Journal of Applied Physiology* 77:2177-2187.

Shoff, S.M., J.A.M. Perlman, K.J. Cruickshanks, R. Klein, B.E. Klein, and L.L. Ritter. 1993. Glycosylated hemoglobin concentrations and vitamin E, vitamin C, and beta-carotene intake in diabetic and nondiabetic older adults. *American Journal of Clinical Nutrition* 58:412-416.

Short, R., D.D. Williams, and D.M. Bowden. 1997. Circulating antioxidants as determinants of the rate of biological aging in pigtailed macaques. *Journal of Gerontology* 52:B26-B38.

Sies, H. 1993. Strategies of antioxidant defence. *European Journal of Biochemistry* 215:213-219.

Sies, H., W. Stahl, and A.R. Sundquist. 1992. Antioxidant functions of vitamins. Vitamin E and C, beta-carotene, and other carotenoids. *Annals of the New York Academy of Sciences.* 669:7-20.

Simic, M.G. 1992. The rate of DNA damage and aging. In: *Free radicals and aging*, I. Emerit and B. Chance (eds.), pp. 20-30. Birkauser Verlag, Basel, Switzerland.

Socci, D.J., B.M. Crandall, and G.W. Arendash. 1995. Chronic antioxidant treatment improves the cognitive performance of aged rats. *Brain Research* 693:88-94.

Sohal, R.S. 1994. Oxidative damage, mitochondrial oxidants and antioxidant defense during aging and in response to food restriction in the mouse. *Mechanisms of Ageing and Development* 74:124-135.

Sohal, R.S., and R. Weindruch. 1996. Oxidative stress, caloric restriction, and aging. *Science* 273:59-63.

Sohal, R.S., H.H. Ku, S. Agarwal, M.J. Forster, and H. Lal. 1994. Oxidative damage, mitochondrial oxidant generation and antioxidant defenses during aging and in response to food restriction in the mouse. *Mechanisms of Ageing and Development* 74:121-133.

Spencer, N.F., M.E. Poynter, S.Y. Im, and R.A. Daynes. 1997. Constitutive activation of NF-kappa B in an animal model of aging. *International Immunology* 9:1581-1588.

Sram, R.J., B. Binkova, J. Topinka, F. Kotesovec, I. Fojtikova, I. Hanel, J. Klaschka, J. Kocisova, M. Prosek, and J. Machalek. 1993. Effects of antioxidant supplementation in an elderly population. *Basic Life Sciences* 61:459-477.

St. Clair, D.K., X.S. Wan, T.D. Oberley, K.E. Muse, and W.H. St. Clair. 1992. Suppression of radiation-induced neoplastic transformation by overexpression of mitochondrial superoxide dismutase. *Molecular Carcinogenesis* 6:238-242.

Stadtman, E.R. 1992. Protein oxidation and aging. *Science* 257:1220-1224.

Stadtman, E.R., P.E. Starke-Reed, and C.N. Oliver. 1992. Protein modification in aging. In: *Free Radicals and Aging*, I. Emerit and B. Chance (eds.), pp. 64-72. Birkhauser Verlag, Basel, Switzerland.

Sumida, S., T. Doi, M. Sakurai, Y. Yosioka, and K. Okamura. 1997. Effect of a single bout of exercise and beta-carotene supplementation on the urinary excretion of 8-hydroxy-deoxyguanosine in humans. *Free Radical Research* 27:607-618.

Taylor, A. 1992. Role of nutrients in delaying cataracts. *Annals of the New York Academy of Sciences* 669:111-123.

Teikari, J.M., L. Laatikainen, J. Virtamo, J. Haukka, M. Rautalahti, K. Liesto, D. Albanes, P. Taylor, and O.P. Heinonen. 1998. Six-year supplementation with alphatocopherol and beta-carotene and age-related maculopathy. *Acta Opthalmologica Scandinavica* 76:224-229.

Thurman, J.E., and Mooradian, A.D. 1997. Vitamin supplementation therapy in the elderly. *Drugs and Aging* 11:433-449.

Varma, S.D., P.S. Devamanoharan, and S.M. Morris. 1995. Prevention of cataracts by nutritional and metabolic antioxidants. *Critical Reviews in Food Science and Nutrition* 35:111-129.

Vasankari, T.J., U.M. Kujala, T.M. Vasankari, T. Vuorimaa, and M. Ahotupa. 1997. Increased serum and LDL lipoprotein antioxidant potential after antioxidant supplementation in endurance athletes. *American Journal of Clinical Nutrition* 65:1052-1056.

Venditti, P., M.C. Piro, G. Axtiaco, and S. DiMeo. 1996. Effect of exercise on tissue antioxidant capacity and heart electrical properties in male and female rats. *European Journal of Applied Physiology and Occupational Physiology* 74:322-329.

Wagner, J.R., C.C. Hu, and B.N. Ames. 1992. Endogenous oxidative damage of deoxycitidine in DNA. *Proceedings of the National Academy of Sciences of the USA* 89:3380-3384.

Wards, J. 1988. Should antioxidant vitamins be routinely recommended for older people? *Drugs and Aging* 12:169-175.

Wards, J. 1994. Free radicals, antioxidants and preventive geriatrics. *Australian Family Physician* 23:1297-1301.

Warren, J.A., R.R. Jenkins, L. Packer, E.H. Witt, and R.B. Armstrong. 1992. Elevated muscle vitamin E does not attenuate eccentric exercise-induced muscle injury. *Journal of Applied Physiology* 72:2168-2175.

West, S., S. Vitale, J. Hallfrisch, B. Munoz, D. Muller, S. Bressler, and N.M. Bressler. 1994. Are antioxidants or supplements protective for age-related macular degeneration? *Archives of Ophthalmology* 112:222-227.

Williams, L.R. 1995. Oxidative stress, age-related neurodegeneration and the potential for neurotrophic treatment. *Cerebrovascular and Brain Metabolism Review* 7:55-73.

Witt, E.H., A.Z. Reznick, C.A. Viguie, P. Sarke-Reed, and L. Packer. 1992. Exercise, oxidative damage and effects of antioxidants manipulation. *Journal of Nutrition* 122:766-773.

Witt, E., A.Z. Reznick, and L. Packer. 1992a. Vitamin E inhibits protein oxidation in skeletal muscle of resting and exercised rats. *Biochemical and Biophysics Research Communication* 189 (2): 801-806.

Xia, E., G. Rao, H. Van Remmen, A.R. Heydari, and A. Richardson. 1995. Activities of antioxidant enzymes in various tissues of male Fischer 344 rats are altered by food restriction. *Journal of Nutrition* 125:195-201.

Yamasaki, T., and B.H. Lau. 1997. Garlic compounds protect vascular endothelial cells from oxidant injury. *Nippon Yakurigaku Zasshi Folia Pharmacologica Japonica* 110 Suppl. 1:138P-141P.

Yoneda, T., M. Hiramatsu, M. Sakainoto, K. Togasaki, M. Komatsu, and K. Yamaguchi. 1995. Antioxidant effects of beta catechin. *Biochemistry and Molecular Biology International* 35:995-1008.

Youngman, L.D. 1993. Protein restriction and calorie restriction compared: Effects on DNA damage, carcinogenesis, and oxidative damage. *Mutation Research* 295:165-179.

Yu, B.P. 1996. Aging and oxidative stress: Modulation by dietary restriction. *Free Radical Biology and Medicine* 21:651-668.

Yu, B.P., C.M. Kang, J.S. Han, and D.S. Kim. 1998. Can antioxidant supplementation slow the aging process? *Biofactors* 7:93-101.

Zemlan, F.P., O.J. Thienhaus, and H.B. Bosmann. 1989. Superoxide dismutase activity in Alzheimer's disease. *Brain Research* 476:160-162.

Chapter 4

Reactive Oxygen Species and Skeletal Muscle Function

Francisco H. Andrade

Department of Neurology, Case Western Reserve University and University Hospitals of Cleveland, Cleveland, Ohio

The aphorism "life is harsh" is all so true, because we exist in an atmosphere that contains 21% oxygen, a fairly reactive gas. Oxygen use is mandatory for so-called aerobic organisms such as ourselves, mainly as the final electron acceptor in mitochondrial oxidative phosphorylation. It is also a source of discomfort and concern, real and imaginary, as anybody taking daily vitamin E and C supplements will attest. Moreover, a few common expressions used to describe our physical state can be darkly amusing, as they seem analogous to the effects of oxygen on the world around us: feeling "rusty," or "rotting" with age.

The study of oxygen and its redox-active derivatives grows ever more inclusive as data accumulate. Emphasis has shifted from the pathology of reactive oxygen species to the study of their biology: How are they harnessed by biological systems to fulfill important cellular functions? Physiologically relevant production of reactive oxygen species is now recognized, and the list of processes that can be regulated by them is growing continuously.

Interest in the role of reactive oxygen intermediates ("free radicals") on skeletal muscle function can be traced to a few seminal studies published in the late 1970s. However, the link between muscle physiology and the study of free radicals, albeit tenuous, goes back further. More than 40 years ago, two reports demonstrated that oxygen toxicity and ionizing radiation cause cellular damage by a common mechanism: the generation of reactive oxygen species (Fenn et al., 1957; Gerschman et al., 1954). One of the co-authors of these studies was Wallace O. Fenn, well recognized in the field of muscle physiology as the discoverer of the property of muscle energetics that disproved the viscoelastic theory of muscle contraction: the Fenn effect.

History aside, this chapter will explore the current understanding of how reactive oxygen species influence the biology of skeletal muscles. From putative sources and cellular targets of reactive oxygen species, it will move on to the effects of free radicals on the function of normal skeletal muscle, their role in exercise and fatigue, and how they may also mediate changes observed with aging and age-related pathologies.

Sources of Reactive Oxygen

The relevance of the study of redox effects on any physiological system rests on the premise that reactive oxygen species are present in the particular cellular milieu. In this case, either free radicals are generated locally within the muscle fibers or they are imported from the surrounding environment. Then, the continuous interaction between reactive oxygen species and the cellular environment determines the cellular redox state (or redox balance). This term refers to the state of the different cellular redox pools, including sulfhydryls (glutathione, cysteine, and other small thiols) and pyridine nucleotides (Chance et al., 1979; Kehrer and Lund, 1994). There is evidence that the cellular redox balance can be harnessed and changed to better suit specific metabolic needs, and that it can be considered as a dynamic condition, fluctuating from a relatively reduced state to relative oxidation, depending on functional requirements (Hwang et al., 1992; Ziegler, 1985). Indeed, for some cellular processes, a particular redox state is required for optimal function; and changes to the oxidized or reduced state are detrimental (figure 4.1) (Aghdasi et al., 1997b; Andrade et al., 1998b). This concept of optimal redox state may be useful in studying the biology of reactive oxygen species.

The list of redox-active compounds is not strictly limited to reactive oxygen species such as superoxide anion, hydroxyl radical, hydrogen peroxide (H_2O_2), and nitric oxide (NO). Important, biologically active compounds are derived from the interactions of the primary reactive oxygen species between themselves (e.g., peroxynitrite) and with cellular components (e.g., nitrosothiols, aldehydes, isoprostanes, glycated proteins). Hence, cellular redox balance shifts in response to overall changes in the content of reactive oxygen species and their biologically active derivatives.

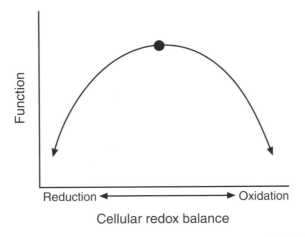

Figure 4.1   A scheme relating function to cellular redox balance or state. The shape of the curve implies that there is a redox state at which function is optimal (●). The position of reduction and oxidation is arbitrary and probably specific to the regulated process.

So, do skeletal muscles produce reactive oxygen species? The answer to this question is an unequivocal yes. Free radical generation by skeletal muscle has been detected and quantified by a wide array of methods: spectrophotometry, fluorescence, chemiluminescence, and spin traps (Balon and Nadler, 1994; Diaz et al., 1993; Jackson et al., 1985; Phung et al., 1994; Reid et al., 1992a).

Certain ubiquitous metabolic pathways are considered the main sources of reactive oxygen species, even if their relative importance in the production of free radicals by skeletal muscle remains unknown: They include mitochondrial oxidative phosphorylation, cytosolic and membrane-bound nicotinamide adenine dinucleotide phosphate (NADPH) oxidoreductases, xanthine dehydrogenase/oxidase, and eicosanoid synthesis.

Mitochondrial Respiration

The mitochondrial electron transport chain that drives oxidative phosphorylation is a group of enzymes and electron carriers whose coordinated activities couple the oxidation of metabolic substrates to the production of adenosine triphosphate (ATP). Oxygen is the final electron acceptor, and it is reduced to water during this process. About 2 to 3% of the oxygen used by the mitochondria "leaks" out incompletely reduced as superoxide anion or H_2O_2 (figure 4.2) (Chance et al., 1979; Naqui et al., 1986). Although the fraction of oxygen leaking out as free radicals is fairly small, the amount of reactive oxygen species generated can be significant if the tissue is metabolically active and uses a lot

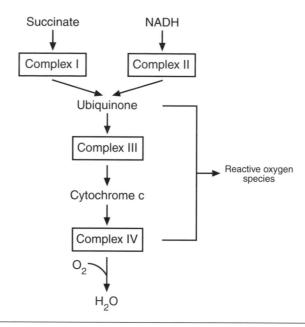

Figure 4.2 Sequence of respiratory electron carriers in mitochondria for the oxidation of succinate and reduced nicotinamide–adenine dinucleotide (NADH)-linked substrates. Probable "leaks" of reactive oxygen species are the reduction of ubiquinone and the catalytic step of complex IV.

of oxygen. Such is the case with liver, brain, heart, and active skeletal muscle. Moreover, it is unclear whether this small degree of mitochondrial uncoupling is fixed or can be altered under physiological conditions. For example, free cytosolic calcium ($[Ca^{2+}]_i$) fluctuates widely during contractions of skeletal muscle. Calcium can move very rapidly into mitochondria, where it activates various dehydrogenases and accelerates oxidative phosphorylation (Bowser et al., 1998; Duchen, 1999; Hansford, 1994). This is an attractive mechanism for coupling ATP supply to demand, but contractile activity also increases the production of reactive oxygen species in skeletal muscle (Davies et al., 1982; Diaz et al., 1993; Kolbeck et al., 1997; Reid et al., 1992a, 1992b). It remains undetermined to what extent this activity-dependent increase in free radical production is caused by higher oxygen consumption or by a bigger fraction of the consumed oxygen being leaked as incompletely reduced species.

Sarcolemmal and Cytosolic Oxidoreductases

Cellular oxidoreductases use NADPH as a source of reducing equivalents during catalysis. Because electrons are shunted from one molecule to another, they may generate reactive oxygen species. The most widely recognized (and currently very popular) of these enzymes is NO synthase (NOS). The free radical NO is an important physiological messenger in a wide variety of tissues (Ignarro, 1990; Moncada et al., 1991). It has recently been shown that mammalian skeletal muscle fibers express the three isoforms of NOS: neuronal (nNOS), endothelial (eNOS), and inducible (iNOS) (Gath et al., 1996; Kobzik et al., 1994, 1995; Silvagno et al., 1996; Thompson et al., 1996). Interestingly, the nNOS expressed in skeletal muscle may actually be a distinct isoform, resulting from alternative splicing of its genetic locus in this tissue (Silvagno et al., 1996). Neuronal NOS is normally found along the sarcolemma of fast skeletal muscle fibers, together with the dystrophin-glycoprotein complex via its linkage to $\alpha 1$-syntrophin (Brenman et al., 1995, 1996). The functional significance of this distribution pattern remains unknown. The eNOS isoform is found in the cytosol, apparently associated with mitochondria (Kobzik et al., 1995). Finally, iNOS expression can be induced in skeletal muscles in the same manner as in other tissues in response to a septic challenge (Hussain et al., 1997; Thompson et al., 1996). Increased NO production by this isoform, expressed during sepsis, has been implicated in skeletal muscle dysfunction (Gath et al., 1996; Hussain et al., 1997; Krause et al., 1998).

Other cytosolic and membrane-associated oxidoreductases may contribute significantly to cellular production of reactive oxygen species. Some may be under hormonal control, whereas others participate in xenobiotic metabolism (Crane et al., 1994; Radjendirane et al., 1998). Surprisingly, skeletal muscle expresses two NADPH:quinone oxidoreductases (NQ1 and NQ2) at levels comparable to those found in liver, which suggests that skeletal muscle has a role in the detoxification of xenobiotics. These enzymes use the reducing equivalents provided by NADPH (NQ1) and dihydronicotinamide riboside (NQ2) to reduce quinones (vitamin K and xenobiotics) and may activate certain hydroquinones to generate reactive oxygen species (Jaiswal, 1994; Radjendirane et al., 1998). Because of the very specific role of these enzymes in detoxification pathways, an important question is why skeletal muscle has such a high content of these enzymes.

Xanthine Oxidase

Xanthine oxidase is a cytosolic enzyme that produces superoxide anion and H_2O_2 during the conversion of hypoxanthine to xanthine. This enzyme is normally present as an NAD$^+$-dependent dehydrogenase, xanthine dehydrogenase, which generates NADH and xanthine as reaction products. Under conditions of increased calcium load, a calcium-dependent protease converts the dehydrogenase to an oxidase that uses molecular oxygen instead of NAD$^+$ as the electron acceptor, with the consequent production of xanthine plus superoxide anion and H_2O_2 (Cross and Jones, 1991). This enzyme is most abundant in endothelial cells but also can be found in skeletal muscle fibers (Apple et al., 1991; McCutchan et al., 1990). It is still unknown whether physiological stresses, such as activity, are sufficient to mediate the conversion from dehydrogenase to oxidase and increase the production of reactive oxygen species.

Eicosanoids

The use of arachidonic acid as a substrate for the generation of small signal transducing molecules results in the production of endoperoxide derivatives by the enzyme cyclooxygenase. These hydroperoxides are usually transformed to more stable eicosanoid derivatives, but most remain redox-active because reactive elements are present within the molecular structure (Cilento and Adam, 1995). Recently, a new series of prostaglandins, the isoprostanes, have been described. These compounds are generated nonenzymatically, secondary to peroxidation of arachidonic acid (Morrow et al., 1990). The arachidonic acid peroxides yield bicycloendoperoxide derivatives, similar to prostaglandin G_2. These are then reduced to F_2-isoprostanes (prostaglandin F_2-like). The production of isoprostanes increases during periods of oxidative stress, and their measurement is an index of lipid peroxidation (Morrow et al., 1992). Furthermore, because the isoprostanes are analogous to prostaglandins and related eicosanoids, they may play biologically important roles as signaling molecules and as mediators of oxidative damage (Morrow and Roberts, 1997).

Although intracellular production of reactive oxygen species has been emphasized so far, the extracellular compartment can also be a source of free radicals and their biologically active derivatives. The sarcolemma allows the passage of NO, superoxide anion, H_2O_2, and other organic peroxides (Kolbeck et al., 1997; Murrant et al., 1999; Reid et al., 1992b). This renders the muscle fibers susceptible to redox changes in the extracellular environment and may allow reactive oxygen species to be used as cell-to-cell signals. Endothelial cells in capillary beds have eNOS and xanthine oxidase (Apple et al., 1991; Dawson et al., 1991; Friedl et al., 1990; Michel and Feron, 1997). NO may be produced in a different location and transported in the circulation bound to hemoglobin or in combination with thiols (Jia et al., 1996; Stamler, 1994). This has interesting implications for the use of reactive oxygen species as humoral signaling factors, affecting function at sites removed from their production locus.

Cellular Targets

The cellular production of reactive oxygen species is checked by specific and nonspecific antioxidant systems. Their biological effects depend on the cellular redox state, a balance of free radical generation and scavenging (as previously discussed), which appears to be a tightly controlled variable under normal conditions. Indeed, the redox state of the intracellular compartments can be adjusted to fit functional and metabolic requirements (Gilbert, 1990; Hwang et al., 1992; Ziegler, 1985).

What defines a redox-sensitive target? Reactive oxygen species can interact with proteins, lipids, carbohydrates, and nucleic acids. The sensitivity of a particular target is defined by two factors: (1) the local redox state, which may vary from the overall cellular balance, and (2) the intrinsic sensitivity of the molecule to oxidation–reduction. The existence of compartments within the cytosol is not a new concept. Common examples are the association of glycolytic enzymes to the triads, the abundance of creatine phosphokinase along the sarcomeric M-line, and the functionally distinct subsarcolemmal and intermyofibrillar mitochondria (Cogswell et al., 1993; Han et al., 1992). This last item also may be relevant in the context of redox regulation considering that mitochondria are an important source of reactive oxygen species. Further examples of possible compartmentalized redox influences are the localization of NOS to the sarcolemma and mitochondria and the oxidized redox state of the endoplasmic reticulum (Hwang et al., 1992; Kobzik et al., 1994, 1995). Although these sources of reactive oxygen species and localized modulation of cellular function have not been definitively linked, their compartmentalization certainly makes that a tantalizing possibility.

The magnitude of the change in redox state determines the biological effect; if the magnitude of change is small or very localized, it will only affect the most sensitive targets, such as proteins containing metal centers (e.g., heme) or accessible thiol groups, essential for activity. The oxidation or reduction of these reactive sites would then have a functional consequence (figure 4.1) (Brigelius, 1985; Gilbert, 1984; Ziegler, 1985). The list of proteins whose function can be modulated in this way is long (see Gilbert, 1984).

This section is limited to those cellular targets of reactive oxygen species that appear to be most relevant to skeletal muscle. Roughly, this list includes calcium handling proteins, myofibrillar proteins, metabolic enzymes, and signal transduction pathways.

Calcium Handling

Oxidants augment calcium release and induce muscle contractures, apparently by increasing the opening probability of calcium release channels in the sarcoplasmic reticulum (ryanodine receptors) (Favero et al., 1995; Stoyanovsky et al., 1997; Stuart et al., 1992; Trimm et al., 1986).This effect is so robust that a proposed model of excitation–contraction coupling requires that critical thiol groups be oxidized for the calcium release channel tetramer to open, and that these critical thiol groups subsequently be reduced for channel closure (Abramson and Salama, 1989; Zaidi et al., 1989). A more likely scenario is that the redox state of the calcium release channels modulates opening probability and channel behavior but does not mediate channel opening per se (Ríos and Pizarro, 1991). Along these lines, recent reports have demonstrated that the ryanodine receptors have a biphasic response to oxidants: an initial increase in opening probability upon oxidation, which turns into a decrease with further oxidation (Aghdasi et al., 1997b). This suggests the existence of an optimal redox state for ryanodine receptor function; the state of redox-sensitive thiol subpopulations modulates channel activity (figure 4.1). There is also evidence of antagonistic interaction between different reactive oxygen species, whereas the functional response of the calcium release channel to the oxidant depends on the order in which the reactive sites are modified (Aghdasi et al., 1997a; Mészàros et al., 1996; Zable et al., 1997).

Calcium uptake into the sarcoplasmic reticulum is mediated by calcium ATPases (calcium "pumps"). The activity of these calcium pumps is inhibited following exposure to oxidants (figure 4.3) (Andrade et al., 1998b; Dinis et al., 1993; Okabe et al., 1982; Scherer and Deamer, 1986). The decrease in catalytic activity can be prevented by reducing agents and correlates with loss of protein thiol content (Scherer and Deamer, 1986), which suggests that some thiol groups are important for calcium-ATPase activity. However, the function of these membrane-spanning proteins also can be disrupted by the loss of membrane organization resulting from significant lipid peroxidation (Dinis et al., 1993). This points to the importance of extrinsic factors on enzyme structure/function and introduces one more mechanism by which reactive oxygen species may exert their influence.

Myofilaments

Interestingly, the effects of redox modification on the function of the contractile proteins have been explored as a consequence of spin probes being used to study actomyosin kinetics (Crowder and Cooke, 1984; Duke et al., 1976; Kwon et al., 1994). Thiol groups on actin and myosin are routinely labeled with spin or fluorescent probes to determine the steps in the interaction between actin and myosin; simultaneously, they have been identified as important determinants of actomyosin ATPase activity (Brooke and Kaiser, 1970; Crowder and Cooke, 1984; Kwon et al., 1994). The thiol groups on actin are fairly resistant to oxidation, which suggests that they are restricted to inaccessible parts of the protein (Liu et al., 1990). On the other hand, the myosin molecule contains over 40 thiol residues, 12 or 13 of which are on each of the two myosin heads

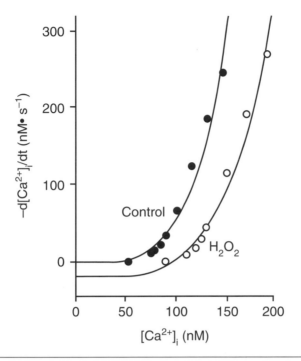

Figure 4.3   Effect of prolonged exposure to H_2O_2 (>15 min) on calcium reuptake. This is studied in skeletal muscle fibers by following the decline in $[Ca^{2+}]_i$ at the end of a contraction (see Andrade et al., 1998b, for details). The rate of $[Ca^{2+}]_i$ decline is plotted against $[Ca^{2+}]_i$ and fitted to a model that estimates the rate of calcium uptake and the passive calcium leak (y-axis intercept) from the sarcoplasmic reticulum. H_2O_2, shown by $\bigcirc$, decreases uptake by 40% and doubles the passive leak.

(Lowey et al., 1969). Two of them, SH_1 and SH_2, have been probed extensively because of their proximity to the actin- and ATP-binding regions (Crowder and Cooke, 1984; Duke et al., 1976). When these thiols are oxidized, myosin ATPase activity decreases significantly, at least in vitro (Crowder and Cooke, 1984; Root and Reisler, 1992). The magnitude of the functional change in vivo may be proportional to the fraction of total thiol groups that becomes oxidized (Crowder and Cooke, 1984).

Tropomyosin is an α-helical coiled-coil dimeric protein that normally blocks the interaction between actin and myosin. Upon calcium binding to troponin, tropomyosin shifts from its resting position on the actin filament, allowing actin to be bound to myosin with the concomitant hydrolysis of ATP and movement. Tropomyosin has an intersubunit disulfide bridge at Cys-190, which can be reduced, enhancing binding to actin and decreasing ATPase activity (Walsh and Wegner, 1980; Williams and Swenson, 1982). By modulating the behavior of tropomyosin, this disulfide bridge potentially may regulate the interaction between actin and myosin.

The regulators of myofilament calcium sensitivity are troponin and the regulatory light chains (Brenner, 1988; Metzger and Moss, 1992). These proteins have significant primary sequence homology as calcium-binding proteins (Collins, 1976). Cardiac troponin C becomes insensitive to calcium when an intramolecular disulfide bridge is formed (Putkey et al., 1993). However, the same effect has not been demonstrated in skeletal muscle troponin C (Potter et al., 1976).

Enzymes

The activity of a number of enzymes requires metal-containing prosthetic groups (iron-sulfur centers, heme) or reactive thiols in the active site (cysteine residues, coenzyme A, α-lipoic acid). These moieties may render enzyme activity susceptible to redox modulation, because their oxidation or reduction can alter the kinetics of the catalyzed reaction. The list of enzymes whose activities can be modulated with oxidants or reductants is very long, and it includes enzymes involved in glycolysis, tricarboxylic acid cycle, and oxidative phosphorylation (Gardner et al., 1995; Gilbert, 1984; Hyslop et al., 1988; Stamler, 1994; Zhang et al., 1990; Ziegler, 1985). The full description of known redox-sensitive enzymes is beyond the scope of this chapter; interested readers may consult the references listed here.

How the cellular redox state is harnessed for redox regulation of enzyme activity under physiological conditions is not fully understood (Brigelius, 1985; Gilbert, 1982). This is better explained by the behavior of NO in skeletal muscle. Activity increases NO production: NO dilates the arterioles and increases blood flow to the muscle; enhances glucose transport to the muscle fibers; and inhibits creatine phosphokinase, glyceraldehyde-3-phosphate dehydrogenase, aconitase, and cytochrome c oxidase (Balon and Nadler, 1997; Cleeter et al., 1994; King et al., 1994; Mohr et al., 1996; Stamler, 1994; Wolosker et al., 1996). From these facts, some conclusions begin to emerge: Muscle activity consumes energy; increased NO production by the active muscle increases the supply of glucose (blood flow and glucose transport), slows its consumption (glyceraldehyde-3-phosphate dehydrogenase, aconitase, cytochrome c oxidase), and inhibits the use of ATP (creatine phosphokinase). NO may then function as a metabolic "brake" that helps couple energy demand with supply.

Integrating the known facts about redox modulation of enzyme function within the context of expected changes in cellular redox state provides a theoretical framework of the physiological importance of reactive oxygen species as regulators of cellular metabolism. The veracity of this theoretical framework awaits further experimental data.

Signal Transduction

Most reactive oxygen species and their redox-active derivatives can cross the sarcolemma (Murrant et al., 1999). Although some of their biological effects can be explained by the direct interaction with reactive moieties on cellular constituents, they can also initiate or mediate signaling pathways. A pharmacological example of this is the way certain oxidizers mimic insulin activity by activating its receptor (Brigelius, 1985; Schmid et al., 1998). Receptors and enzymes associated with specific signal transduction systems can respond to reactive oxygen species via the same mechanism described in the preceding section, that is, the presence of redox-sensitive sites on these molecules.

This argument is easily made for NO, because it is well known that most of its biological actions are through a guanosine cyclic $3',5'$-monophosphate (cGMP)-dependent signaling system. NO stimulates soluble guanylyl cyclases, a family of cytosolic, heme- and copper-containing heterodimeric enzymes, increasing cGMP concentration. The second messenger cGMP, in turn, activates a number of protein kinases, phosphodiesterases, and ion channels, the actual effectors of the NO signal (Schmidt and Walter, 1994; Schmidt et al., 1993; Stamler, 1994). For example, the changes induced by NO on the contractile function of skeletal muscle are at least partially due to a cGMP-mediated response and not to direct redox effects on contractile proteins (Kobzik et al., 1994).

The role of other reactive oxygen species in signal transduction is not quite as well characterized. It is known that oxidants stimulate membrane receptors, protein kinases, and channels and activate the redox-sensitive transcription factors AP-1 and nuclear factor-κB (Devary et al., 1993; Simon et al., 1998; Whisler et al., 1995). However, this could represent a generic stress response to cellular damage, not specific to the initiating reactive oxygen species. In this sense, free radicals are a noxious signal, and the cellular response aims to limit or repair damage (Kletzien et al., 1994).

Evidence that reactive oxygen species may initiate or participate in signaling pathways comes mostly from other tissues and cell types. Free radicals, lipid peroxidation, and advanced end glycation products can stimulate extracellular signal-regulated kinases (ERK) and other members of the mitogen-activated protein (MAP) kinase family of growth factor signal transduction systems (Abe et al., 1998; Chen et al., 1995; Kunapuli et al., 1998; Milligan et al., 1998). More importantly, some studies suggest that the production of reactive oxygen species is a required step in certain signaling pathways (Joneson and Bar-Sagi, 1998; Sundaresan et al., 1995). Supposedly, these free radicals act on "redox switches" downstream to transmit the signal and initiate a cellular response (Lander et al., 1997; Xie et al., 1999). These are mostly redox-sensitive cysteine residues that modulate the activity of signaling enzymes and transcription factors when oxidized or reduced (Flohé et al., 1997; Lander et al., 1995; Peng et al., 1995).

From this background, an interesting question develops: Could reactive oxygen species be the link between muscle activity and muscle growth and adaptation? Recent studies have demonstrated that the response to the cytokine tumor necrosis factor α (TNF-α) is mediated by reactive oxygen species: Antioxidants and NOS inhibitors block the response to TNF-α, whereas NO donors and superoxide anion generators mimic it (Buck and Chojkier, 1996; Li et al., 1998). Whether this is part of a common mechanism for redox modulation of signal transduction and control of gene expression in skeletal muscle is unknown. Furthermore, important issues remain untested, such as the source of the regulatory reactive oxygen species and its control, the specificity of the response, and the identity of the redox targets.

Normal Muscle Function

Recent studies have suggested that reactive oxygen species may participate in the basic processes of contractile function such as excitation–contraction coupling. As mentioned previously, the function of calcium release channels and pumps in the sarcoplasmic reticulum can be influenced by oxidants and reductants. Oxidizers induce calcium release from isolated sarcoplasmic reticulum vesicles and stimulate the activity of single calcium release channels reconstituted in lipid bilayers (Abramson et al., 1995; Favero et al., 1995; Stuart et al., 1992). These effects seem to be fully reversible or preventable with reducing agents (Favero et al., 1995; Salama et al., 1992; Zable et al., 1997). From these data, investigators proposed that excitation–contraction coupling requires a change in the redox state of the calcium release channels: When the sarcolemma depolarizes, the voltage sensor on the T-tubule membrane oxidizes particular sulfhydryl groups on the calcium release channels present in the terminal cisternae of the sarcoplasmic reticulum, allowing channel opening and calcium efflux to the cytosol. This process is terminated when ubiquitous cytosolic reducing compounds, such as glutathione and cysteine, restore the calcium release channels to their original redox state (Salama et al., 1992). Whether this mechanism is active in whole skeletal muscle fibers is controversial, because it may not be fast enough to explain the fast kinetics of calcium release by the sarcoplasmic reticulum (Ríos and Pizarro, 1991). Nonetheless, available experimental data are supportive: Brief exposures to H_2O_2 increase twitch forces and prolong the time to peak tension and the half-relaxation time. In contrast, the antioxidant enzyme catalase has the opposite effects: shortening time to peak tension and half-relaxation time and decreasing twitch and submaximal tetanic forces. These data suggest that catalase acts as an external sink for endogenous H_2O_2 and drives the intracellular redox balance toward a more reduced state, inhibiting calcium release from the sarcoplasmic reticulum and decreasing force production (Reid et al., 1993b). Alternatively, H_2O_2 increases $[Ca^{2+}]_i$ and submaximal force by slowing the activity of the calcium pumps on the sarcoplasmic reticulum (figure 4.3) (Scherer and Deamer, 1986). Or, the change in force is independent of changes in $[Ca^{2+}]_i$ and is caused by alterations in the myofibrillar calcium sensitivity (figure 4.4) (Wilson et al., 1991).

Recent studies have shown that the sensitivity of contractile function to changes in cellular redox balance does not necessarily depend on altered calcium kinetics. Brief exposures to H_2O_2 and dithiothreitol (DTT, a reducing agent) do not change tetanic $[Ca^{2+}]_i$ but rather increase and decrease force,

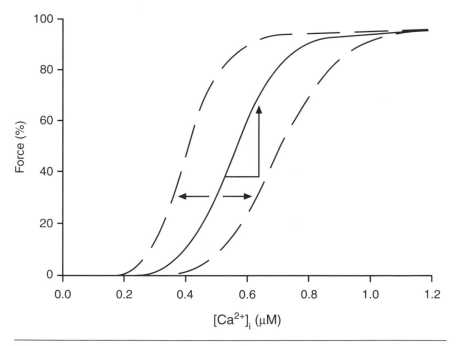

Figure 4.4   Idealized force–$[Ca^{2+}]_i$ relationship (solid line). Force can be modulated by increasing myofibrillar calcium sensitivity (less calcium required to activate the myofilaments, dashed line to the left), by decreasing myofibrillar calcium sensitivity (dashed line to the right), and by increasing $[Ca^{2+}]_i$ (upward shift along the solid line).

respectively. Prolonged exposures to H_2O_2 decrease force, without altering tetanic $[Ca^{2+}]_i$. The time-dependence of the H_2O_2 response is illustrated in figure 4.5. The late-phase decrease in force with H_2O_2 can be completely reversed with DTT. Even longer exposures at higher H_2O_2 concentrations are needed to alter the calcium reuptake by the sarcoplasmic reticulum and increase tetanic $[Ca^{2+}]_i$, as evidenced in the last three time points in figure 4.5. The divergent results obtained with H_2O_2 and DTT and the time dependence of the H_2O_2 effects strongly suggest that the cellular targets of these reagents have optimal redox states (figure 4.1). In other words, the function of certain determinants of contractile activity depends on the redox state of their component proteins. Therefore, calcium release and reuptake are fairly insensitive to redox modulation, because fibers have to be exposed to high concentrations or for prolonged periods before sarcoplasmic reticulum function is affected (figures 4.3 and 4.5). On the other hand, myofibrillar calcium sensitivity appears to be exquisitely susceptible to variations of the cellular redox balance (figure 4.5) (Andrade et al., 1998b; Brotto and Nosek, 1996; Posterino and Lamb, 1996).

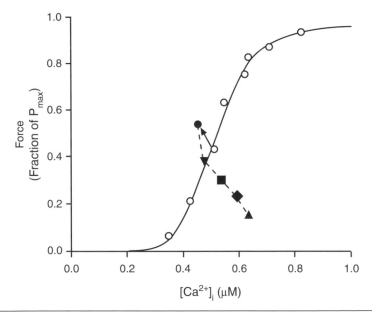

Figure 4.5 Force–$[Ca^{2+}]_i$ relationship from a single skeletal muscle fiber under control conditions (○ and solid line). The black symbols represent $[Ca^{2+}]_i$ and force during 40 Hz tetani at selected times during incubation with 100 μM H_2O_2 (2 min, ●; 6 min, ▼; 7 min, ■; 8 min, ◆; 10 min, ▲).

Adapted, by permission, from F.H. Andrade, et al., 1998, "Effect of hydrogen peroxide and dithiothreitol on contractile function of single skeletal muscle fibres from the mouse," *Journal of Physiology* 509:565-575.

Nitric oxide also modulates normal skeletal muscle function. One of the original studies on the presence of NO synthase in skeletal muscle found that inhibition of NO synthesis increased force production by rat diaphragm muscle bundles, whereas NO donors reduced force (Kobzik et al., 1994). The decreased force in response to NO was only partially explained by cGMP-dependent signaling, because simultaneous inhibition of guanylyl cyclase did not completely abolish the response to the NO donors. The remaining force deficit was then attributed to NO-induced changes of the redox state of contractile proteins (Kobzik et al., 1994). In skinned skeletal muscle fibers, NO donors decrease maximal force, myofibrillar calcium sensitivity (Ca_{50}), maximum shortening velocity (V_0), and myosin ATPase activity. These effects were explained as due to NO-induced alterations in the kinetics of the interaction between myosin and actin, presumably in response to the oxidation of reactive sulfhydryl groups on the contractile proteins, particularly myosin (Galler et al., 1997; Perkins et al., 1997). Factors inherent to more isolated systems (i.e., skinned fibers and sarcoplasmic reticulum vesicles, see following) may influence the experimental results, such as the use of the reductant DTT during their preparation (e.g., 10 mM in Perkins et al., 1997). This chemical would shift the native redox state of the contractile proteins to an abnormally reduced state, with unpredictable effects on function. Moreover, prolonged exposure to redox agents ($\geq$20 min in Perkins et al., 1997) may cause nonspecific oxidation of thiol groups throughout the fiber, resulting in artifactual alterations in function (Crowder and Cooke, 1984). It has already been shown that the redox-sensitive behavior of isolated calcium release channels depends on the order in which the experimental reagents are tested, which suggests there are subpopulations of sulfhydryl groups in the channel tetramer, each influencing function in a particular manner (Aghdasi et al., 1997a, 1997b).

The effects of nitric oxide on force production and $[Ca^{2+}]_i$ also have been studied in single skeletal muscle fibers, a more intact experimental system. Nitric oxide donors reduced myofibrillar calcium sensitivity, whereas $[Ca^{2+}]_i$ transients increased during submaximal tetani (figure 4.6). V_0, the rate of force redevelopment, and maximal force production at saturating $[Ca^{2+}]_i$ were unchanged on exposure to the NO donor S-nitroso-N-acetylcysteine (SNAC). These findings suggest that the decrease in myofibrillar calcium sensitivity is not caused by a reduction in the number of recruited cross-bridges or altered cross-bridge kinetics. The effects of NO on single fiber function are also cGMP-independent, which suggests direct interactions with target proteins, either by oxidation of sulfhydryl residues or by nitrosylation of nucleophilic centers on the contractile elements themselves (Andrade et al., 1998c).

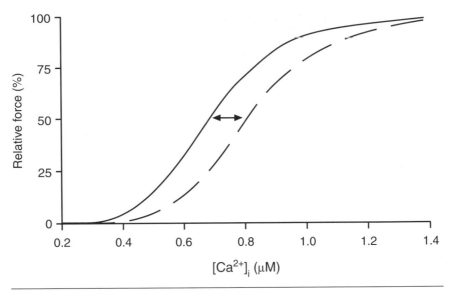

Figure 4.6 Force–[Ca^{2+}]$_i$ relationship from a single skeletal muscle fiber under control conditions (solid line) and during incubation with the NO donor SNAC (dashed line). The NO donor decreased myofibrillar calcium sensitivity, shifting the relationship to the right of control. This effect was fully reversible upon washout (arrow).

Adapted, by permission, from F.H. Andrade, et al., 1998, "Effect of nitric oxide on single skeletal muscle fibres from the mouse," *Journal of Physiology* 509:577-586.

Little is known about the effect of NO on calcium release by the sarcoplasmic reticulum. NO donors have been reported both to decrease and to increase the open probability of isolated calcium release channels (Mészàros et al., 1996; Stoyanovsky et al., 1997). These conflicting findings could be explained by the biphasic response of isolated calcium release channels to NO, which reflects distinct thiol subpopulations in the channel tetramer and reinforces the concept of optimal redox state (Aghdasi et al., 1997a, 1997b). An important caveat in the interpretation of these reports is that the experimental preparations include a reducing agent such as dithiothreitol or β-mercaptoethanol, which effectively change the native redox state of the system under study. Results on intact single fibers indicate that NO donors and NO$_2^-$ have no effect on the passive calcium leak from the sarcoplasmic reticulum, which would be equivalent to the open probability measured in isolated calcium release channels. The increase in tetanic [Ca^{2+}]$_i$ detected in this study could partially be due to a small increase in voltage-activated calcium release, sufficient to largely offset the concurrent decrease in myofibrillar calcium sensitivity and maintain force unchanged (Andrade et al., 1998c).

Muscle Fatigue and Injury

Interest in the influence of reactive oxygen species on skeletal muscle function has concentrated on their putative role in the development of fatigue and injury (Andrade et al., 1998a; Anzueto et al., 1992, 1994; Davies et al., 1982; Krause et al., 1998). Muscle fatigue can be defined as the loss of the capacity to develop force and/or velocity, as a result of muscular activity under load, which is reversible with rest (NHLBI Workshop, 1990). Fatigue due to changes in intrinsic muscle components is the reversible decline in muscle performance not explained by changes in neural drive or neuromuscular mechanisms. A multitude of factors have been proposed and examined to explain this type of fatigue, such as increased intracellular sodium content; decreased intracellular potassium; decreased intracellular pH; changes in myofilament calcium sensitivity; increased intracellular levels of inorganic phosphate; loss of calcium from the sarcoplasmic reticulum; depletion of metabolic substrates such as glycogen, phosphocreatine, or ATP; and others. The reader is directed to recent reviews on this topic for further discussion (Allen et al., 1995; Fitts, 1994; Westerblad et al., 1991, 1998).

The changes in muscle performance and biochemistry seen with increased contractile activity seem to overlap with evidence of cellular injury (Fridén and Lieber, 1992; Hoppeler, 1986; Newham et al., 1983). There is no general agreement on the difference between fatigue and injury, because both processes are reversible in normal subjects. The commonsense approach is that it takes longer to recover from injury (hours to days) than from fatigue (minutes to hours). Still, there is sufficient overlap to suggest that fatigue and injury represent grades of a continuous scale; that is, these two terms describe the severity of the cellular changes in response to activity. Unaccustomed muscular exertion results in discomfort and pain, which may persist for several days (the "weekend warrior" syndrome). After all, exercise is followed by acute structural and biochemical changes in muscle, which depend on the type and intensity of the activity and the subjects' physical condition. It has been known for some time that strenuous contractile activity increases the release of cytosolic enzymes from skeletal muscle, a hallmark of muscle damage (Clarkson et al., 1986; Highman and Altland, 1963; Jones et al., 1983; Newham et al., 1986). Moreover, activity-related morphological changes include mitochondrial swelling and rupture, T-tubular dilation and occlusion, sarcolemmal disruption, myofibrillar lysis, Z-band streaming, and evidence of necrosis and inflammation some hours or days after the exercise bout (Armstrong, 1984; Hoppeler, 1986).

As evidence of muscle damage with exercise accumulated, some studies showed that vitamin E (α-tocopherol) seemed to accelerate muscle recovery.

Results were not always consistent, but it was suggested that the beneficial effects of vitamin E could be due to its antioxidant properties (Helgheim et al., 1979; Lawrence et al., 1975; Shephard et al., 1974). Soon after, Dillard and coworkers found an unexpected increase in pentane, an index of lipid peroxidation, in expired air of exercising human subjects (Dillard et al., 1978). These investigators concluded that exercise induced increased lipid peroxidation in vivo and that an antioxidant, vitamin E, significantly reduced it at rest and during exercise. Later, two studies in rats showed increased indices of lipid peroxidation in liver and hindlimb muscles after exhaustive exercise (Brady et al., 1979; Gee and Tappel, 1981). It was then proposed that reactive oxygen species, probably leaked during mitochondrial respiration, were responsible for damage to skeletal muscles during exercise (Davies et al., 1981; Dillard et al., 1978).

In a pivotal study, Davies and coworkers demonstrated loss of mitochondrial respiratory control, impairment of sarcoplasmic and endoplasmic reticular integrity, and increased levels of lipid peroxidation products in rat hindlimb muscles and liver after exhaustive exercise. These findings correlated with a 3-fold increase in the content of reactive oxygen species in these tissues measured with electron spin resonance spectroscopy (Davies et al., 1982). Another study found parallel increases in serum creatine kinase (leaked from muscle fibers, an index of muscle injury) and free radical electron spin resonance signals after contractile activity (Jackson et al., 1985). These results have been confirmed by other studies on humans, dogs, and rodents. Evidence of increased lipid peroxidation after intense exercise coincides with increases in the plasma levels of creatine kinase and lactate dehydrogenase, decreases in the activity of skeletal muscle mitochondrial enzymes, loss of muscle force development, and even morphological changes of affected muscles (Barclay and Hansel, 1991; Duthie et al., 1990; Kanter et al., 1988; Lawler et al., 1993; Lovlin et al., 1987; Zerba et al., 1990). There is also evidence of increased levels of oxidized proteins and ribonucleic acid after exercise, with vitamin E preventing these changes, at least partially (Reznick et al., 1992; Witt et al., 1992).

Other researchers have attempted to modulate the impairment induced by oxidative stress by altering the redox milieu of skeletal muscles. Barclay and Hansel reported that development of fatigue was accelerated by the addition of xanthine oxidase, an enzyme that generates superoxide anion. On the other hand, the rate of development of fatigue was slowed by the infusion of dimethyl sulfoxide (DMSO, a nonspecific hydroxyl radical scavenger), allopurinol (an inhibitor of xanthine oxidase), or desferoxamine (an iron chelator). These data support the notion that reactive oxygen species participate in the development of fatigue (Barclay and Hansel, 1991). Similarly, Zerba and coworkers, by administering the antioxidant enzyme superoxide dismutase, attenuated the immediate loss of contractile function and the severity of the delayed injury in mouse muscles after eccentric contraction (Zerba et al., 1990). This finding suggests a dual role for reactive oxygen species: mediating both the initial fatigue and the delayed, probably inflammatory, consequences of acute exercise.

The production of reactive oxygen species by skeletal muscle increases with activity (Balon and Nadler, 1994; Diaz et al., 1993; Kolbeck et al., 1997; Reid et al., 1992a, 1992b). Moreover, prolonged contractile activity depletes cellular antioxidants and induces lipid peroxidation, protein oxidation, and deoxyribonucleic acid (DNA) damage (Alessio, 1993; Davies et al., 1982; Duthie et al., 1990; Radák et al., 1999, Sumida et al., 1989; Witt et al., 1992). However, the link between activity-induced oxidative stress and fatigue is still rather indirect. When antioxidants have been provided to experimental subjects, the indices of oxidative stress have been improved, but endurance generally has not changed (Helgheim et al., 1979; Khanna et al., 1999; Sastre et al., 1992; Witt et al., 1992). This argues against a causative role for reactive oxygen species in fatigue, unless the antioxidant interventions were not sufficient to sustain muscular activity for longer periods. The exception to this has been the use of N-acetylcysteine, a thiol-containing amino acid derivative that can also be a precursor in the synthesis of glutathione. N-acetylcysteine has been shown to increase endurance in humans, rabbits, rats, and isolated muscles (Diaz et al., 1994; Reid et al., 1994; Shindoh et al., 1990; Supinski et al., 1995). Glutathione is readily oxidized during contractile activity (Anzueto et al., 1992; Gohil et al., 1988; Lew et al., 1985). Glutathione is part of a redox system that uses NADPH to maintain an intracellular reducing environment (Kehrer and Lund, 1994). The disruption of the glutathione redox cycle accelerates the appearance of muscle dysfunction when contractile activity is increased in vivo. Therefore, the activity of the glutathione redox cycle appears to delay the onset of contractile dysfunction induced by increased mechanical loads. Both the cellular content of glutathione and the ability to use it through its redox cycle are necessary to protect skeletal muscle against the deleterious effects of oxidative stress induced by repeated contractions: Chemical depletion of cellular glutathione and inhibition of gluthatione peroxidase or glutathione reductase activity have the same consequences on skeletal muscle function (Andrade et al., 1998a; Morales et al., 1993, 1994).

In view of the evidence accumulated so far, further mechanistic studies on the role of glutathione in the maintenance of muscle function are warranted. Nevertheless, the more general question concerning the participation of reactive oxygen species in the development of muscle fatigue remains nebulous. Because of the multiplicity of sources and effects, their study will continue to be a complex undertaking.

Aging

The loss of function due to aging is particularly obvious in the musculoskeletal system. Apparently, the cumulative effects of cellular and genetic damage are important factors in aging. The production of reactive oxygen species throughout life is one of these time-dependent insults that damage cellular components. Starting from the premise that aerobic cells generate reactive oxygen species and that antioxidant defenses are not fully efficient, the oxidative stress hypothesis of aging postulates that aging occurs as oxidative damage accumulates in tissues and organs. Certain postmitotic, mitochondria-rich tissues, such as brain, heart, and skeletal muscle, show the highest degree of oxidative damage with age (Sohal and Weindruch, 1996; Wei et al., 1996). Coincidentally, these tissues are also targets of age-related disorders in which oxidative stress has been implicated (Beal, 1996; Cortopassi and Wong, 1999; Di Monte et al., 1992; Kopsidas et al., 1998; Schapira and Cooper, 1992).

Diminished motor capacity in the elderly is a complex phenomenon, because motor activity results from the interplay of central and peripheral elements plus the influence of humoral factors and behavioral patterns such as activity levels (Larsson, 1998; Li et al., 1997; Phillips et al., 1996). Nonetheless, important intrinsic changes take place in skeletal muscle as a consequence of age. The activity of mitochondrial and other metabolic enzymes decreases with age, and contractile function is lost, perhaps in a muscle-specific manner (Hansford and Castro, 1982; Ji et al., 1990; Powers et al., 1992). Aging induces a relative uncoupling between the sarcolemma and the sarcoplasmic reticulum, rendering the release of calcium upon depolarization less efficient and impairing contractile function (Mayhew et al., 1998; Renganathan and Delbono, 1998; Renganathan et al., 1997). Skeletal muscle also undergoes age-related oxidative stress as evidenced by increased lipid peroxidation and oxidative damage to DNA and proteins (Ji et al., 1990; Lass et al., 1998; Mecocci et al., 1999; Ohkuwa et al., 1997; Schöneich et al., 1999; Viner et al., 1999). The loss of cytochrome c oxidase activity in muscle fibers from old subjects is at least partially explained by mitochondrial DNA mutations, presumably due to the long-term deleterious effects of reactive oxygen species generated during mitochondrial respiration (Kadenbach et al., 1995; Müller-Höcker et al., 1993, 1996). A recently published report demonstrated that skeletal muscle from aged rats is in a relative state of oxidative stress, with increased generation of reactive oxygen species (Bejma and Ji, 1999). Furthermore, aging increases the activity of antioxidant enzymes such as superoxide dismutase, catalase, glutathione peroxidase, and glutathione reductase in at least some skeletal muscles (Ji et al., 1990; Ji, 1993; Leeuwenburgh et al., 1994; Oh-ishi et al., 1996). All these findings point to skeletal muscle damage induced by age-dependent oxidative stress: damage to cellular components, loss of enzyme function, and up-regulation of antioxidant defenses.

Conclusion

Happily, we have learned to live with oxygen. The evolutionary wisdom of incorporating a widely available, albeit reactive, compound into our metabolic pathways is evident by the homology in metabolic pathways of most, if not all, aerobic organisms. Oxygen use is a critical step in the production of our energy currency (ATP), and we use its more reactive forms and by-products to regulate cellular function and growth. At the same time, oxygen may be the prime reason why our lives are finite: Living in a "toxic" environment has detrimental consequences and may cause disease and accelerate aging.

The concept of cellular redox state or balance as originally proposed considers reactive oxygen species as normal components of the physiological milieu, their production and use to be purposefully regulated to maintain homeostasis (Reid, 1996). Some of the reports summarized here present evidence of biphasic responses to reactive oxygen species. As mentioned previously, this suggests that there may be an optimal redox state for function: Deviations to a grossly "reduced" or "oxidized" state would be detrimental (figure 4.1). This "mental model" of redox modulation of function seems to work for skeletal muscle. At least, it provides us with a first approximates the effects of reactive oxygen species on skeletal muscle function and how they may participate in the development of fatigue and injury.

Where do we go from here? Current knowledge is little more than a list of sources of reactive oxygen species and their putative targets. Important pieces are still missing from this puzzle: primary sites and control of production, cellular distribution of free radicals, and signal transduction. Although the presence of reactive oxygen species has been demonstrated in skeletal muscle during fatigue, aging, and some diseases, their mechanistic participation in each one of these processes remains to be fully elucidated. Moreover, production, buffering, and susceptibility to free radicals may not necessarily be the same in all muscle fibers. Therefore, the coherent picture provided by the idea of redox homeostasis is a good starting point for the study of the biology of reactive oxygen species in skeletal muscle.

References

Abe, M.K., S. Kartha, A.Y. Karpova, J. Li, P.T. Liu, W.-L. Kuo, and M.B. Hershenson. 1998. Hydrogen peroxide activates extracellular signal-regulated kinase via protein kinase C, raf-1, and MEK1. *American Journal of Respiratory Cell and Molecular Biology* 18:562-569.

Abramson, J.J., and G. Salama. 1989. Critical sulfhydryls regulate calcium release from sarcoplasmic reticulum. *Journal of Bioenergetics and Biomembranes* 21:283-294.

Abramson, J.J., A.C. Zable, T.G. Favero, and G. Salama. 1995. Thimerosal interacts with the Ca^{2+} release channel ryanodine receptor from skeletal muscle sarcoplasmic reticulum. *Journal of Biological Chemistry* 270:29644-29647.

Aghdasi, B., M.B. Reid, and S.L. Hamilton. 1997a. Nitric oxide protects the skeletal muscle Ca^{2+} release channel from oxidation induced activation. *Journal of Biological Chemistry* 272:25462-25467.

Aghdasi, B., J.-Z. Zhang, Y. Wu, M.B. Reid, and S.L. Hamilton. 1997b. Multiple classes of sulfhydryls modulate the skeletal muscle Ca^{2+} release channel. *Journal of Biological Chemistry* 272:3739-3748.

Alessio, H.M. 1993. Exercise-induced oxidative stress. *Medicine and Science in Sports and Exercise* 25:218-224.

Allen, D.G., J. Lännergren, and H. Westerblad. 1995. Muscle cell function during prolonged activity: Cellular mechanisms of fatigue. *Experimental Physiology* 80:497-527.

Andrade, F.H., A. Anzueto, W. Napier, S. Levine, R.A. Lawrence, S.G. Jenkinson, and L.C. Maxwell. 1998a. Effects of selenium deficiency on diaphragmatic function after resistive loading. *Acta Physiologica Scandinavica* 162:141-148.

Andrade, F.H., M.B. Reid, D.G. Allen, and H. Westerblad. 1998b. Effect of hydrogen peroxide and dithiothreitol on contractile function of single skeletal muscle fibres from the mouse. *Journal of Physiology* 509:565-575.

Andrade, F.H., M.B. Reid, D.G. Allen, and H. Westerblad. 1998c. Effect of nitric oxide on single skeletal muscle fibres from the mouse. *Journal of Physiology* 509:577-586.

Anzueto, A., F.H. Andrade, L.C. Maxwell, S.M. Levine, R.A. Lawrence, W.J. Gibbons, and S.G. Jenkinson. 1992. Resistive breathing activates the glutathione redox cycle and impairs performance of rat diaphragm. *Journal of Applied Physiology* 72:529-534.

Anzueto, A., J.M. Brassard, F.H. Andrade, R.A. Lawrence, L.C. Maxwell, S.M. Levine, and S.G. Jenkinson. 1994. Effects of hyperoxia on rat diaphragm function. *Journal of Applied Physiology* 77:63-68.

Apple, F.S., J.E. Hyde, A.M. Ingersoll-Stroubos, and A. Theologides. 1991. Geographic distribution of xanthine oxidase, free radical scavengers, creatine kinase, and lactate dehydrogenase enzyme systems in rat heart and skeletal muscles. *American Journal of Anatomy* 192:319-323.

Armstrong, R.B. 1984. Mechanisms of exercise-induced delayed onset muscular soreness: A brief review. *Medicine and Science in Sports and Exercise* 16:529-538.

Balon, T.W., and J.L. Nadler. 1994. Nitric oxide release is present from incubated skeletal muscle preparations. *Journal of Applied Physiology* 77:2519-2521.

Balon, T.W., and J.L. Nadler. 1997. Evidence that nitric oxide increases glucose transport in skeletal muscle. *Journal of Applied Physiology* 82:359-363.

Barclay, J.K., and M. Hansel. 1991. Free radicals may contribute to oxidative skeletal muscle fatigue. *Canadian Journal of Physiology and Pharmacology* 69:279-284.

Beal, M.F. 1996. Mitochondria, free radicals, and neurodegeneration. *Current Opinion in Neurobiology* 6:661-666.

Bejma, J., and L.L. Ji. 1999. Aging and acute exercise enhance free radical generation in rat skeletal muscle. *Journal of Applied Physiology* 87:465-470.

Bowser, D.N., T. Minamikawa, P. Nagley, and D.A. Williams. 1998. Role of mitochondria in calcium regulation of spontaneously contracting cardiac muscle cells. *Biophysical Journal* 75:2004-2014.

Brady, P.S., L.J. Brady, and D.E. Ullrey. 1979. Selenium, vitamin E and the response to swimming stress in the rat. *Journal of Nutrition* 109:1103-1109.

Brenman, J.E., D.S. Chao, S.H. Gee, A.W. McGee, S.E. Craven, D.R. Santillano, Z. Wu, F. Huang, H. Xia, M.F. Peters, S.C. Froehner, and D.S. Bredt. 1996. Interaction of nitric oxide synthase with the postsynaptic density protein PSD-95 and α1-syntrophin mediated by PDZ domains. *Cell* 84:757-767.

Brenman, J., D. Chao, H. Xia, K. Aldape, and D. Bredt. 1995. Nitric oxide synthase complexed with dystrophin and absent from skeletal muscle sarcolemma in Duchenne muscular dystrophy. *Cell* 82:743-752.

Brenner, B. 1988. Effect of Ca^{2+} on cross-bridge turnover kinetics in skinned single rabbit psoas fibers: Implications for regulation of muscle contraction. *Proceedings of the National Academy of Sciences of the USA* 85:3265-3269.

Brigelius, R. 1985. Mixed disulfides: Biological functions and increase in oxidative stress. In: *Oxidative Stress*, H. Sies (ed.), pp. 243-272. Academic Press, London.

Brooke, M.H., and K.K. Kaiser. 1970. Three "myosin adenosine triphosphatase" systems: The nature of their pH lability and sulfhydryl dependence. *Journal of Histochemistry and Cytochemistry* 18:670-672.

Brotto, M.A.P., and T.M. Nosek. 1996. Hydrogen peroxide disrupts Ca^{2+} release from the sarcoplasmic reticulum of rat skeletal muscle fibers. *Journal of Applied Physiology* 81:731-737.

Buck, M., and M. Chojkier. 1996. Muscle wasting and dedifferentiation induced by oxidative stress in a murine model of cachexia is prevented by inhibitors of nitric oxide synthesis and antioxidants. *EMBO Journal* 15:1753-1765.

Chance, B., H. Sies, and A. Boveris. 1979. Hydroperoxide metabolism in mammalian organs. *Physiological Reviews* 59:527-605.

Chen, Q., N. Olashaw, and J. Wu. 1995. Participation of reactive oxygen species in the lysophosphatidic acid-stimulated mitogen-activated protein kinase activation pathway. *Journal of Biological Chemistry* 270:28499-28502.

Cilento, G., and W. Adam. 1995. From free radicals to electronically excited species. *Free Radical Biology and Medicine* 19:103-114.

Clarkson, P.M., W.C. Byrnes, K.M. McCormick, L.P. Turcotte, and J.S. White. 1986. Muscle soreness and serum creatine kinase activity following isometric, eccentric, and concentric exercise. *International Journal of Sports Medicine* 7:152-155.

Cleeter, M.W.J., J.M. Cooper, V.M. Darley-Usmar, S. Moncada, and A.H.V. Schapira. 1994. Reversible inhibition of cytochrome c oxidase, the terminal enzyme of the mitochondrial respiratory chain, by nitric oxide. *FEBS Letters* 345:50-54.

Cogswell, A.M., R.J. Stevens, and D.A. Hood. 1993. Properties of skeletal muscle mitochondria isolated from subsarcolemmal and intermyofibrillar regions. *American Journal of Physiology* 264:C383-C389.

Collins, J.H. 1976. Homology of myosin DTNB light chain with alkali light chains, troponin C and parvalbumin. *Nature* 259:699-700.

Cortopassi, G.A., and A. Wong. 1999. Mitochondrial in organismal aging and degeneration. *Biochimica et Biophysica Acta* 1410:183-193.

Crane, F.L., I.L. Sun, R.A. Crowe, F.J. Alcain, and H. Löw. 1994. Coenzyme Q_{10}, plasma membrane oxidase and growth control. *Molecular Aspects of Medicine* 15:s1-s11.

Cross, A.R., and O.T.G. Jones. 1991. Enzymic mechanisms of superoxide production. *Biochimica et Biophysica Acta* 1057:281-298.

Crowder, M.S., and R. Cooke. 1984. The effect of myosin sulfhydryl modification on the mechanics of fiber contraction. *Journal of Muscle Research and Cell Motility* 5:131-146.

Davies, K.J.A., L. Packer, and G.A. Brooks. 1981. Biochemical adaptation of mitochondria, muscle, and whole animal respiration to endurance training. *Archives of Biochemistry and Biophysics* 209:539-554.

Davies, K.J.A., A.T. Quintanilha, G.A. Brooks, and L. Packer. 1982. Free radicals and tissue damage produced by exercise. *Biochemical and Biophysical Research Communications* 107:1198-1205.

Dawson, T.M., D.S. Bredt, M.H.P. Fotuhi, and S.H. Snyder. 1991. Nitric oxide synthase and neuronal NADPH diaphorase are identical in brain and peripheral tissues. *Proceedings of the National Academy of Sciences of the USA* 88:7797-7801.

Devary, Y., C. Rosette, J.A. DiDonato, and M. Karin. 1993. NF-κB activation by ultraviolet light not dependent on a nuclear signal. *Science* 261:1442-1444.

Di Monte, D.A., P. Chan, and M.S. Sandy. 1992. Glutathione in Parkinson's disease: A link between oxidative stress and mitochondrial damage? *Annals of Neurology* 32:S111-S115.

Diaz, P.T., E. Brownstein, and T.L. Clanton. 1994. Effects of N-acetylcysteine on in vitro diaphragm function are temperature dependent. *Journal of Applied Physiology* 77:2424-2439.

Diaz, P.T., Z.-W. She, B. Davis, and T.L. Clanton. 1993. Hydroxylation of salicylate by the in vitro diaphragm: Evidence for hydroxyl radical production during fatigue. *Journal of Applied Physiology* 75:540-545.

Dillard, C.J., R.E. Litov, W.M. Savin, E.E. Dumelin, and A.L. Tappel. 1978. Effects of exercise, vitamin E, and ozone on pulmonary function and lipid peroxidation. *Journal of Applied Physiology* 45:927-932.

Dinis, T.C.P., L.M. Almeida, and V.M.C. Madeira. 1993. Lipid peroxidation in sarcoplasmic reticulum membranes: Effect on functional and biophysical properties. *Archives of Biochemistry and Biophysics* 301:256-264.

Duchen, M.R. 1999. Contributions of mitochondria to animal physiology: From homeostatic sensor to calcium signalling and cell death. *Journal of Physiology* 516:1-17.

Duke, J., R. Takashi, K. Ue, and M.F. Morales. 1976. Reciprocal reactivities of specific thiols when actin binds to myosin. *Proceedings of the National Academy of Sciences of the USA* 73:302-306.

Duthie, G.G., J.D. Robertson, R.J. Maughan, and P.C. Morrice. 1990. Blood antioxidant status and erythrocyte lipid peroxidation following distance running. *Archives of Biochemistry and Biophysics* 282:78-83.

Favero, T.G., A.C. Zable, and J.J. Abramson. 1995. Hydrogen peroxide stimulates the Ca^{2+} release channel from skeletal muscle sarcoplasmic reticulum. *Journal of Biological Chemistry* 270:25557-25563.

Fenn, W.O., R. Gerschman, D.L. Gilbert, D.E. Terwillinger, and F.V. Cothran. 1957. Mutagenic effects of high oxygen tension on E. coli. *Proceedings of the National Academy of Sciences of the USA* 43:1027-1031.

Fitts, R.H. 1994. Cellular mechanisms of muscle fatigue. *Physiological Reviews* 74:49-94.

Flohé, L., R. Brigelius-Flohé, C. Saliou, M.G. Traber, and L. Packer. 1997. Redox regulation of NF-kappa B activation. *Free Radical Biology and Medicine* 22:1115-1126.

Fridén, J., and R.L. Lieber. 1992. Structural and mechanical basis of exercise-induced muscle injury. *Medicine and Science in Sports and Exercise* 24:521-530.

Friedl, H.P., D.J. Smith, G.O. Till, P.D. Thomson, D.S. Louis, and P.A. Ward. 1990. Ischemia-reperfusion in humans. Appearance of xanthine oxidase activity. *American Journal of Pathology* 136:491-495.

Galler, S., K. Hilber, and A. Göbesberger. 1997. Effects of nitric oxide on force-generating proteins of skeletal muscle. *Pflügers Archiv* 434:242-245.

Gardner, P.R., I. Raineri, L.B. Epstein, and C.W. White. 1995. Superoxide radical and iron modulate aconitase activity in mammalian cells. *Journal of Biological Chemistry* 270:13399-13405.

Gath, I., E.I. Closs, U. Gödtel-Armbrust, S. Schmitt, M. Nakane, I. Wessler, and U. Förstermann. 1996. Inducible NO synthase II and neuronal NO synthase I are constitutively expressed in different structures of guinea pig skeletal muscle: Implications for contractile function. *FASEB Journal* 10:1614-1620.

Gee, D.L., and A.L. Tappel. 1981. The effect of exhaustive exercise on expired pentane as a measure of in vivo lipid peroxidation in the rat. *Life Sciences* 28:2425-2429.

Gerschman, R., D.L. Gilbert, S.W. Nye, P. Dwyer, and W.O. Fenn. 1954. Oxygen poisoning and X-irradiation: A mechanism in common. *Science* 119:623-626.

Gilbert, H.F. 1982. Biological disulfides: The third messenger? *Journal of Biological Chemistry* 257:12086-12091.

Gilbert, H.F. 1984. Redox control of enzyme activities by thiol/disulfide exchange. *Methods in Enzymology* 107:330-351.

Gilbert, H.F. 1990. Molecular and cellular aspects of thiol-disulfide exchange. *Advances in Enzymology* 63:69-172.

Gohil, K., C. Viguie, W.C. Stanley, G.A. Brooks, and L. Packer. 1988. Blood glutathione oxidation during human exercise. *Journal of Applied Physiology* 64:115-119.

Han, J.W., R. Thieleczek, M. Varsanyi, and L.M.J. Heilmeyer. 1992. Compartmentalized ATP synthesis in skeletal muscle triads. *Biochemistry* 31:377-384.

Hansford, R.G., and F. Castro. 1982. Age-linked changes in the activity of enzymes of the tricarboxylate cycle and lipid oxidation, and of carnitine content, in muscles of the rat. *Mechanisms of Ageing and Development* 19:191-201.

Hansford, R.G. 1994. Role of calcium in respiratory control. *Medicine and Science in Sports and Exercise* 26:44-51.

Helgheim, I., Ø. Hetland, S. Nilsson, F. Ingjer, and S.B. Strömme. 1979. The effects of vitamin E on serum enzyme levels following heavy exercise. *European Journal of Applied Physiology* 40:283-289.

Highman, B., and P.D. Altland. 1963. Effects of exercise and training on serum enzyme and tissue changes in rats. *American Journal of Physiology* 205:162-166.

Hoppeler, H. 1986. Exercise-induced ultrastructural changes in skeletal muscle. *International Journal of Sports Medicine* 7:187-204.

Hussain, S.N., A. Giaid, Q. El Dawiri, D. Sakkal, R. Hattori, and Y. Guo. 1997. Expression of nitric oxide synthases and GTP cyclohydrolase I in the ventilatory and limb muscles during endotoxemia. *American Journal of Respiratory Cell and Molecular Biology* 17:173-180.

Hwang, C., A.J. Sinskey, and H.F. Lodish. 1992. Oxidized redox state of glutathione in the endoplasmic reticulum. *Science* 257:1496-1502.

Hyslop, P.A., D.B. Hinshaw, W.A.J. Halsey, I.U. Schraufstätter, R.D. Sauerheber, R.G. Spragg, J.H. Jackson, and C.G. Cochrane. 1988. Mechanisms of oxidant-mediated cell injury. The glycolytic and mitochondrial pathways of ADP phosphorylation are major intracellular targets inactivated by hydrogen peroxide. *Journal of Biological Chemistry* 263:1665-1675.

Ignarro, L.J. 1990. Biosynthesis and metabolism of endothelium-derived nitric oxide. *Annual Review of Pharmacology and Toxicology* 30:535-560.

Jackson, M.J., R.H.T. Edwards, and M.C.R. Symons. 1985. Electron spin resonance studies of intact mammalian skeletal muscle. *Biochimica et Biophysica Acta* 847:185-190.

Jaiswal, A.K. 1994. Human NAD(P)H:quinone oxidoreductase-2. Gene structure, activity, and tissue-specific expression. *Journal of Biological Chemistry* 269:14502-14508.

Ji, L.L. 1993. Antioxidant enzyme response to exercise and aging. *Medicine and Science in Sports and Exercise* 25:225-231.

Ji, L.L., D. Dillon, and E. Wu. 1990. Alteration of antioxidant enzymes with aging in rat skeletal muscle and liver. *American Journal of Physiology* 258:R918-R923.

Jia, L., C. Bonaventura, J. Bonaventura, and J.S. Stamler. 1996. S-nitrosohaemoglobin: A dynamic activity of blood involved in vascular control. *Nature* 380:221-226.

Jones, D.A., M.J. Jackson, and R.H.T. Edwards. 1983. Release of intracellular enzymes from an isolated mammalian skeletal muscle preparation. *Clinical Science* 65:193-201.

Joneson, T., and D. Bar-Sagi. 1998. A rac1 effector site controlling mitogenesis through superoxide production. *Journal of Biological Chemistry* 273:17991-17994.

Kadenbach, B., C. Münscher, V. Frank, J. Müller-Höcker, and J. Napiwotzki. 1995. Human aging is associated with stochastic somatic mutations of mitochondrial DNA. *Mutation Research* 338:161-172.

Kanter, M.M., G.R. Lesmes, L.A. Kaminsky, J.L. Ham-Saeger, and N.D. Nequin. 1988. Serum creatine kinase and lactate dehydrogenase changes following an eighty kilometer race. Relationship to lipid peroxidation. *European Journal of Applied Physiology* 57:60-63.

Kehrer, J.P., and L.G. Lund. 1994. Cellular reducing equivalents and oxidative stress. *Free Radical Biology and Medicine* 17:65-75.

Khanna, S., M. Atalay, D.E. Laaksonen, M. Gul, S. Roy, and C.K. Sen. 1999. α-Lipoic acid supplementation: Tissue glutathione homeostasis at rest and after exercise. *Journal of Applied Physiology* 86:1191-1196.

King, C.E., M.J. Melinyshyn, J.D. Mewburn, S.E. Curtis, M.J. Winn, S.M. Cain, and C.K. Chapler. 1994. Canine hindlimb blood flow and O_2 uptake after inhibition of EDRF/NO synthesis. *Journal of Applied Physiology* 76:1166-1171.

Kletzien, R.F., P.K.W. Harris, and L.A. Foellmi. 1994. Glucose-6-phosphate dehydrogenase: A "housekeeping" enzyme subject to tissue-specific regulation by hormones, nutrients, and oxidant stress. *FASEB Journal* 8:174-181.

Kobzik, L., M.B. Reid, D.S. Bredt, and J.S. Stamler. 1994. Nitric oxide in skeletal muscle. *Nature* 372:546-548.

Kobzik, L., B. Stringer, J.-L. Balligand, M.B. Reid, and J.S. Stamler. 1995. Endothelial type nitric oxide synthase in skeletal muscle fibers: Mitochondrial relationships. *Biochemical and Biophysical Research Communications* 211:375-381.

Kolbeck, R.C., Z.-W. She, L.A. Callahan, and T.M. Nosek. 1997. Increased superoxide production during fatigue in the perfused rat diaphragm. *American Journal of Respiratory and Critical Care Medicine* 156:140-145.

Kopsidas, G., S.A. Kovalenko, J.M. Kelso, and S.W. Linnane. 1998. An age-associated correlation between cellular bioenergy decline and mtDNA rearrangements in human skeletal muscle. *Mutation Research* 421:27-36.

Krause, K.M., M.R. Moody, F.H. Andrade, A.A. Taylor, L. Kobzik, and M.B. Reid. 1998. Peritonitis causes diaphragm weakness in rats. *American Journal of Respiratory and Critical Care Medicine* 157:1277-1282.

Kunapuli, P., J.A. Lawson, J.A. Rokach, J.L. Meinkoth, and G.A. FitzGerald. 1998. Prostaglandin F_{2a} (PGF_{2a}) and the isoprostane, 8,12-iso-isoprostane F_{2a}-III, induce cardiomyocyte hypertrophy. *Journal of Biological Chemistry* 273:22442-22452.

Kwon, H., P.M.D. Hardwicke, J.H. Collins, X. Zhao, and A.G. Szent-Györgyi. 1994. Myosin filament ATPase is enhanced by intramolecular cross-linked actin. *Journal of Muscle Research and Cell Motility* 15:555-562.

Lander, H.M., D.P. Hajjar, B.L. Hempstead, U.A. Mirza, B.T. Chait, S. Campbell, and L.A. Quilliam. 1997. A molecular redox switch on p21[ras]. *Journal of Biological Chemistry* 272:4323-4326.

Lander, H.M., J.S. Ogiste, K.K. Teng, and A. Novogrodsky. 1995. p21[ras] as a common signaling target of reactive free radicals and cellular redox stress. *Journal of Biological Chemistry* 270:21195-21198.

Larsson, L. 1998. The age-related motor disability: Underlying mechanisms in skeletal muscle at the motor unit, cellular and molecular level. *Acta Physiologica Scandinavica* 163:S27-S29.

Lass, A., B.H. Sohal, R. Weindruch, M.J. Forster, and R.S. Sohal. 1998. Caloric restriction prevents age-associated accrual of oxidative damage to mouse skeletal muscle mitochondria. *Free Radical Biology and Medicine* 25:1089-1097.

Lawler, J.M., S.K. Powers, T. Visser, H. Van Dijk, M.J. Kordus, and L.L. Ji. 1993. Acute exercise and skeletal muscle antioxidant and metabolic enzymes: Effect of fiber type and age. *American Journal of Physiology* 265:R1344-R1350.

Lawrence, J.D., R.C. Bower, W.P. Riehl, and J.L. Smith. 1975. Effects of α-tocopherol acetate on the swimming endurance of trained swimmers. *American Journal of Clinical Nutrition* 28:205-208.

Leeuwenburgh, C., R. Fiebig, R. Chandwaney, and L.L. Ji. 1994. Aging and exercise training in skeletal muscle: Responses of glutathione and antioxidant enzyme systems. *American Journal of Physiology* 267:R439-R445.

Lew, H., S. Pyke, and A. Quintanilha. 1985. Changes in the glutathione status of plasma, liver and muscle following exhaustive exercise in rats. *FEBS Letters* 185:262-266.

Li, X., S.M. Hughes, G. Salviati, A. Teresi, and L. Larsson. 1997. Thyroid hormone effects on contractility and myosin composition of soleus muscle and single fibres from young and old rats. *Journal of Physiology* 494:555-567.

Li, Y.-P., R.J. Schwartz, I.D. Waddell, B.R. Holloway, and M.B. Reid. 1998. Skeletal muscle myocytes undergo protein loss and reactive oxygen-mediated NF-κB activation in response to tumor necrosis factor α. *FASEB Journal* 12:871-880.

Liu, D.F., D. Wang, and A. Stracher. 1990. The accessibility of the thiol groups on G- and F-actin of rabbit muscle. *Biochemical Journal* 266:453-459.

Lovlin, R., W. Cottle, I. Pyke, M. Kavanagh, and A.N. Belcastro. 1987. Are indices of free radical damage related to exercise intensity? *European Journal of Applied Physiology* 56:313-316.

Lowey, S., H.S. Slayter, A.G. Weeds, and H. Baker. 1969. Substructure of the myosin molecule. I. Subfragments of myosin by enzymic degradation. *Journal of Molecular Biology* 42:1-29.

Mayhew, M., M. Renganathan, and O. Delbono. 1998. Effectiveness of caloric restriction in preventing age-related changes in rat skeletal muscle. *Biochemical and Biophysical Research Communications* 251:95-99.

McCutchan, H.J., J.R. Schwappach, E.G. Enquist, D.L. Walden, L.S. Terada, O.K. Reiss, J.A. Leff, and J.E. Repine. 1990. Xanthine oxidase-derived H_2O_2 contributes to reperfusion injury of ischemic skeletal muscle. *American Journal of Physiology* 258:H1415-H1419.

Mecocci, P., G. Fano, S. Fulle, U. MacGarvey, L. Shinobu, M.C. Polidori, A. Cherubini, J. Vecchiet, U. Senin, and M.F. Beal. 1999. Age-dependent increases in oxidative damage to DNA, lipids, and proteins in human skeletal muscle. *Free Radical Biology and Medicine* 26:303-308.

Mészàros, L., I. Minarovic, and A. Zahradnikova. 1996. Inhibition of the skeletal muscle ryanodine receptor calcium release channel by nitric oxide. *FEBS Letters* 380:49-52.

Metzger, J.M., and R.L. Moss. 1992. Myosin light chain 2 modulates calcium-sensitive cross-bridge transitions in vertebrate skeletal muscle. *Biophysical Journal* 63:460-468.

Michel, T., and O. Feron. 1997. Nitric oxide synthases: Which, where, how, and why? *Journal of Clinical Investigation* 100:2146-2152.

Milligan, S.A., M.W. Owens, and M.B. Grisham. 1998. Differential regulation of extracellular signal-regulated kinase and nuclear factor-κB signal transduction pathways by hydrogen peroxide and tumor necrosis factor. *Archives of Biochemistry and Biophysics* 352:255-262.

Mohr, S., J.S. Stamler, and B. Brune. 1996. Posttranslational modification of glyceraldehyde-3-phosphate dehydrogenase by S-nitrosylation and subsequent NADH attachment. *Journal of Biological Chemistry* 271:4209-4214.

Moncada, S., R.M.J. Palmer, and E.A. Higgs. 1991. Nitric oxide: Physiology, pathophysiology, and pharmacology. *Pharmacological Reviews* 43:109-142.

Morales, C.F., A. Anzueto, F. Andrade, J. Brassard, S.M. Levine, L.C. Maxwell, R.A. Lawrence, and S.G. Jenkinson. 1994. Buthionine sulfoximine treatment impairs rat diaphragm function. *American Journal of Respiratory and Critical Care Medicine* 149:915-919.

Morales, C.F., A. Anzueto, F. Andrade, S.M. Levine, L.C. Maxwell, R.A. Lawrence, and S.G. Jenkinson. 1993. Diethylmaleate produces diaphragmatic impairment after resistive breathing. *Journal of Applied Physiology* 75:2406-2411.

Morrow, J.D., J.A. Awad, T. Kato, K. Takahashi, K.F. Badr, L.J. Roberts II, and R.F. Burk. 1992. Formation of novel non-cyclooxygenase-derived prostanoids (F_2-isoprostanes) in carbon tetrachloride hepatotoxicity. *Journal of Clinical Investigation* 90:2502-2507.

Morrow, J.D., T.M. Harris, and L.J. Roberts II. 1990. Noncyclooxygenase oxidative formation of a series of novel prostaglandins: Analytical ramifications for measurement of eicosanoids. *Analytical Biochemistry* 184:1-10.

Morrow, J.D., and L.J. Roberts. 1997. The isoprostanes: Unique bioactive products of lipid peroxidation. *Progress in Lipid Research* 36:1-21.

Müller-Höcker, J., S. Schäfer, T.A. Link, S. Possekel, and C. Hammer. 1996. Defects of the respiratory chain in various tissues of old monkeys: A cytochemical-immunocytochemical study. *Mechanisms of Ageing and Development* 86:197-213.

Müller-Höcker, J., P. Seibel, K. Schneiderbanger, and B. Kadenbach. 1993. Different in situ hybridization patterns of mitochondrial DNA in cytochrome *c* oxidase-deficient extraocular muscle fibres in the elderly. *Virchows Archiv A: Pathological Anatomy and Histopathology* 422:7-15.

Murrant, C.L., F.H. Andrade, and M.B. Reid. 1999. Exogenous reactive oxygen and nitric oxide alter redox state in skeletal muscle fibers. *Acta Physiologica Scandinavica* 166:111-121.

Naqui, A., B. Chance, and E. Cadenas. 1986. Reactive oxygen intermediates in biochemistry. *Annual Review of Biochemistry* 55:137-166.

Newham, D.J., G. McPhail, K.R. Mills, and R.H.T. Edwards. 1983. Ultrastructural changes after concentric and eccentric contractions of human muscle. *Journal of the Neurological Sciences* 61:109-122.

Newham, D.J., D.A. Jones, S.E.J. Tolfree, and R.H.T. Edwards. 1986. Skeletal muscle damage: A study of isotope uptake, enzyme efflux and pain after stepping. *European Journal of Applied Physiology* 55:106-112.

NHLBI Workshop. 1990. Respiratory muscle fatigue. Report of the Respiratory Muscle Fatigue Workshop Group. *American Review of Respiratory Disease* 142:474-480.

Oh-ishi, S., K. Toshinai, T. Kizaki, S. Haga, K. Fuduka, N. Nagata, and H. Ohno. 1996. Effects of aging and/or training on antioxidant enzyme system in diaphragm of mice. *Respiration Physiology* 105:195-202.

Ohkuwa, T., Y. Sato, and M. Naoi. 1997. Glutathione status and reactive oxygen generation in tissues of young and old exercised rats. *Acta Physiologica Scandinavica* 159:237-244.

Okabe, E., E. Hiyama, M. Oyama, C. Odajima, H. Ito, and Y.W. Cho. 1982. Free radical damage to sarcoplasmic reticulum of masseter muscle by arachidonic acid and prostaglandin G_2. *Pharmacology* 25:138-148.

Peng, H.-B., P. Libby, and J.K. Liao. 1995. Induction and stabilization of I_Ba by nitric oxide mediates inhibition of NF-κB. *Journal of Biological Chemistry* 270:14214-14219.

Perkins, W.J., Y.-S. Han, and G.C. Sieck. 1997. Skeletal muscle force and actomyosin ATPase activity reduced by nitric oxide donor. *Journal of Applied Physiology* 83:1326-1332.

Phillips, S.K., K.M. Rook, N.C. Siddle, S.A. Bruce, and R.C. Woledge. 1996. Muscle weakness in women occurs at an earlier age than in men, but strength is preserved by hormone replacement therapy. *Clinical Science* 84:85-98.

Phung, C.D., J.A. Ezieme, and J.F. Turrens. 1994. Hydrogen peroxide metabolism in skeletal muscle mitochondria. *Archives of Biochemistry and Biophysics* 315:479-482.

Posterino, G.S., and G.D. Lamb. 1996. Effects of reducing agents and oxidants on excitation-contraction coupling in skeletal muscle fibres of rat and toad. *Journal of Physiology* 496:809-825.

Potter, J.D., J.C. Seidel, P. Leavis, S.S. Lehrer, and J. Gergely. 1976. Effect of Ca^{2+} binding on troponin C. Changes in spin label mobility, extrinsic fluorescence, and sulfhydryl reactivity. *Journal of Biological Chemistry* 251:7551-7556.

Powers, S.K., J. Lawler, D. Criswell, F.K. Lieu, and S. Dodd. 1992. Alterations in diaphragmatic oxidative and antioxidant enzymes in the senescent Fischer 344 rat. *Journal of Applied Physiology* 72:2317-2321.

Putkey, J.A., D.G. Dotson, and P. Mouawad. 1993. Formation of inter- and intramolecular disulfide bonds can activate cardiac troponin C. *Journal of Biological Chemistry* 268:6827-6830.

Radák, Z., J. Pucsok, S. Mecseki, T. Csont, and P. Ferdinandy. 1999. Muscle soreness-induced restriction in force generation is accompanied by increased nitric oxide content and DNA damage in human skeletal muscle. *Free Radical Biology and Medicine* 26:1059-1063.

Radjendirane, V., P. Joseph, Y.-H. Lee, S. Kimura, A.J.P. Klein-Szanto, F.J. Gonzalez, and A.K. Jaiswal. 1998. Disruption of the DT diaphorase (NQO1) gene in mice leads to increased menadione toxicity. *Journal of Biological Chemistry* 273:7382-7389.

Reid, M.B. 1996. Reactive oxygen and nitric oxide in skeletal muscle. *News in Physiological Sciences* 11:114-119.

Reid, M.B., G.J. Grubwieser, D.S. Stokic, S.M. Koch, and A.A. Leis. 1993a. Development and reversal of fatigue in human tibialis anterior. *Muscle and Nerve* 16:1239-1245.

Reid, M.B., K.E. Haack, K.M. Franchek, P.A. Valber, L. Kobzik, and M.S. West. 1992a. Reactive oxygen in skeletal muscle I. Intracellular oxidant kinetics and fatigue in vitro. *Journal of Applied Physiology* 73:1797-1804.

Reid, M.B., F.A. Khawli, and M.R. Moody. 1993b. Reactive oxygen in skeletal muscle III. Contractility of unfatigued muscle. *Journal of Applied Physiology* 75:1081-1087.

Reid, M.B., T. Shoji, M.R. Moody, and M.L. Entman. 1992b. Reactive oxygen in skeletal muscle II. Extracellular release of free radicals. *Journal of Applied Physiology* 73:1805-1809.

Reid, M.B., D.S. Stokic, S.M. Koch, F.A. Khawli, and A.A. Leis. 1994. N-Acetylcysteine inhibits muscle fatigue in humans. *Journal of Clinical Investigation* 94:2468-2474.

Renganathan, M., and O. Delbono. 1998. Caloric restriction prevents age-related decline in skeletal muscle dihydropyridine receptor and ryanodine receptor expression. *FEBS Letters* 434:346-350.

Renganathan, M., M.L. Messi, and O. Delbono. 1997. Dihydropyridine receptor-ryanodine receptor uncoupling in aged skeletal muscle. *Journal of Membrane Biology* 157:247-253.

Reznick, A.Z., E. Witt, M. Matsumoto, and L. Packer. 1992. Vitamin E inhibits protein oxidation in skeletal muscle of resting and exercised rats. *Biochemical and Biophysical Research Communications* 189:801-806.

Ríos, E., and G. Pizarro. 1991. Voltage sensor of excitation-contraction coupling in skeletal muscle. *Physiological Reviews* 71:849-908.

Root, D.D., and E. Reisler. 1992. Cooperativity of thiol-modified myosin filaments. ATPase and motility assays of myosin function. *Biophysical Journal* 63:730-740.

Salama, G., J.J. Abramson, and G.K. Pike. 1992. Sulphydryl reagents trigger Ca^{2+} release from the sarcoplasmic reticulum of skinned rabbit psoas fibres. *Journal of Physiology* 454:389-420.

Sastre, J., M. Asensi, E. Gascó, F.V. Pallardó, J.A. Ferrero, T. Furukawa, and J. Viña. 1992. Exhaustive physical exercise causes oxidation of glutathione status in blood: Prevention by antioxidant administration. *American Journal of Physiology* 263:R992-R995.

Schapira, A.H.V., and J.M. Cooper. 1992. Mitochondrial function in neurodegeneration and ageing. *Mutation Research* 275:133-143.

Scherer, N.M., and D.W. Deamer. 1986. Oxidative stress impairs the function of sarcoplasmic reticulum by oxidation of sulfhydryl groups in the Ca^{2+}-ATPase. *Archives of Biochemistry and Biophysics* 246:589-601.

Schmid, E., J. El Benna, D. Galter, G. Klein, and W. Dröge. 1998. Redox priming of the insulin receptor β-chain associated with altered tyrosine kinase activity and insulin responsiveness in the absence of tyrosine autophosphorylation. *FASEB Journal* 12:863-870.

Schmidt, H.H.H.W., S.M. Lohmann, and U. Walter. 1993. The nitric oxide and cGMP signal transduction system: Regulation and mechanism of action. *Biochimica et Biophysica Acta* 1178:153-175.

Schmidt, H.H.H.W., and U. Walter. 1994. NO at work. *Cell* 78:919-925.

Schöneich, C., R.I. Viner, D.A. Ferrington, and D.J. Bigelow. 1999. Age-related chemical modification of the skeletal muscle sarcoplasmic reticulum Ca-ATPase of the rat. *Mechanisms of Ageing and Development* 107:221-231.

Shephard, R.J., R. Campbell, P. Pimm, D. Stuart, and G.R. Wright. 1974. Vitamin E, exercise, and recovery from physical activity. *European Journal of Applied Physiology* 33:119-126.

Shindoh, C., A. DiMarco, A. Thomas, P. Manubay, and G. Supinski. 1990. Effect of N-acetylcysteine on diaphragm fatigue. *Journal of Applied Physiology* 68:2107-2113.

Silvagno, F., H. Xia, and D.S. Bredt. 1996. Neuronal nitric-oxide synthase-μ, an alternatively spliced isoform expressed in differentiated skeletal muscle. *Journal of Biological Chemistry* 271:11204-11208.

Simon, A.R., U. Rai, B.L. Fanburg, and B.H. Cochran. 1998. Activation of the JAK-STAT pathway by reactive oxygen species. *American Journal of Physiology* 275:C1640-C1652.

Sohal, R.S., and R. Weindruch. 1996. Oxidative stress, caloric restriction, and aging. *Science* 273:59-63.

Stamler, J.S. 1994. Redox signaling: Nitrosylation and related target interactions of nitric oxide. *Cell* 78:931-936.

Stoyanovsky, D., T. Murphy, P.R. Anno, Y.-M. Kim, and G. Salama. 1997. Nitric oxide activates skeletal and cardiac ryanodine receptors. *Cell Calcium* 21:19-29.

Stuart, J., I.N. Pessah, T.G. Favero, and J.J. Abramson. 1992. Photooxidation of skeletal muscle sarcoplasmic reticulum induces rapid calcium release. *Archives of Biochemistry and Biophysics* 292:512-521.

Sumida, S., K. Tanaka, H. Kitao, and F. Nakadomo. 1989. Exercise-induced lipid peroxidation and leakage of enzymes before and after vitamin E supplementation. *International Journal of Biochemistry* 21:835-838.

Sundaresan, M., Z.X. Yu, V.J. Ferrans, K. Irani, and T. Finkel. 1995. Requirement for generation of H_2O_2 for plateled-derived growth factor signal transduction. *Science* 270:296-299.

Supinski, G.S., D. Stofan, R. Ciufo, and A. DiMarco. 1995. N-Acetylcysteine administration and loaded breathing. *Journal of Applied Physiology* 79:340-347.

Thompson, M., L. Becker, D. Bryant, G. Williams, D. Levin, L. Margraf, and B.P. Giroir. 1996. Expression of the inducible nitric oxide synthase gene in diaphragm and skeletal muscle. *Journal of Applied Physiology* 81:2415-2420.

Trimm, J.L., G. Salama, and J.J. Abramson. 1986. Sulfhydryl oxidation induces rapid calcium release from sarcoplasmic reticulum vesicles. *Journal of Biological Chemistry* 261:16092-16098.

Viner, R.I., D.A. Ferrington, T.D. Williams, D.J. Bigelow, and C. Schöneich. 1999. Protein modification during biological aging: Selective tyrosine nitration of the SERCA2a isoform of the sarcoplasmic reticulum Ca^{2+}-ATPase in skeletal muscle. *Biochemical Journal* 340:657-669.

Walsh, T.P., and A. Wegner. 1980. Effect of the state of oxidation of cysteine 190 of tropomyosin on the assembly of the actin-tropomyosin complex. *Biochimica et Biophysica Acta* 626:79-87.

Wei, Y.H., S.H. Kao, and H.C. Lee. 1996. Simultaneous increase of mitochondrial DNA deletions and lipid peroxidation in human aging. *Annals of the New York Academy of Sciences* 786:24-43.

Westerblad, H., D.G. Allen, J.D. Bruton, F.H. Andrade, and J. Lännergren. 1998. Mechanisms underlying the reduction of isometric force in skeletal muscle fatigue. *Acta Physiologica Scandinavica* 162:253-260.

Westerblad, H., J.A. Lee, J. Lännergren, and D.G. Allen. 1991. Cellular mechanisms of fatigue in skeletal muscle. *American Journal of Physiology* 261:C195-C209.

Whisler, R.L., M.A. Goyette, I.S. Grants, and Y.G. Newhouse. 1995. Sublethal levels of oxidant stress stimulate multiple serine/threonine kinases and suppress protein phosphatases in Jurkat T cells. *Archives of Biochemistry and Biophysics* 319:23-25.

Williams, D.L. Jr., and C.A. Swenson. 1982. Disulfide bridges in tropomyosin. *European Journal of Biochemistry* 127:495-499.

Wilson, G.J., C.G. Dos Remedios, D.G. Stephenson, and D.A. Williams. 1991. Effects of sulphydryl modification on skinned rat skeletal muscle fibres using 5,5'-dithiobis(2-nitrobenzoic acid). *Journal of Physiology* 437:409-430.

Witt, E.H., A.Z. Reznick, C.A. Viguie, P. Starke-Reed, and L. Packer. 1992. Exercise, oxidative damage and effects of antioxidant manipulation. *Journal of Nutrition* 122:766-773.

Wolosker, H., R. Panizzutti, and S. Engelender. 1996. Inhibition of creatine kinase by S-nitrosoglutathione. *FEBS Letters* 392:274-276.

Xie, Z., P. Kometiani, J. Liu, J. Li, J.I. Shapiro, and A. Askari. 1999. Intracellular reactive oxygen species mediate the linkage of Na^+/K^+-ATPase to hypertrophy and its marker genes in cardiac myocytes. *Journal of Biological Chemistry* 274:19323-19328.

Zable, A.C., T.G. Favero, and J.J. Abramson. 1997. Glutathione modulates ryanodine receptor from skeletal muscle sarcoplasmic reticulum. *Journal of Biological Chemistry* 272:7069-7077.

Zaidi, N.F., C.F. Lagenaur, J.J. Abramson, I. Pessah, and G. Salama. 1989. Reactive disulfides trigger Ca^{2+} release from sarcoplasmic reticulum via an oxidation reaction. *Journal of Biological Chemistry* 264:21725-21736.

Zerba, E., T.E. Komorowski, and J.A. Faulkner. 1990. Free radical injury to skeletal muscles of young, adult, and old mice. *American Journal of Physiology* 258:C429-C435.

Zhang, Y., O. Marcillat, C. Giulivi, L. Ernster, and K.J.A. Davies. 1990. The oxidative inactivation of mitochondrial electron transport chain components and ATPase. *Journal of Biological Chemistry* 265:16330-16336.

Ziegler, D.M. 1985. Role of reversible oxidation-reduction of enzyme thiols-disulfides in metabolic regulation. *Annual Review of Biochemistry* 54:305-329.

Chapter 5

Adaptation of the Heart to Ischemic Stress: Role of Nitric Oxide, Exercise, and Aging in Myocardial Preconditioning

Peter Ferdinandy
Department of Biochemistry, University of Szeged, Szeged, Hungary

Zoltán Szilvássy
Department of Pharmacology, University of Debrecen, Debrecen, Hungary

Arpad Tosaki
Department of Pharmacology, University of Debrecen, Debrecen, Hungary

Ischemic stress is the most common type of stress that affects the heart. Cardiac ischemia occurs when the blood flow to the heart is insufficient to fuel the metabolic demand of the myocardium. As a consequence, the pump function of the heart deteriorates, the rhythmic contraction is disturbed (i.e., arrhythmia), and, in cases of extreme and long-lasting ischemia, irreversible tissue damage may occur, which is labeled myocardial necrosis or infarction. In humans, this complex pathological situation is termed ischemic heart disease. Ischemic heart disease is a major cause of mortality in developed societies. Ischemic heart disease that develops due to sustained hypertension is responsible for about 50% of morbidity and mortality in the elderly in these societies. Procedures that rapidly restore blood flow (i.e., reperfusion of the ischemic zone of the myocardium) can reduce mortality by approximately half. However, the mortality benefit diminishes if treatment is administered late (Braunwald, 1985). Although early reperfusion is the only way to salvage the ischemic myocardium, reperfusion itself causes arrhythmias and reversible functional deterioration of the heart. Therefore, cardioprotective agents that can improve myocardial function, decrease the incidence of arrhythmias, lessen necrotic tissue mass, and delay the onset of necrosis during ischemia/ reperfusion are of great importance. Because previous attempts to attenuate or delay the consequences of ischemic injury with pharmacological tools have been largely unsuccessful, new ways to protect the ischemic/reperfused myocardium are being explored. Ischemic preconditioning of myocardium is a well-described adaptive response in which brief exposure to ischemia markedly enhances the heart's ability to withstand a subsequent ischemic insult. The underlying molecular mechanisms of this phenomenon have been investigated extensively in the hope of identifying new, rational approaches to therapeutic protection of the ischemic myocardium. In this chapter, we review the roles of nitric oxide exercise, aging, and aging-related diseases in the preconditioning phenomenon, the heart's remarkable capability to adapt to ischemia.

Myocardial Adaptation to Ischemia

Myocardium may adapt to ischemia in several ways. In the long term, repetitive ischemic insults over months or years may result in angiogenic adaptation, in which coronary collateral vessels are developed to provide a better blood supply to the ischemic region from adjacent, nonischemic regions. Chronic low-grade ischemia over prolonged periods can induce a biochemical, energy-sparing adaptation of the myocardium called *hibernation*. More than a decade ago, Murry et al. (1986) showed that the heart is able to adapt to ischemic stress by an endogenous cardioprotective mechanism termed *ischemic preconditioning* (for reviews, see Baxter and Yellon, 1994; Parratt, 1995). Ischemic preconditioning, which falls within a spectrum of adaptive responses to ischemia, represents the ability of myocardium to adapt to sublethal ischemic stress in the short term so that it is more resistant to a subsequent, potentially injurious period of ischemia. Although the effectiveness of preconditioning is markedly attenuated in some disease states (for review, see Ferdinandy et al., 1998c), preconditioning confers a remarkable cardioprotective effect in a variety of species, including humans. The cardioprotective effect of preconditioning shows two phases. The early phase (classical preconditioning) is manifested within minutes after the ischemic stress and has a duration of less than 3 h; the late phase (second window of protection) is characterized by a slower onset and a duration of up to 72 h (see figure 5.1). In both phases of preconditioning, necrotic tissue mass is reduced, cardiac performance following ischemia/reperfusion is improved, and both the incidence and severity of arrhythmias are reduced (for reviews, see Baxter and Yellon, 1994; Parratt and Szekeres, 1995). Although other tissues and organs may display preconditioning-like adaptations, particularly skeletal muscle, small intestine, and neuronal tissue (see Millar et al., 1996), in this chapter we limit the discussion to adaptation in cardiac tissue. The powerful nature of preconditioning has fueled research into the cellular mechanisms of protection, with the hope that these might provide physiological templates to develop new preconditioning-based therapies for individuals at risk of myocardial ischemia. Although initially described as a laboratory phenomenon, preconditioning can be induced in human myocardium and may even occur as a natural feature of ischemic syndromes in humans (Kloner and Yellon, 1994; Lawson, 1995).

A full discussion of the mechanisms thought to underlie preconditioning is beyond the scope of this chapter, but this subject is covered in greater depth elsewhere (Ferdinandy et al., 1998c; Millar et al., 1996; Parratt, 1995). Despite intensive research in the last decade, the biochemical mechanism of ischemic preconditioning remains a subject of debate. The discrepancies are generally attributed to species differences; different preconditioning stimuli such as no-flow ischemia, low-flow ischemia, or rapid pacing models (Ferdinandy et al., 1995b); and different study end points, such as myocardial function, arrhythmias, or infarct size (for review, see Baxter et al., 1996; Parratt, 1994;

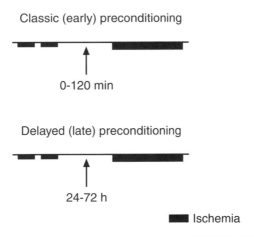

Figure 5.1 Experimental approach to early and late preconditioning. Ischemic preconditioning describes adaptation of the heart to one or more brief periods of ischemia so that the myocardium is more resistant to a subsequent period of ischemia. Experimentally, preconditioning can be induced by one or more single periods of ischemia (or other stimuli such as rapid cardiac pacing or myocardial stretch) lasting 3 to 5 min. The end points used to assess the protective effect of preconditioning include infarct size, arrhythmias, and functional parameters following a long-lasting "test ischemia."

Walker and Yellon, 1992). A wealth of evidence exists that both classic preconditioning and delayed preconditioning are initiated during brief ischemia by the actions of diffusible mediators, especially adenosine, bradykinin, nitric oxide, and catecholamines. In other words, although ischemia and reperfusion result in a release and accumulation of several substances that lead to injury, some of these factors trigger the development of an adaptive response (see figure 5.2). These triggers of adaptation may be sufficient individually to initiate adaptive responses, or their actions may be augmented by the generation of reactive oxygen species during preconditioning and ischemia/reperfusion. Complex signaling cascades are activated, which are the subject of considerable controversy, particularly with regard to classic preconditioning. There is evidence that kinases such as protein kinase C (PKC) must be activated for the adaptation to occur. An alternative pathway proposes stimulation of guanosine cyclic $3',5'$-monophosphate (cGMP) synthesis and the generation of nitric oxide. The distal effectors of both the early and late forms of protection are unclear, although pharmacological evidence implicates an important role of the adenosine triphosphate (ATP)-sensitive K^+ channel (KATP) in both classic and delayed preconditioning. Also, a great deal of evidence proves that accumulation of the inducible member of the 70 kDa heat-shock protein family (HSP70) in response to ischemic stress in the hearts of different species and in different cell lines confers long-lasting protection against a variety of stressors including ischemia/reperfusion (for reviews, see Black and Lucchesi, 1993; Mestril and Dillmann, 1995; Yellon et al., 1993).

Natural Preconditioning: Physical Exercise

Hypercholesterolemia, diabetes mellitus, and aging are risk factors for myocardial infarction probably partly because they diminish the heart's capability to adapt to ischemia (see following discussion). Therefore, clinical application of preconditioning before ischemic heart disease develops would be advantageous. How can preconditioning be applied to prevent heart disease? Drug-induced preconditioning is an obvious choice (for review, see Parratt, 1995); however, a natural way to induce preconditioning would be more beneficial. Traditional experimental triggers for preconditioning include low-flow ischemia and different types of no-flow ischemia induced by coronary occlusion or global ischemia, conditions that lead to low oxygen supply (for review, see Kloner et al., 1995). The naturally occurring "synonyms" of these triggers are coronary stenosis, spasm, or complete coronary occlusion, which all represent advanced-stage ischemic heart disease. Both classic preconditioning and late preconditioning can be induced experimentally by rapid cardiac pacing (Ferdinandy et al., 1997a; Szekeres et al., 1993; Szilvássy et al., 1994a) and

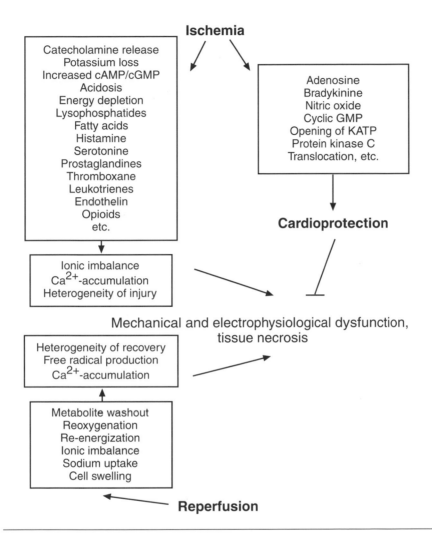

Figure 5.2   The most important biochemical events occurring during ischemia and reperfusion. Ischemia and reperfusion result in the release and accumulation of several substances that lead to injury; however, some of these factors also trigger or mediate an adaptive response to ischemia, which confers cytoprotection on the heart. cAMP, adenosine 3′,5′-cyclic monophosphate; cGMP, cyclic guanosine 3′,5′-monophosphate; KATP, adenosine triphosphate–sensitive K+ channel.

myocardial stretch (Ovize et al., 1994), both of which result in high oxygen demand. This can be mimicked in healthy subjects by physical exercise, which leads to significant tachycardia and expansion of plasma volume, thus stretching the myocardium. In accordance, recent studies showed that both cardiac pacing and exercise induced preconditioning in humans (Capecchi et al., 1997; Correa and Schaefer, 1997; Maybaum et al., 1996).

Thus, "natural preconditioning" includes active and passive preconditioning (Ferdinandy et al., 1995c). Active preconditioning induced by tachycardia may be provoked by physical training, whereas passive preconditioning induced by decreased oxygen supply is a life threatening consequence of ischemic heart disease. We previously reported that in rat hearts, the mechanism of active preconditioning induced by rapid ventricular pacing may be different from that of passive preconditioning induced by no-flow ischemia (Ferdinandy et al., 1995c).

It is well known that regular exercise decreases cardiovascular mortality, but it is unknown whether this is due to the preconditioning phenomenon or to other factors. Therefore, the biochemical mechanism of active, training- or pacing-induced preconditioning may warrant greater pharmacological and clinical interest, because this preconditioning can be safely and naturally induced in healthy subjects to prevent ischemic heart disease, including myocardial infarction.

Nitric Oxide and Myocardial Preconditioning

In the normal heart, nitric oxide is synthesized from L-arginine by Ca^{2+}-dependent nitric oxide synthases in cardiac myocytes, in vascular and endocardial endothelium (nitric oxide synthase [NOS] III), and in specific cardiac neurons (NOS I); nitric oxide plays an important role in regulating coronary circulation and cardiac contractile function. Basal release of nitric oxide acts as an important coronary vasodilator, inhibits aggregation of platelets, and depresses mitochondrial respiration. These effects of nitric oxide help maintain sufficient blood supply to the heart to fuel the metabolic needs of the contracting cardiac myocytes.

It is well known that acute physical exercise increases nitric oxide formation in the vasculature, including the coronary arteries, possibly due to increased shear stress (Jungersten et al., 1997; Lamontagne et al., 1991; Poveda et al., 1997). In addition, basal synthesis is elevated in trained subjects, which may explain the beneficial effects of physical exercise on cardiovascular health (Jungersten et al., 1997).

Aging reduces the capability of the coronary vasculature to relax in response to some endothelium-dependent vasodilators, possibly by reduced synthesis of nitric oxide in the coronary vascular bed (for review, see Marin, 1995). Coronary vascular aging may also affect the sympathetic regulation of coronary arterial tone by attenuating relaxation to catecholamines via α_2-adrenoceptors (for review, see Marin, 1995).

Nitric Oxide in Cardiac Ischemia/Reperfusion

The role of nitric oxide in myocardial ischemia and reperfusion is controversial. Numerous studies have shown that nitric oxide synthase inhibitors improved myocardial function following ischemia/reperfusion injury; therefore, the researchers considered nitric oxide to be a mediator of ischemia/reperfusion injury in these studies (Matheis et al., 1992; Schulz and Wambolt, 1995; Wang and Zweier 1996; Yasmin et al., 1997). Other studies, however, described the detrimental effect of blockade of nitric oxide synthesis and the cardioprotective effect of nitric oxide donors (Beresewicz et al., 1995; Curtis and Pabla, 1997; Ferdinandy et al., 1995a; Hasebe et al., 1993; Lefer et al., 1993; Linz et al., 1992). Yasmin et al. (1997) showed the protective action of both nitric oxide synthase inhibitors and nitric oxide donors. These discrepancies may be attributed to the researchers using different doses of nitric oxide synthase inhibitors to modulate nitric oxide metabolism without measuring actual changes in nitric oxide synthase activities, nitric oxide, or cGMP concentrations during ischemia/reperfusion. Other studies, however, proved that myocardial ischemia increases the activity of Ca^{2+}-dependent nitric oxide synthase (Depre et al., 1997) and leads to accumulation of nitric oxide, which might contribute to ischemia/reperfusion injury (Zweier et al., 1995a). Depre et al. (1997) and Depre and Huc (1994) demonstrated that global ischemia in the rabbit perfused heart increased cardiac nitric oxide synthase activity and cGMP content. Based on indirect pharmacological evidence, ischemia-induced increase in nitric oxide synthesis has been suggested (Kitakaze et al., 1995; Node et al., 1996). Zweier et al. (1995a), using electron spin resonance technique to directly measure cardiac nitric oxide concentration, reported that global myocardial ischemia markedly increases myocardial nitric oxide content, which leads to ischemia/reperfusion injury in the rat isolated heart. Our laboratory confirmed this finding (Ferdinandy et al., 1998a). Zweier at al. (1995b) showed that N^G-nitro-L-arginine methyl ester (up to 1 µM) only partially inhibited ischemia-induced nitric oxide accumulation, while improving the recovery of postischemic myocardial function. Using radio-labeled $^{15}NO^{2-}$, Zweier et al. (1995b) demonstrated that the ischemic heart favors the reduction of nitrite to nitric oxide due to low pH. These results show that increased nitric oxide concentration during ischemia/reperfusion contributes to myocardial injury, and that a non-NOS-dependent formation of nitric oxide might play a role, at least in part, in the accumulation of nitric oxide during myocardial ischemia (for review, see Weitzberg and Lundberg, 1998). Whatever the mechanism of the accumulation of nitric oxide during ischemia and reperfusion, the evidence suggests that this accumulation contributes to ischemia/reperfusion injury. The mechanism by which nitric oxide deteriorates myocardial function is not known. Recent studies suggest, however, that the harmful effects of nitric oxide in the heart and in the vasculature are due not to nitric oxide itself but to peroxynitrite, a reaction product of nitric oxide and superoxide (Schulz et al., 1997; Yasmin et al., 1997; for reviews, see Szabó, 1996; Beckman and Koppenol, 1996).

Nitric Oxide in Myocardial Ischemic Preconditioning

Nitric oxide has been proposed to play a key role in both the early (Ferdinandy et al., 1996; Vegh et al., 1992) and the late phases of preconditioning (Bolli et al., 1997a, 1997b). For example, Vegh et al. (1992) demonstrated that 10 mg/kg N^G-nitro-L-arginine methyl ester, a blocker of nitric oxide synthase, administered before and after preconditioning abolished the antiarrhythmic effect of preconditioning in a coronary occlusion model in anesthetized dogs. However, Lu et al. (1995) reported that 10 mg/kg N^G-mono-methyl-L-arginine or N^G-nitro-L-arginine methyl ester did not affect the antiarrhythmic effect of preconditioning in anesthetized rats with coronary occlusion/reperfusion. Weselcouch et al. (1995) showed that N^G-nitro-L-arginine methyl ester (30 µM) in rat hearts perfused in Langendorff mode did not interfere with the effect of preconditioning on postischemic myocardial function and lactate dehydrogenase release. In these studies, surprisingly, the different nitric oxide synthase inhibitors interfered with neither preconditioning nor the outcome of ischemia/reperfusion without preconditioning (although N^G-mono-methyl-L-arginine is approximately 10-fold less potent than N^G-nitro-L-arginine methyl-ester; Rees et al., 1990). Because nitric oxide generation was not determined and only a single dose of nitric oxide synthase inhibitor was used in these studies, it is difficult to interpret these negative results. Using electron spin resonance technique to measure cardiac nitric oxide content, we have previously shown that a decrease in basal cardiac nitric oxide content due to N^G-nitro-L-arginine treatment, hypercholesterolemia, or selective depletion of neurotransmitters from cardiac sensory neurons decreases pacing-induced preconditioning in isolated working rat hearts (Ferdinandy et al., 1996, 1997a, 1997b). We have also shown that if basal nitric oxide content was decreased after the preconditioning protocol, the effect of pacing-induced preconditioning on myocardial stunning remained intact (Ferdinandy et al., 1997b). Changes in cardiac nitric oxide content during preconditioning followed by ischemia/reperfusion have not been investigated; however, this might elucidate the role of nitric oxide in the triggering and mediating mechanisms of ischemic preconditioning. In another study from our laboratory, electron spin resonance spectroscopy was used to measure cardiac nitric oxide content at the end of a 30-min period of global normo-thermic ischemia in isolated rat hearts; the results showed that precondition-ing, similar to the effects of nitric oxide synthase inhibitors, attenuates the accumulation of nitric oxide during ischemia and reperfusion (Csonka et al., 1999).

Preconditioning and Aging

It is well known that the aged heart is less resistant to ischemia/reperfusion than the adult heart. Most experiments that examined aging were carried out using old but healthy laboratory animals. Therefore, most of the present knowl-edge about aging is based on studies in which the diseases commonly associ-ated with aging were not taken into account. However, most of the elderly population suffers from diseases such as hypertension, hypercholesterolemia, and diabetes, which colors the effects of aging. When studying the effects of aging on the heart, it is important to distinguish between the effects of aging itself and the effects of aging combined with some aging-related diseases. Fol-lowing is a list of the most common age-related diseases that can significantly modify the ability of the aging heart to resist ischemic stress.

* Hypercholesterolemia
* Diabetes mellitus
* Adverse effects of pharmacological maneuvers (e.g. development of nitrate tolerance, oral antidiabetic therapy, insulin resistance)
* Hypertrophy of the ventricles due to hypertension
* Failing myocardium

Effect of Aging on the Ischemic Tolerance of the Myocardium

It is well known that elderly people exhibit high risk of myocardial infarction and that hearts from aged animals are more susceptible to ischemia/reperfusion injury (for review, see Keller and Feit, 1995). The biochemical background of this phenomenon has been extensively investigated, and researchers have found that aging alters intracellular Ca^{2+}-channel activity. Ca^{2+}-channel activity de-creases in the mature heart but increases in the aged heart, resulting in in-creased intracellular Ca^{2+} accumulation (Matsuda et al., 1997). Enhanced coro-nary endothelial dysfunction (Amrani et al., 1996a) and intracellular acidosis (Stewart et al., 1995), and decreased formation of nitric oxide (Amrani et al., 1996b), high-energy phosphates (Ramani et al., 1996), adenosine (Ramani et al., 1996), and the natural antioxidants coenzymes Q_{10} and Q_9 (Muscari et al.,

1995) may all contribute to the increased susceptibility of the aging heart to ischemic stress.

Despite its great clinical significance, the heart's ability to adapt to ischemic stress in aging models has been investigated in very few studies. Abete et al. (1996) reported the loss of classical preconditioning in rats, whereas Burns et al. (1996) found that preconditioning was preserved in an aged sheep model of preconditioning. Abete et al. (1997, 1998) compared angina-induced protection against myocardial infarction in adult and elderly patients. The researchers found that although the presence of angina before acute myocardial infarction seemed to protect against in-hospital outcomes in adults, this effect was less obvious in elderly patients. These studies suggested that the protection afforded by angina in adult patients may involve classic preconditioning, which seems to be lost in senescent individuals. Late preconditioning has not been studied yet in aging models.

The biochemical background of the lost preconditioning effect in the elderly is not known. Recent investigations show that some adaptive biochemical mechanisms such as heat-shock response are diminished in aged hearts (Locke and Tanguay, 1996; Nitta et al., 1994). Because late preconditioning is based at least in part on expression of heat-shock proteins (Baxter et al., 1996), one may speculate if the protective effect of late preconditioning deteriorates with aging. HSP70 induction is an obvious pathway for therapeutic intervention, either by genetic or pharmacological manipulations, against diminished adaptive responses in aging.

Effect of Aging-Associated Diseases on the Ischemic Tolerance of the Myocardium

The majority of studies on preconditioning have been devoted to reconfirming the ability of the heart of an otherwise healthy animal to develop tolerance to ischemic challenges and analyzing mechanisms activated in the principally healthy heart in response to ischemia of whatever origin. Despite the incontestable clinical significance of preconditioning, very few studies have focused on its effect in hearts of diseased animals suffering from abnormalities relevant to coronary artery disease. Nevertheless, clinical studies have shown that the presence of several conditions such as hypercholesterolemia, diabetes, aging, smoking, chronic medication with oral antidiabetics, and development of nitrate tolerance all increase mortality rate due to myocardial infarction (for review, see Kannel and Mcgee, 1979; Roberts, 1995). Therefore, we suggested that these conditions may interfere with the biochemical pathways of the endogenous adaptive response to ischemia. Consequently, because the hearts of diabetic rats and hypercholesterolemic rabbits (two independent risk factors of coronary disease) were found to lose their adaptive capability to ischemia, at least in part, preconditioning was proposed to be a "healthy heart phenomenon" (Szilvássy et al., 1995; Tosaki et al., 1996). Study of preconditioning in complex pathological animal models will reveal more about the underlying biochemical mechanisms of preconditioning and will promote clinical application of preconditioning to prevent ischemic heart disease in high-risk populations. Development of new drugs to restore the protective effect of preconditioning in several diseases often related to aging will be of great importance in future cardiovascular research. The following part of this chapter focuses on the present knowledge and future perspectives of preconditioning in the diseased heart.

Preconditioning in the Hypertrophied and Failing Myocardium

Systemic arterial hypertension is a recognized risk factor for the development of ischemic heart disease. Long-standing hypertension also may be associated with the development of left ventricular hypertrophy. Left ventricular hypertrophy may be regarded as a structural adaptive response to sustained pressure overload; that is, the increasing mass of myocardium is able to perform more powerful contractions to work against the increasing systemic blood pressure. This condition, however, has severe prognostic implications, independent of or combined with other risk factors. Because of the higher occurrence of hypertension in the elderly, the incidence of left ventricular hypertrophy also increases with aging. Large cohort studies such as the Framingham Heart Study show that hypertensives with left ventricular hypertrophy have substantially increased risk of mortality from all cardiovascular causes including ischemic heart disease (Frohlich et al., 1992). Besides involving myocyte hypertrophy, left ventricular hypertrophy involves complex alterations of myocardial and vascular architecture such as fibrosis and inadequate microvascular perfusion as well as marked electrophysiological and biochemical changes. Eventually, left ventricular hypertrophy may deteriorate from a "compensated" state to serious impairment of the pump function of the left ventricle, ultimately resulting in cardiac failure when the heart is unable to cope with the high systemic blood pressure. Experimental evidence shows that the hypertrophied myocardium is at greater risk from ischemia/reperfusion insults even during the compensated state, although the risk may be greater still when failure supervenes. The high prevalence of left ventricular hypertrophy and cardiac failure has led several investigators to examine the preconditioning response in these conditions, because any preconditioning therapy would have to effectively address these conditions.

The first study examining the preconditioning response in left ventricular hypertrophy was described by Speechly-Dick et al. (1994) using a rat model of hypertension induced by deoxycorticosterone (DOCA) and salt administration. DOCA-salt rapidly elevates blood pressure without elevating plasma renin-angiotensin system activity. After 4 wk of DOCA-salt, hypertension was established, and a 30% enlargement of the left ventricle was present without any frank signs of cardiac failure. The classic preconditioning response could be induced in these animals, indicated by a marked limitation of infarct size after a sustained coronary artery occlusion in vivo. Boutros and Wang (1995) examined the effects of preconditioning in hypertrophy associated with a genetic form of hypertension in the spontaneously hypertensive rat. Preconditioning enhanced postischemic contractile recovery in hypertrophied hearts, as did exogenously applied adenosine. Pantos et al. (1996) examined classic preconditioning in the hypertrophied hearts of rats after 5 wk of abdominal aortic stenosis. The researchers examined preconditioning during Langendorff perfusion, finding that postischemic recovery of contractile function was improved by preconditioning in the hypertrophied hearts. Randall et al. (1997) studied the effect of preconditioning in a transgenic hypertensive rat, overexpressing the renin gene ([mREN-2]27 rat). Again the preconditioning response assessed by postischemic contractile recovery of isolated hearts was preserved in this model of hypertrophy. In contrast to these four studies, Moolman et al. (1997) reported that ischemic preconditioning does not protect hypertrophied myocardium from genetically hypertensive rats. The reason for the discordant finding is not clear; it is possible that the animals used in their study were senescent or were developing heart failure. Tajima et al. (1997) showed that the expression of heat-shock proteins is attenuated in hypertrophied hearts, which may suggest a decreased capability of these hearts to adapt to stress. Hence the majority of studies of hypertensive hearts undertaken to date have described the preservation of cardioprotective classic preconditioning mechanisms when there is no evidence of cardiac failure. However, a recent report hinted that, in cardiac failure, the preconditioning response may be attenuated. Dekker et al. (1998) studied perfused papillary muscles from rabbits with combined pressure and volume overload-induced cardiac failure. The end points used in that study to assess responses to ischemia, namely the time to onset of intracellular calcium rise and ischemic contracture, were delayed by preconditioning in normal myocardium but were not affected by preconditioning in the failing myocardium. However, these end points cannot be related easily to any functional index of injury, and more detailed examination of this question is required. Also, the effect of late preconditioning in the hypertrophied and failing myocardium has not been studied yet.

Interaction Between Hypercholesterolemia/Atherosclerosis and Preconditioning

Hyperlipidemia; atherosclerosis; and their major clinical sequel, ischemic heart disease, are the leading causes of death in the developed world. The reduced capability of hyperlipidemic/atherosclerotic individuals' hearts to adapt to repetitive ischemic challenges is clinically indicated by the syndrome *angina pectoris*. Moreover, several studies suggest that individuals with coronary artery disease who experience frequent ischemic episodes related to either exercise (Kent et al., 1982) or mental stress (Jain et al., 1995) have a worse prognosis than those without such episodes. On the other hand, in clinical practice the terms "warm-up angina" or "walk-through angina," similar to the term "ischemic preconditioning" in animal studies, have been used to describe the heart's adaptive capability to ischemia. That the vast majority of individuals with stable or unstable angina pectoris suffer from coronary artery disease suggests that the diseased hearts of individuals with coronary sclerosis do disclose the preconditioning phenomenon. As an experimental approach to this controversy, an earlier study from our laboratory with chronically instrumented conscious rabbits (Szilvássy et al., 1995) demonstrated that protection by classic preconditioning against myocardial stunning and electrophysiological changes was lost when the rabbits developed hypercholesterolemia and atherosclerosis because of exposure to dietary cholesterol. When the hypercholesterolemic and atherosclerotic conscious rabbits exhibiting loss of the preconditioning phenomenon were reexposed to a normal diet, serum cholesterol levels normalized in parallel with recapture of the preconditioning phenomenon in the presence of a constant degree of intimal lesions; this indicates that atherosclerosis per se without hypercholesterolemia does not significantly interact with preconditioning (Szilvássy et al., 1995). This finding was subsequently confirmed in isolated hearts of rats that were exposed to dietary cholesterol overload but did not develop significant amounts of atherosclerotic lesions in the large conductance vessels and in the coronary vasculature (Ferdinandy et al., 1997b). Benzuly et al. (1994) also concluded that abnormal vasoreactivity due to atherosclerosis returns to normal during the course of atherosclerosis regression, a process attributed to early resorption of lipids from the arterial wall that occurs before any detectable changes in the mass of atherosclerotic lesions. Alternatively, in the very early stage of dietary atherosclerosis (4-wk exposure of the rabbits to cholesterol-enriched diet), with no evidence of intimal lesions, dietary hypercholesterolemia completely blocked the preconditioning effect (Szilvássy et al., 1995). This observation is similar to Pitkanen et al.'s (1996) finding in young clinical patients with familial hypercholesterolemia, where a strongly significant decrease in coronary flow reserve precedes atherosclerotic plaque formation. These findings suggest that the ability of the in situ heart of conscious rabbits to adapt to repetitive challenges of transient global myocardial ischemia can be modulated by dietary manipulations regardless of the presence of intimal lesions (i.e., atherosclerosis). The experimental observation that this short-term adaptive mechanism is blocked by hypercholesterolemia but not atherosclerosis seems consistent with the clinical experience that individuals with coronary artery stenosis exhibit the preconditioning phenomenon (Deutsch et al., 1990; Ottani et al., 1995, Tzivoni and Maybaum, 1997). These observations emphasize the role of antihyperlipidemic therapy not only in preventing but also in treating ischemic heart disease caused by hyperlipidemia/atherosclerosis (Quyyumi et al., 1997).

The mechanism of the lost classic preconditioning in hypercholesterolemia is not completely understood. In previous studies with conscious rabbits, we have demonstrated that the anti-ischemic effect conferred by preconditioning with rapid pacing is associated with an increase in cardiac cGMP concentration (Szilvássy et al., 1994a). Pharmacological inhibition of nitric oxide synthesis also has been shown to block preconditioning in dogs and isolated working rat hearts (Ferdinandy et al., 1996; Vegh et al., 1992). These results suggest that the protection provided by preconditioning may be related to formation of nitric oxide from endothelial cells (Parratt, 1995), neurons (Ferdinandy et al., 1997a), and/or cardiac myocytes. Nitric oxide stimulates formation of cGMP (Moncada et al., 1991), with four results: (1) reduction of the influx of calcium through L-type calcium channels, (2) stimulation of cGMP-sensitive phosphodiesterase, a mechanism that explains the reduced adenosine $3',5'$-cyclic monophosphate (cAMP) level in parallel with high cGMP levels in the preconditioned rabbit hearts (Szilvássy et al., 1994a), (3) reduction of norepinephrine release from cardiac sympathetic nerves (Sandhu et al., 1996; Schwarz et al., 1995), and (4) reduction of oxygen consumption (Parratt, 1995). Functional defects have long been identified in endothelium-dependent vasorelaxation in hypercholesterolemia and atherosclerosis (Flavahan, 1992), because vasodilation was found impaired in response to several different endothelium-dependent vasodilators such as bradykinin, muscarinic agonists, substance P, 5-hydroxytryptamine, ATP, thrombin, histamine, and A23187. Reduced generation of nitric oxide is an attractive hypothesis to explain reduced endothelium-dependent vasodilation due to hypercholesterolemia. Jacobs et al. (1990) suggested that inactivation of nitric oxide by oxidized low density lipoprotein is also involved in this mechanism. Increasing evidence has accumulated that a high-cholesterol diet impairs nitric oxide–cGMP signaling in endothelial and nonendothelial cells as well (Deliconstatinos et al., 1995; Lefer et al., 1993; Simonet et al., 1993), which raises the question whether the loss of ischemic preconditioning in a high-cholesterol diet is caused by impaired nitric oxide metabolism in the heart. Accordingly, Szekeres et al. (1997) demonstrated that the cGMP content of hearts of hyperlipidemic/atherosclerotic rabbits was almost three times lower than that seen in healthy animals' hearts. Moreover, unlike the normal heart, the hypercholesterolemic/atherosclerotic one is unable to produce increased amounts of cGMP in response to ischemia. In addition, an increase in cardiac cAMP concentration with much more severe electrophysiological and hemodynamic changes caused by ischemia with or without preconditioning is seen in hypercholesterolemic animals (Szekeres et al., 1997; Szilvássy et al., 1995). Another study from our laboratory showed a correlation between reduced cardiac nitric oxide content and a loss of preconditioning in hearts of hypercholesterolemic rats (Ferdinandy ct al., 1997b). Nevertheless, according to our recent finding (Ferdinandy et al., 1998b), farnesol, a major polyprenyl product, restored the preconditioning phenomenon but did not restore nitric oxide synthesis in hearts from hypercholesterolemic rats. This finding may suggest that the nitric oxide–cGMP pathway is activated in the initial phase of the biochemical cascade of preconditioning, and that the excess exogenous cholesterol that inhibits formation of several polyprenyl derivatives, including farnesol via inhibition of hydroxymethyl-glutaryl-coenzyme-A (HMG-CoA) reductase enzyme (Casey, 1992), leads to signal transduction failure. Recent evidence suggests that prenylation of several proteins, including nuclear protein lamin B and a number of guanosine 5′-triphosphate (GTP)-binding regulatory proteins (G proteins), which control a wide spectrum of signal transduction pathways, is a prerequisite for their normal physiological function (Casey, 1992; Finegold et al., 1990). These findings raise the question whether using HMG-CoA inhibitors as an antihyperlipidemic therapy interferes with preconditioning, as does hypercholesterolemia itself.

It is not known if late preconditioning works under pathological conditions such as hypercholesterolemia and atherosclerosis. Currently, the literature includes only one paper (from our laboratory) dealing with this problem. According to our result, chronically instrumented conscious rabbits subjected to a cholesterol-enriched diet over 8 wk exhibited loss of the short-term (classic) preconditioning phenomenon, but the long-term protection was preserved; however, stronger preconditioning stimuli were necessary to induce late preconditioning as compared to the nonhyperlipidemic animals. Moreover, development of the long-term protection in hypercholesterolemic/atherosclerotic animals was independent of the cGMP system (Szekeres et al., 1997). Therefore, it is conceivable that the presence of the short-term or classic preconditioning might not be a prerequisite for late preconditioning.

In conclusion, according to our present knowledge, high-cholesterol-diet–induced impairment of the nitric oxide–cGMP pathway and inhibition of polyprenyl biosynthesis may impair signal transduction processes, thereby leading to the loss of preconditioning. Further studies are needed to examine the interaction of hyperlipidemia with other well-known mediators of preconditioning such as adenosine. However, it seems that late preconditioning might be used to prevent ischemic heart disease even in hypercholesterolemia and atherosclerosis.

Preconditioning and Nitrate Therapy

Organic nitrates have been the mainstay therapy of angina pectoris for more than a century. During the last decade, much has been learned about the mechanism of action of this pharmacological class of drugs. One of the fascinating results of research on the pharmacology of nitrates, especially in context of endogenous cardioprotection, is that these agents elicit their cardiovascular effects by releasing nitric oxide (for review, see Harrison and Bates, 1993). If we assume that nitric oxide is a key mediator of ischemic preconditioning, it is plausible that nitrate therapy could be used to effect ischemic preconditioning (Parratt, 1995). Indeed, nitroglycerin at doses that marginally decreased blood pressure attained an anti-ischemic effect of approximately the same magnitude as did preconditioning in conscious rabbits (Szilvássy et al., 1997). The administration of nitrates in ischemic heart disease caused by hypercholesterolemia/atherosclerosis is therefore of particular interest, because these conditions are characterized by a deficiency in nitric oxide, as discussed previously. Nevertheless, the major problem with continuous administration of nitrates is the development of vascular tolerance, which often attenuates the therapeutic effect as well (Bassenge and Zanzinger, 1992). In our laboratory, we have shown that, besides vascular sites of action, the anti-ischemic effect of nitroglycerin involves a direct myocardial mechanism independent of the development of vascular tolerance to nitroglycerin in isolated working rat hearts (Ferdinandy et al., 1995a). This observation was confirmed by Szilvássy et al. (1997), who showed that, besides the potent direct anti-ischemic effect of nitroglycerin, the loss of the preconditioning phenomenon was seen in conscious rabbits made tolerant to the hypotensive effect of the organic nitrate. Moreover, neither nitroglycerin nor application of preconditioning stimuli increased cardiac cGMP content in the tolerant state (Szilvássy et al., 1997). Szilvássy et al. concluded that nitroglycerin might elicit cardioprotection without involvement of cGMP, whereas maneuvers that make the heart unable to produce increased amounts of cGMP in response to an ischemic challenge also render the heart unable to adapt to repetitive ischemic insults (Szilvássy et al., 1994b). Indeed, a recent study from our laboratory showed that cardiac cGMP content does not necessarily reflect the actual cardiac nitric oxide concentration (Csont et al., 1998).

To avoid rapid development of nitrate tolerance, intermittent nitrate therapy is used in clinical practice. Intermittent nitrate therapy effectively prevents nitrate tolerance and the loss of the preconditioning effect; however, during nitrate-free periods, the heart is more vulnerable to a single ischemic episode in conscious rabbits (Szilvássy et al., 1997). This finding emphasizes the complexity of the interaction of nitrates with ischemic heart disease and encourages further studies to develop an optimal method of nitrate therapy.

In spite of its great clinical significance, there are no experimental data available on the interaction of nitrate therapy with late preconditioning.

Preconditioning in Diabetes

Epidemiological studies have clearly shown that both insulin-dependent (IDDM) and non-insulin-dependent diabetics (NIDDM, or elderly diabetics) are more prone to develop myocardial infarctions and postinfarction complications that occur independently of coronary atherosclerosis (Stone et al., 1989). Although numerous clinical data show that diabetes is a risk factor for ischemic heart disease, the literature is inconsistent regarding the susceptibility of hearts from diabetic animals to ischemia/reperfusion injury. Because insulin not only regulates the balance of metabolic fuels received by the heart but also regulates metabolism and myocardial perfusion by acting on various intracellular regulatory proteins and messenger systems (for review, see Stanley et al., 1997), it is plausible that the complex metabolic disorder of diabetes may interfere with the biochemical pathways of preconditioning. Tremendous data are available on the effect of experimental diabetes on the outcome of a single ischemic episode, and several excellent reviews discuss whether hearts from diabetic animals are either more sensitive or less sensitive to transient ischemia (for review, see Feuvray and Lopaschuk, 1997; Paulson, 1997). It is surprising, however, that only very few experimental studies have addressed whether diabetes interacts with ischemic preconditioning. The first study to examine the preconditioning response in experimental streptozotocin-induced diabetes was reported by Liu et al. (1993). Diabetic rat hearts were found to be more resistant to myocardial infarction in vivo than normal control hearts, and classic preconditioning conferred additional protection; however, rats were selected for ischemia/reperfusion 2 wk after streptozotocin. Tosaki et al. (1996) examined the evolution of streptozotocin-induced diabetic response to ischemia/reperfusion and preconditioning in isolated rat hearts. They found that, in the early phase of experimental diabetes (2 wk), the diabetic heart is more resistant to ischemia/reperfusion, although this protection was not seen in the 4- and 6-wk diabetics; moreover, the 8-wk diabetic group exhibited a worse outcome from ischemia/reperfusion injury. In contrast to nondiabetic hearts, four cycles of preconditioning did not afford protection in 4- or 8-wk diabetics, showing that classical preconditioning against myocardial stunning and ischemia-induced cellular ion imbalance is not effective in chronic diabetes. Further studies are needed to elucidate the interaction of diabetes and preconditioning in other species and to examine if diabetes interferes with the development of late preconditioning.

Beyond the possible detrimental effect of diabetes per se on the preconditioning phenomenon, the picture is further colored by the feasible influence of antidiabetic drugs on cardiac response to repetitive ischemic challenges. Drug treatments of NIDDM include sulfonylureas, biguanides, alfa-glucosidase inhibitors, and insulin (for review, see Scheen, 1997). Unfortunately, it is not known which class of drugs best retards progression of the disease and prevents late complications of diabetes such as neuropathy, angiopathy, hypertension, or ischemic heart disease. The United Kingdom Prospective Diabetes Study (UKPDS) is expected to answer this important question (UKPDS 13, 1995; UKPDS 16, 1995).

Recently, the use of sulfonylurea derivatives has raised many concerns (Engler and Yellon, 1996). Sulfonylureas blockade ATP-sensitive potassium channels in pancreatic island beta-cells (for review, see Panten et al., 1996). Because these agents are not specific for beta-cell sulfonylurea receptors (SUR1), which together with an inwardly rectifying K^+-channel make up the functional ATP-sensitive potassium channels; sulfonylurea derivatives block extrapancreatic ATP-sensitive potassium channels including sarcolemmal ATP-sensitive potassium channels via binding to sarcolemmal sulfonylurea receptor (SUR2) (for review, see Nichols et al., 1996). Blockade of cardiac ATP-sensitive potassium channels worsens the outcome of the heart from ischemia/reperfusion injury (Ferdinandy et al., 1995b) and abolishes both classical preconditioning and late preconditioning (Ferdinandy et al., 1995c; Mei et al., 1996; Qian et al., 1996). In addition to interacting with ischemia/reperfusion and preconditioning, sulfonylurea derivatives may abolish the therapeutic effect of some frequently used anti-ischemic and hypotensive drugs such as nicorandil (Auchampach et al., 1992; Mitani et al., 1991;) and nitroglycerine (Csont et al., 1999).

Insulin therapy also shows some unfavorable effects on the cardiovascular system. The most important adverse effect of insulin therapy is body-weight gain that requires hyperinsulinemia to maintain blood glucose levels within a normal range. Hyperinsulinemia associated with insulin resistance is suspected to promote hyperlipidemia and atherosclerosis in diabetic individuals (Stout, 1990). The interaction between hyperlipidemia/atherosclerosis was discussed previously. Thus, we think that dietary therapeutic manipulations should be emphasized to restore glucose homeostasis in NIDDM individuals.

Bimoclomol, a novel derivative of hydroxylamine [2-hydroxy-3-(1-piperidinyl)-propoxyl]-3-pyridine-carboximidoil-chloride, is currently under development to alleviate diabetic complications. We recently showed that nontoxic concentrations of the compound provide wide cytoprotective effect by inducing HSP70 in different cells and tissues, including a marked anti-ischemic and antiarrhythmic effect on the ischemic/reperfused myocardium. Therefore, Bimoclomol may be the first candidate of a new class of drugs with HSP70-inducing effect (Vigh et al., 1997) that provides a wide cytoprotective action in the myocardium during ischemic stress.

Conclusion

Ischemic heart disease is responsible for about 50% of morbidity and mortality in the elderly. Myocardial ischemic preconditioning, an endogenous adaptive mechanism of the heart to stress, confers a remarkable cardioprotection against ischemia/reperfusion injury in a variety of species, including humans. The naturally occurring triggers of preconditioning are coronary stenosis, spasm, or complete coronary occlusion, which all represent an advanced-stage ischemic heart disease. However, preconditioning can be elicited physiologically in healthy subjects through physical exercise, and therefore exercise may be used to prevent ischemic heart disease. The biochemical mechanism of preconditioning includes modulation of nitric oxide metabolism. The effectiveness of preconditioning is markedly attenuated in aging and in some disease states frequently associated with aging (e.g., hypercholesterolemia, diabetes, nitrate tolerance, etc.). The mechanism by which aging and the aforementioned diseases attenuate the cardioprotective effect of preconditioning is not completely understood, but their interaction with nitric oxide–dependent cellular signaling pathways is suspected. Further studies on the mechanism of preconditioning in both healthy and diseased myocardium will lead to development of drugs that will enhance the elderly heart's ability to adapt to ischemic stress, thereby improving quality of life in the elderly.

References

Abete, P., P. Ferrara, S. Bianco, C. Calabrese, C. Napoli, F. Cacciatore, N. Ferrara, and F. Rengo. 1998. Age-related effects of acidosis in isolated cardiac muscle. *Journals of Gerontology Series A, Biological Sciences and Medical Sciences* 53:B42-B48.

Abete, P., N. Ferrara, F. Cacciatore, A. Madrid, S. Bianco, C. Calabrese, C. Napoli, P. Scognamiglio, O. Bollella, A. Cioppa, G. Longobardi, and F. Rengo. 1997. Angina-induced protection against myocardial infarction in adult and elderly patients: A loss of preconditioning mechanism in the aging heart? *Journal of the American College of Cardiology* 30:947-954.

Abete, P., N. Ferrara, A. Cioppa, P. Ferrara, S. Bianco, C. Calabrese, F. Cacciatore, G. Longobardi, and F. Rengo. 1996. Preconditioning does not prevent postischemic dysfunction in aging heart. *Journal of the American College of Cardiology* 27:1777-1786.

Auchampach, J.A., I. Cavero, and G.J. Gross. 1992. Nicorandil attenuates myocardial dysfunction associated with transient ischemia by opening ATP–dependent potassium channels. *Journal of Cardiovascular Pharmacology* 20:765-771.

Amrani, M., A.H. Chester, J. Jayakumar, and M.H. Yacoub. 1996a. Aging reduces postischemic recovery of coronary endothelial function. *Journal of Thoracic and Cardiovascular Surgery* 111:238-245.

Amrani, M., A.T. Goodwin, C.C. Gray, and M.H. Yacoub. 1996b. Ageing is associated with reduced basal and stimulated release of nitric oxide by the coronary endothelium. *Acta Physiologica Scandinavica* 157:79-84.

Baxter, G.F., M.S. Marber, and D.M. Yellon. 1996. Myocardial stress response, cytoprotective proteins and the second window of protection against infarction. In: *Myocardial Preconditioning,* C.L. Wainwright and J.R. Parratt (eds.), pp. 233-250. R.G. Landes. Austin, Texas.

Baxter, G.F., and D.M. Yellon. 1994. Ischaemic preconditioning of myocardium: A new paradigm for clinical cardioprotection. *British Journal of Clinical Pharmacology* 38:381-387.

Bassenge, E., and J. Zanzinger. 1992. Nitrates in different vascular beds, nitrate tolerance, and interactions with endothelial function. *American Journal of Cardiology* 70:23B-29B.

Beckman, J.S., and W.H. Koppenol. 1996. Nitric oxide, superoxide, and peroxynitrite: The good, the bad and ugly. *American Journal of Physiology* 271:C1424-C1437.

Benzuly, K.H., R.C. Padgett, S. Kaul, D. Piegors, M.L. Armstrong, and D.D. Heistad. 1994. Functional improvement precedes structural regression of atherosclerosis. *Circulation* 89:1810-1818.

Beresewicz, A., E. Karwatowska-Prokopczuk, B. Lewartowski, and K. Cedro-Ceremuzynska. 1995. A protective role of nitric oxide in isolated ischaemic reperfused rat heart. *Cardiovascular Research* 30:1001-1008.

Black, S.C., and B.R. Lucchesi. 1993. Heat shock proteins and the ischemic heart: An endogenous protective mechanism. *Circulation* 87:1048-1051.

Bolli, R., Z.A. Bhatti, X.L. Tang, Y.M. Qiu, Q. Zhang, Y. Guo, and A.K. Jadoon. 1997a. Evidence that late preconditioning against myocardial stunning in conscious rabbits is triggered by the generation of nitric oxide. *Circulation Research* 81:42-52.

Bolli, R., S. Manchikalapudi, X.L. Tang, H. Takano, Y. Qiu, Y. Guo, Q. Zhang, and A.K. Jadoon. 1997b. The protective effect of late preconditioning against myocardial stunning in conscious rabbits is mediated by nitric oxide synthase. Evidence that nitric oxide acts both as a trigger and as a mediator of the late phase of ischemic preconditioning. *Circulation Research* 81:1094-1107.

Boutros, A., and J. Wang. 1995. Ischemic preconditioning, adenosine and bethanechol protect spontaneously hypertensive isolated rat hearts. *Journal of Pharmacology and Experimental Therapeutics* 275:1148-1156.

Braunwald, E.K. 1985. Myocardial reperfusion: A double-edged sword? *Journal of Clinical Investigation* 76:1713-1719.

Burns, P.G., I.B. Krunkenkamp, C.A. Calderone, R.J. Kirvaitis, G.R. Guadette, and S. Levitsky. 1996. Is the preconditioning response conserved in senescent myocardium? *Annals of Thoracic Surgery* 61:925-929.

Capecchi, P.L., F.L. Pasini, G. Cati, M. Colafati, A. Acciavatti, L. Ceccatelli, S. Petri, A. de Lalla, and T. Di Perri. 1997. Experimental model of short-time exercise-induced preconditioning in POAD patients. *Angiology* 48:469-480.

Casey, P.J. Biochemistry of protein prenylation. 1992. *Journal of Lipid Research* 33:1731-1740.

Correa, S.D., and S. Schaefer. 1997. Blockade of KATP channels with glibenclamide does not abolish preconditioning during demand ischemia. *American Journal of Cardiology* 79:75-78.

Csonka, C., Z. Szilvássy, F. Fülöp, T. Páli, R. Schulz, A. Tosaki, I.E. Blasig, and P. Ferdinandy. 1999. Classic preconditioning decreases the harmful accumulation of nitric oxide during ischemia and reperfusion in rat hearts. *Circulation* 100: 2260-2266.

Csont, T., T. Páli, Z. Szilvássy, and P. Ferdinandy. 1998. Lack of correlation between myocardial nitric oxide and cyclic guanosine monophosphate content in both nitrate-tolerant and nontolerant rats. *Biochemical Pharmacology* 56:1139-1144.

Csont, T., Z. Szilvássy, F. Fülöp, S. Nedeianu, T. Páli, A. Tosaki, L. Dux, and P. Ferdinandy. 1999. Direct myocardial anti-ischaemic effect of GTN in both nitrate-tolerant and nontolerant rats: a cGMP-independent activation of K_{ATP}. *British Journal of Pharmacology* 128: 1427-1434.

Curtis, M.J., and R. Pabla. 1997. Nitric oxide supplementation or synthesis block—Which is the better approach to treatment of heart disease? *Trends in Pharmacological Sciences* 18:239-244.

Deliconstatinos, G., V. Villioyou, and J.C. Stavrides. 1995. Modulation of particulate nitric oxide synthase activity and peroxynitrite synthesis in cholesterol enriched endothelial cell membranes. *Biochemical Pharmacology* 49:1589-1600.

Dekker, L.R., H. Rademaker, J.T. Vermeulen, T. Opthof, R. Coronel, J.A. Spaan, and M.J. Janse. 1998. Cellular uncoupling during ischemia in hypertrophied and failing rabbit ventricular myocardium: Effects of preconditioning. *Circulation* 97:1724-1730.

Depre, C., L. Fierain, and L. Hue. 1997. Activation of nitric oxide synthase by ischaemia in the perfused heart. *Cardiovascular Research* 33:82-87.

Depre, C., and L. Hue. 1994. Cyclic GMP in the perfused rat heart. Effect of ischaemia, anoxia and nitric oxide synthase inhibitor. *FEBS Letters* 345:241-245.

Deutsch, E., M. Berger, W.G. Kussmaul, J.W. Hirshfield, H.C. Herman, and W.L. Laskey. 1990. Adaptation to ischemia during percutaneous transluminal coronary angioplasty: Clinical, haemodynamic, and metabolic features. *Circulation* 82:2044-2051.

Engler, R.L., and D.M. Yellon. 1996. Sulfonylurea K_{ATP} blockade in type II diabetes and preconditioning in cardiovascular disease—Time for reconsideration. *Circulation* 94:2297-2301.

Ferdinandy, P., Y. Appelbaum, C. Csonka, I.E. Blasig, and A. Tosaki. 1998a. Role of nitric oxide and TPEN, a potent metal chelator, in ischaemic and reperfused rat isolated hearts. *Clinical and Experimental Pharmacology and Physiology* 25:496-502.

Ferdinandy, P., C. Csonka, T. Csont, Z. Szilvássy, and L. Dux. 1998b. Rapid pacing-induced preconditioning is recaptured by farnesol treatment in hearts of cholesterol-fed rats: Role of polyprenyl-derivatives and nitric oxide. *Molecular and Cellular Biochemistry* 186:27-34.

Ferdinandy, P., T. Csont, C. Csonka, M. Torok, M. Dux, J. Nemeth, L.I. Horvath, L. Dux, Z. Szilvássy, and G. Jancso. 1997a. Capsaicin-sensitive local sensory innervation is involved in pacing-induced preconditioning in rat hearts: Role of nitric oxide and CGRP? *Naunyn-Schmiedebergs Archives of Pharmacology* 356:356-363.

Ferdinandy, P., Z. Szilvássy, N. Balogh, C. Csonka, T. Csont, M. Koltai, and L. Dux. 1996. Nitric oxide is involved in active preconditioning in isolated working rat hearts. *Annals of the New York Academy of Sciences* 793:489-493.

Ferdinandy, P., Z. Szilvássy, and G.F. Baxter. 1998c. Myocardial stress adaptation in disease states: Is preconditioning a healthy heart phenomenon? *Trends in Pharmacological Sciences* 19:223-229.

Ferdinandy, P., Z. Szilvássy, T. Csont, C. Csonka, E. Nagy, M. Koltai, and L. Dux. 1995a. Nitroglycerin-induced direct protection of the ischaemic myocardium in isolated working hearts of rats with vascular tolerance to nitroglycerin. *British Journal of Pharmacology* 115:1129-1131.

Ferdinandy, P., Z. Szilvássy, M.T. Droy-Lefaix, T. Tarrade, and M. Koltai. 1995b. K_{ATP} channel modulation in working rat hearts with coronary occlusion: Effects of cromakalim, cicletanine, and glibenclamide. *Cardiovascular Research* 30:781-787.

Ferdinandy, P., Z. Szilvássy, L.I. Horvath, T. Csont, C. Csonka, E. Nagy, R. Szentgyörgyi, I. Nagy, M. Koltai, and L. Dux. 1997b. Loss of pacing-induced preconditioning in rat hearts: Role of nitric oxide and cholesterol-enriched diet. *Journal of Molecular and Cellular Cardiology* 29:3321-3333.

Ferdinandy, P., Z. Szilvássy, M. Koltai, and L. Dux. 1995c. Ventricular overdrive pacing-induced preconditioning and no-flow ischemia-induced preconditioning in isolated working rat hearts. *Journal of Cardiovascular Pharmacology* 25:97-104.

Feuvray, D., and G.D. Lopaschuk. 1997. Controversies on the sensitivity of the diabetic heart to ischemic injury: The sensitivity of the diabetic heart to ischemic injury is decreased. *Cardiovascular Research* 34:113-120.

Finegold, A.A., W.R. Schafer, J. Rine, M. Whiteway, and F. Tamanoi. 1990. Common modifications of trimeric G proteins and ras protein: Involvement of polyisoprenylation. *Science* 249:165-169.

Flavahan, N.A. 1992. Atherosclerosis or lipoprotein-induced endothelial dysfunction. Potential mechanisms underlying reduction in EDRF/nitric oxide activity. *Circulation* 85:1927-1938.

Frohlich, E.D., C. Apstein, A.V. Chobanian, R.B. Devereux, H.P. Dustan, V. Dzau, F. Fauad-Tarazi, M.J. Horan, M. Marcus, and B. Massie. 1992. The heart in hypertension. *New England Journal of Medicine* 327:998-1008.

Hasebe, N., Y.-T. Shen, and S.F. Vatner. 1993. Inhibition of endothelium-derived relaxing factor enhances myocardial stunning in conscious dogs. *Circulation* 88:2862-2871.

Harrison, D.G., and J.N. Bates. 1993. The nitrovasodilators. New ideas about old drugs. *Circulation* 87:1461-1467.

Jacobs, M., F. Plane, and F.R. Bruckdorfer. 1990. Native and oxidized low-density lipoproteins have different inhibitory effects on endothelium-derived relaxing factor in the rabbit aorta. *British Journal of Pharmacology* 100:21-26.

Jain, D., M. Burg, R. Soufer, and B.L. Zaret. 1995. Prognostic implications of mental stress-induced silent left ventricular dysfunction in patients with stable angina pectoris. *American Journal of Cardiology* 76:31-35.

Jungersten, L., A. Ambring, B. Wall, and A. Wennmalm. 1997. Both physical fitness and acute exercise regulate nitric oxide formation in healthy humans. *Journal of Applied Physiology* 82:760-764.

Kannel, W.B., and D.L. Mcgee. 1979. Diabetes and cardiovascular disease: The Framingham study. *Journal of the American Medical Association* 241:2035-2038.

Keller, N.M., and F. Feit. 1995. Atherosclerotic heart disease in the elderly. *Current Opinion in Cardiology* 10:427-433.

Kent, K.M., D.R. Rosing, C.J. Ewels, L. Lipson, R. Bonow, and S.E. Epstein. 1982. Prognosis of asymptomatic or mildly symptomatic patients with coronary artery disease. *American Journal of Cardiology* 49:1823-1830.

Kitakaze, M., K. Node, K. Komamura, T. Minamino, M. Inoue, M. Hori, and T. Kamada. 1995. Evidence for nitric oxide generation in cardiomyocytes: Its augmentation by hypoxia. *Journal of Molecular and Cellular Cardiology* 27:2149-2154.

Kloner, R.A., K. Przyklenk, P. Whittaker, and S. Hale. 1995. Preconditioning stimuli and inadvertent preconditioning. *Journal of Molecular and Cellular Cardiology* 27:743-747.

Kloner, R.A., and D. Yellon. 1994. Does ischemic preconditioning occur in patients? *Journal of the American College of Cardiology* 24:1133-1142.

Lamontagne, D., U. Pohl, and R. Busse. 1991. The mechanical activity of the heart stimulates endothelium-derived relaxing factor release in the coronary vascular bed. *Journal of Cardiovascular Pharmacology* 17:S98-S94.

Lawson, C.S. 1995. Preconditioning in man: Progress and prospects. *Journal of Molecular and Cellular Cardiology* 27:961-967.

Lefer, D.J., K. Nakanishi, W.E. Johnston, and J. Vinten-Johansen. 1993. Antineutrophil and myocardial protecting actions of a novel nitric oxide donor after acute myocardial ischemia and reperfusion in dogs. *Circulation* 88:2337-2350.

Lefer, A.M., M.R. Siegfriedand, and X. Ma. 1993. Protection of ischemia-reperfusion injury by sydnonimine NO donors via inhibition of neutrophil-endothelium interaction. *Journal of Cardiovascular Pharmacology* 22(Suppl. 7):S27-S33.

Linz, W., G. Wiemer, and B. Schölkens. 1992. ACE-inhibition induces NO-formation in cultured bovine endothelial cells and protects isolated ischemic rat hearts. *Journal of Molecular and Cellular Cardiology* 24:909-919.

Liu, Y., J.D. Thornton, M.V. Cohen, J.M. Downey, and S.W. Schaffer. 1993. Streptozotocin-induced non-insulin-dependent diabetes protects the heart from infarction. *Circulation* 88:1273-1278.

Locke, M., and R.M. Tanguay. 1996. Diminished heat shock response in the aged myocardium. *Cellular Stress and Chaperones* 1:251-260.

Lu, H.R., P. Remeysen, and F. De Clerck. 1995. Does the antiarrhythmic effect of ischemic preconditioning in rats involve the L-arginine nitric oxide pathway? *Journal of Cardiovascular Pharmacology* 25:524-530.

Marin, J. 1995. Age-related changes in vascular responses: A review. *Mechanisms of Ageing and Development* 79:71-114.

Matheis, G., M.P. Sherman, G.D. Buckberg, D.M. Haybron, H.H. Young, and L.J. Ignarro. 1992. Role of L-arginine-nitric oxide pathway in myocardial reoxygenation injury. *American Journal of Physiology* 262:616-620.

Matsuda, H., J.D. McCully, and S. Levitsky. 1997. Developmental differences in cytosolic calcium accumulation associated with global ischemia: Evidence for differential intracellular calcium channel receptor activity. *Circulation* 96(Suppl. 9):2-8.

Maybaum, S., M. Ilan, J. Mogilevsky, and D. Tzivoni. 1996. Improvement in ischemic parameters during repeated exercise testing: A possible model for myocardial preconditioning. *American Journal of Cardiology* 78:1087-1091.

Mei, D.A., G.T. Elliott, and G.J. Gross. 1996. K_{ATP} channels mediate late preconditioning against infarction produced by monophosphoryl lipid A. *American Journal of Physiology* 271:H2723-H2729.

Mestril, R., and W.H. Dillmann. 1995. Heat shock proteins and protection against myocardial ischemia. *Journal of Molecular and Cellular Cardiology* 27:45-52.

Millar, C.G., G.F. Baxter, and C. Thiemermann. 1996. Protection of the myocardium by ischaemic preconditioning: Mechanisms and therapeutic implications. *Pharmacology and Therapeutics* 69:143-151.

Mitani, A., K. Sakamoto, K. Fukamachi, M. Sakamoto, K. Kurisu, Y. Tsuruhara, F. Fukumura, A. Nakashima, and K. Tokunaga. 1991. Effects of glibenclamide and nicorandil on cardiac function during ischemia and reperfusion in isolated perfused rat hearts. *American Journal of Physiology* 261:H1864-H1871.

Moncada, S., R.M.J. Palmer, and E.A. Higgs. 1991. Nitric oxide: Physiology, pathophysiology and pharmacology. *Pharmacological Reviews* 43:109-142.

Moolman, J.A., S. Genade, E. Tromp, L.H. Opie, and A. Lochner. 1997. Ischaemic preconditioning does not protect hypertrophied myocardium against ischaemia. *South African Medical Journal* 87 (Suppl. 3):C151-C156.

Murry, C., R. Jennings, and K. Reimer. 1986. Preconditioning with ischemia: A delay of lethal cell injury in ischemic myocardium. *Circulation* 74:1124-1136.

Muscari, C., L. Biagetti, C. Stefanelli, E. Giordano, C. Guarnieri, and C.M. Caldarera. 1995. Adaptive changes in coenzyme Q biosynthesis to myocardial reperfusion in young and aged rats. *Journal of Molecular and Cellular Cardiology* 27:283-289.

Nichols, C.G., E.N. Makhina, W.L. Pearson, Q. Sha, and A.N. Lopatin. 1996. Inward rectification and implications for cardiac excitability. *Circulation Research* 78:1-7.

Nitta, Y., K. Abe, M. Aoki, I. Ohno, and S. Isoyama. 1994. Diminished heat shock protein 70 mRNA induction in aged rat hearts after ischemia. *American Journal of Physiology* 267:H1795-H1803.

Node, K., M. Kitakaze, H. Kosaka, K. Komamura, T. Minamino, M. Inoue, M. Tada, M. Hori, and T. Kamada. 1996. Increased release of NO during ischemia reduces myocardial contractility and improves metabolic dysfunction. *Circulation* 93:356-364.

Ottani, F., M. Galvani, D. Ferrini, F. Sorbello, P. Limonetti, D. Pantoli, and F. Rusticali. 1995. Prodromal angina limits myocardial infarct size: A role for ischemic preconditioning. *Circulation* 91:291-297.

Ovize, M., R.A. Kloner, and K. Przyklenk. 1994. Stretch preconditions canine myocardium. *American Journal of Physiology* 266:H137-H146.

Pantos, C.I., C.H. Davos, H.C. Carageorgiou, D.V. Varonos, and D.V. Cokkinos. 1996. Ischaemic preconditioning protects against myocardial dysfunction caused by ischaemia in isolated hypertrophied rat hearts. *Basic Research in Cardiology* 91:444-449.

Panten, U., M. Schwanstecher, and C. Schwanstecher. 1996. Sulfonylurea receptors and mechanism of sulfonylurea action. *Experimental and Clinical Endocrinology and Diabetes* 104:1-9.

Parratt, J.R. 1994. Protection of the heart by ischaemic preconditioning: Mechanisms and possibilities for pharmacological exploitation. *Trends in Pharmacological Sciences* 15:19-25.

Parratt, J.R. 1995. Possibilities for the pharmacological exploitation of ischaemic preconditioning. *Journal of Molecular and Cellular Cardiology* 27:991-1000.

Parratt, J.R., and L. Szekeres. 1995. Delayed protection of the heart against ischaemia.
Trends in Pharmacological Sciences 16:351-355.

Paulson, D.J. 1997. The diabetic heart is more sensitive to ischemic injury. *Cardiovascular
Research* 34:104-112.

Pitkanen, O.P., O.T. Raitakari, H. Ninikoski, P. Nuutila, H. Iida, M. Voipio-Pulkki, R.
Harkonen, U. Wegelius, T. Ronnemaa, J. Viikari, and J. Knuuti. 1996. Coronary flow
reserve is impaired in young men with familial hypercholesterolemia. *Journal of the
American College of Cardiology* 28:1705-1711.

Poveda, J.J., A. Riestra, E. Salas, M.L. Cagigas, C. Lopez-Somoza, J.A. Amado, and J.R.
Berrazueta. 1997. Contribution of nitric oxide to exercise-induced changes in healthy
volunteers: Effects of acute exercise and long-term physical training. *European Journal
of Clinical Investigation* 27:967-971.

Qian, Y.Z., J.E. Levasseur, K.I. Yoshida, and R.C. Kukreja. 1996. K_{ATP} channels in rat heart:
Blockade of ischemic and acetylcholine-mediated preconditioning by glibenclamide.
American Journal of Physiology 271:H23-H28.

Quyyumi, A.A., N. Dakak, D. Mulcahy, N.P. Andrews, S. Husain, J.A. Panza, and R.O.
Cannon. 1997. Nitric oxide activity in the atherosclerotic human coronary circulation.
Journal of the American College of Cardiology 29:308-317.

Ramani, K., W.D. Lust, T.S. Whittingham, and E.J. Lesnefsky. 1996. ATP catabolism and
adenosine generation during ischemia in the aging heart. *Mechanisms of Ageing and
Development* 89:113-124.

Randall, M.D., S.M. Gardiner, and T. Bennett. 1997. Enhanced cardiac preconditioning in
the isolated heart of the transgenic ((mREN-2)27) hypertensive rat. *Cardiovascular Research* 33:400-409.

Rees, S.A., R.M.J. Palmer, R. Schulz, H.F. Hodson, and S. Moncada. 1990. Characterization of three inhibitors of endothelial nitric oxide synthase in vitro and in vivo. *British
Journal of Pharmacology* 101:746-752.

Roberts, W.C. 1995. Preventing and arresting coronary atherosclerosis. *American Heart
Journal* 130:580-600.

Sandhu, R., R. Diaz, V. Thomas, and G. Wilson. 1996. Effect of ischemic preconditioning
of the myocardium on cyclic AMP. *Circulation Research* 78:137-147.

Scheen, A.J. 1997. Drug treatment of non-insulin-dependent diabetes mellitus in the 1990s.
Achievements and future developments. *Drugs* 54:355-368.

Schulz, R., K.L. Dodge, G.D. Lopaschuk, and A.S. Clanachan. 1997. Peroxynitrite impairs
cardiac contractile function by decreasing cardiac efficiency. *American Journal of Physiology* 272:H1212-H1219.

Schulz, R., and R. Wambolt. 1995. Inhibition of nitric oxide synthesis protects the isolated
working rabbit heart from ischaemia-reperfusion injury. *Cardiovascular Research* 30:432-
439.

Schwarz, P., R. Diem, N.J. Dun, and V. Förstermann. 1995. Endogenous and exogenous
nitric oxide inhibits norepinephrine release from rat heart sympathetic nerves. *Circulation Research* 77:841-848.

Simonet, S., D.E. Porro, J. Bailliencourt, J.J. Descombes, P. Mennecier, M. Laubie, and T.J.
Verbeuren. 1993. Hypoxia causes an abnormal contractile response in the atheroscle-

rotic rabbit aorta. Implication of reduced nitric oxide and cGMP production. *Circulation Research* 72:616-630.

Speechly-Dick, M.E., G.F. Baxter, and D.M. Yellon. 1994. Ischaemic preconditioning protects hypertrophied myocardium. *Cardiovascular Research* 28:1025-1029.

Stanley, W.C., G.D. Lopaschuk, and J.G. McCormack. 1997. Regulation of energy substrate metabolism in the diabetic heart. *Cardiovascular Research* 34:25-33.

Stewart, L.C., R.A. Kelly, D.E. Atkinson, and J.S. Ingwall. 1995. pH heterogeneity in aged hypertensive rat hearts distinguishes reperfused from persistently ischemic myocardium. *Journal of Molecular and Cellular Cardiology* 27:321-333.

Stone, P.H., J.E. Muller, and T. Hartwell. 1989. The effect of diabetes mellitus on prognosis and left ventricular function after acute myocardial infarction: Contribution of both coronary disease and diastolic left ventricular dysfunction to the adverse prognosis. *Journal of the American College of Cardiology* 14:49-57.

Stout, R.W. 1990. Insulin and atheroma: 20-year perspective. *Diabetes Care* 13:631-651.

Szabó, C. 1996. The pathophysiological role of peroxynitrite in shock, inflammation, and ischemia-reperfusion injury. *Shock* 6:79-88.

Szekeres, L., J.G. Papp, Z. Szilvássy, E. Udvary, and A. Vegh. 1993. Moderate stress by cardiac pacing may induce both short term and long term cardioprotection. *Cardiovascular Research* 27:593-596.

Szekeres, L., Z. Szilvássy, P. Ferdinandy, I. Nagy, S. Karcsu, and S. Csati. 1997. Delayed cardiac protection against harmful consequences of stress can be induced in experimental atherosclerosis in rabbits. *Journal of Molecular and Cellular Cardiology* 29:1977-1983.

Szilvássy, Z., P. Ferdinandy, P. Bor, I. Jakab, J. Lonovics, and M. Koltai. 1994a. Ventricular overdrive pacing-induced anti-ischaemic effect: A conscious rabbit model of preconditioning. *American Journal of Physiology* 266:H2033-H2041.

Szilvássy, Z., P. Ferdinandy, P. Bor, I. Jakab, I. Nagy, J. Lonovics, and M. Koltai. 1994b. Loss of preconditioning in rabbits with vascular tolerance to preconditioning. *British Journal of Pharmacology* 112:999-1001.

Szilvássy, Z., P. Ferdinandy, J. Szilvássy, I. Nagy, S. Karcsu, J. Lonovics, L. Dux, and M. Koltai. 1995. The loss of pacing-induced preconditioning in atherosclerotic rabbits: Role of hypercholesterolaemia. *Journal of Molecular and Cellular Cardiology* 27:2559-2569.

Szilvássy, Z., P. Ferdinandy, I. Nagy, I. Jakab, and M. Koltai. 1997. The effect of continuous versus intermittent treatment with transdermal nitroglycerin on pacing-induced preconditioning in conscious rabbits. *British Journal of Pharmacology* 121:491-496.

Tajima, M., S. Isoyama, Y. Nitta, and K. Abe. 1997. Attenuation of heat shock protein expression by coronary occlusion in hypertrophied hearts. *American Journal of Physiology* 273:H526-H533.

Tosaki, A., D.T. Engelman, R.M. Engelman, and D.K. Das. 1996. The evolution of diabetic response to ischemia/reperfusion and preconditioning in isolated working rat hearts. *Cardiovascular Research* 31:526-536.

Tzivoni, D., and S. Maybaum. 1997. Attenuation of severity of myocardial ischemia during repeated daily ischaemic episodes. *Journal of the American College of Cardiology* 30:119-124.

United Kingdom Prospective Diabetes Study Group. 1995. 13: Relative efficacy of randomly allocated diet, sulfonylurea, insulin or metformin in patients with newly diagnosed non-insulin-dependent diabetes followed for three years. *British Medical Journal* 44:1249-1258.

United Kingdom Prospective Diabetes Study Group. 1995. 16: Overview of 6 years' therapy of type II diabetes, a progressive disease. *Diabetes* 44:1249-1258.

Vegh, A., L. Szekeres, and J. Parratt. 1992. Preconditioning of the ischaemic myocardium: Involvement of the L-arginine nitric oxide pathway. *British Journal of Pharmacology* 107:648-652.

Vigh, L., P.N. Literati, I. Horvath, Z. Török, G. Balogh, A. Glatz, E. Kovacs, I. Boros, P. Ferdinandy, B. Farkas, L. Jaszlits, A. Jednakovics, L. Koranyi, and B. Maresca. 1997. Bimoclomol: A nontoxic, hydroxylamine derivative with stress protein-inducing activity and cytoprotective effects. *Nature Medicine* 3:1150-1154.

Walker, D.M., and D.M. Yellon. 1992. Ischaemic preconditioning: From mechanisms to exploitation. *Cardiovascular Research* 26:734-739.

Wang, P.H., and J.L. Zweier. 1996. Measurement of nitric oxide and peroxynitrite generation in the postischemic heart—Evidence for peroxynitrite-mediated reperfusion injury. *Journal of Biological Chemistry* 271:29223-29230.

Weitzberg, E., and J.O.N. Lundberg. 1998. Nonenzymatic nitric oxide production in humans. *Nitric Oxide* 2:1-7.

Weselcouch, E.O., A.J. Baird, P. Sleph, and G.J. Grover. 1995. Inhibition of nitric oxide synthesis does not affect ischemic preconditioning in isolated perfused rat hearts. *American Journal of Physiology* 268:H242-H249.

Yasmin, W., K.D. Strynadka, and R. Schulz, R. 1997. Generation of peroxinitrite contributes to ischemia-reperfusion injury in isolated rat hearts. *Cardiovascular Research* 33:422-432.

Yellon, D.M., D.S. Latchman, and M.S. Marber. 1993. Stress proteins—An endogenous route to myocardial protection: Fact or fiction. *Cardiovascular Research* 27:158-161.

Zweier, J.L., P. Wang, and P. Kuppusamy. 1995a. Direct measurement of nitric oxide generation in the ischemic heart using electron paramagnetic resonance spectroscopy. *Journal of Biological Chemistry* 270:304-307.

Zweier, J.L., P. Wang, A. Samuilov, and P. Kuppusamy. 1995b. Enzyme-independent formation of nitric oxide in biological tissues. *Nature Medicine* 1:804-809.

Chapter 6

Oxidative Modification of Proteins and DNA

Zsolt Radák

Laboratory of Exercise and Physiology, School of Sport Sciences, Semmelweis University Budapest, Budapest, Hungary

Sataro Goto

Department of Biochemistry, School of Pharmaceutical Sciences, Toho University, Japan

Deoxyribonucleic acid (DNA) is an information macromolecule consisting of four deoxyribonucleotides that contain the bases adenine, cytosine, guanine, and thymine. Genetic information is stored by specific pairing of nucleotides and copied accordingly into newly synthesized DNA. The replication of DNA is a very tightly controlled process with superb accuracy. However, very rarely the replication machinery skips or adds a nucleotide or bases are modified. Such modifications would produce mutation if the repair system failed. Many types of mutation have been identified, from "simple" exchanges of nucleotide bases that alter the coding sequence for one or more amino acids (missense mutation) to insertion or deletion of one or more nucleotides (frame-shift mutation). These modifications, when transcribed into messenger ribonucleic acid (mRNA) and subsequently translated, produce an improper protein sequence and induce structural or functional changes that render proteins defective at their physiological tasks (dysfunctional) or confer a new function (gain of dysfunction). These structurally and functionally altered proteins may compromise structural integrity or the metabolic or replicative capabilities of the cell by triggering a variety of diseases.

Oxidative Damage of DNA

Amino acid sequences of cellular proteins are encoded in either nuclear DNA (nDNA) or the mitochondrial DNA (mtDNA) double helix. The human mtDNA has 16,569 base pairs, encoding about 10% of mitochondrial proteins, which involve seven subunits of the reduced form of nicotinamide adenine dinucleotide (NADH)-ubiquinone reductase system, cytochrome *b* apoprotein of ubiquinone-cytochrome *c* reductase, three subunits of the cytochrome *c* oxidase, and two subunits of adenosine triphosphate (ATP) synthetase (Ozawa, 1997). These enzymes pass the electrons through complex I (NADH dehydrogenase), complex II (succinate dehydrogenase), complex III (ubiquinol: cytochrome *c*), and complex IV (cytochrome oxidase) to oxygen to produce water, as the end-product of metabolism. Only a relatively small amount (5-15%) of the whole cell protein content is located in mitochondria, and a significant portion of mitochondrial proteins are encoded by nDNA. Although only about 7% of nDNA is ever expressed and, therefore, important for life, the expression of whole mtDNA is vital to maintain a suitable ATP level by oxidative phosphorylation.

Beckman and Ames (1997) suggested that the estimated free radical generation of mitochondria during physiological activities could be even 10-fold less than was estimated earlier (Chance et al., 1979). Nevertheless, free radicals continuously damage macromolecules that involve DNA, the steady-state oxidation of which appears to be very significant; the extent of oxidative adducts is several orders of magnitude larger than those of nonoxidative adducts (Ames et al., 1993). For example, the physiological concentration of 8-hydroxy-2′-deoxyguanosine (8-OHdG) with the 0.3-1.5 8-OH dG/dG-5 might be a result of about a million oxidative attacks on each cell DNA per day (Beckman and Ames, 1998). Free radical attack, one of the most destructive, involves hydroxyl radicals derived from the Fenton reaction that attack DNA bases and add HO• to the electron-rich double bonds in purine and pyrimidine (Henle and Linn, 1997). Despite the abundance of oxidative DNA modifications, there is a very significant concern about the detection of these adducts, because artifactual oxidation may occur during sample preparation or determination. Mostly because of the validity of the available techniques, oxidized purines, formamidopyrimidines, 7-methylguanine, 3-methyladenine, O6-methylguanine, 8-nitroguanine, and 8-OHdG have received widespread attention. Although 8-OHdG represents only about 5% of the total DNA oxidative damage (Richter et al., 1988), significant research has been directed toward this adduct because it has a very high mutagenic potential. It induces G (guanine):C (cytosine) to T (thymine):A (adenosine) transversion during DNA replication (Kuchino et al., 1987), and a link has been established between its accumulation and different physiological disorders (Ames et al., 1993). Richter and his associates (1988) reported that mtDNA has significantly higher extent of oxidative damage than nDNA. This could be because mitochondria consume >90% of the cell's oxygen, and the mitochondrial electron transport chain is one of the major sources of free radical generation. mtDNA is located in the inner membrane, at the site of free radical production. In addition, mitochondria are not equipped with nucleotide excision repair and recombinational DNA repair systems (Clayton et al., 1974), and mtDNA does not have a protective histone coat like nDNA. Finally, mtDNA turns over approximately five times faster in the cardiac muscle of rats than does nDNA (Gross et al., 1969), and this rapid turnover might result in replications of unrepaired mutations and concomitant shifts of mtDNA genotype. These observations might explain why mtDNA receives higher oxidative damage than nDNA. Free radical-induced damage of mtDNA could lead to defects in the function of encoded proteins, resulting in a decreased state 3/state 4 ratio and an age-related decrease of complexes I, II, and IV (Ozawa, 1997).

Postmitotic tissues such as skeletal muscle do not possess the high repair capacity found in more mitotically active tissues. But skeletal muscle is able to respond readily to increased oxidative demand by the proliferation of mitochondria. The metabolic stress due to a defect in mitochondrial respiration might lead to massive increases in the proliferation of mitochondria or development of ragged red-fibers (large number of abnormal mitochondria) as a result of a metabolic feedback loop, in which cells attempt to compensate for a reduced oxidative potential. The proliferation of mitochondria with mutant mtDNA would not resolve the metabolic stress but rather would increase it. The damage of mtDNA in skeletal muscle, therefore, has serious consequences in a variety of diseases and in aging. Some diseases, such as Kearn-Sayre syndrome, Parkinson's disease, mitochondrial myopathy, myotonic dystrophy, myoclonic epilepsy, and neurogenic muscle weakness could be the result of accelerated aging and could be associated with accumulation of DNA damage.

The Effect of Physical Exercise on DNA Damage

It is well known that regular physical exercise significantly enhances health and quality of life, despite the paradox that exercise elevates free radical production. Moreover, it seems probable that the radical-generating effect of exercise contributes significantly to health. Regular exercise causes the body to adapt to exercise-induced stress, which might result in hypertrophy, better cardiac function, lower resting heart rate, improved glucose uptake, and better endurance. That the large increase in oxygen uptake during physical exercise is associated with enhanced free radical formation might be a key element for the adaptive response. Interestingly, both aerobic and anaerobic exercise increases the rate of free radical formation. However, the radical-generating pathways are different, as discussed later. The activity of antioxidant enzymes depends on the rate of radical generation; for example, the increase in superoxide formation stimulates superoxide dismutase expression and activity. Therefore, it is generally accepted that regular exercise increases the activity of the antioxidant enzyme system. However, regular exercise–induced free radical generation and the related oxidative damage also stimulate the repair systems for oxidative damage in lipid, protein, and DNA. The oxidative defense and repair mechanisms for oxidative damage are very plastic and inducible processes. Using regular exercise to induce moderate oxidative stress is somewhat similar to administrating small amounts of certain carcinogens or weak x-ray irradiation to evoke adaptation and enhance resistance. Extensive studies have revealed that, indeed, treatment with a low level of free radicals increases resistance against oxidative stress. For instance, hydrogen peroxide exposure up-regulates transcriptions of catalase and glutathione peroxidase (Wiese et al., 1995). Therefore, we propose that up-regulation of antioxidant and repair mechanisms by regular exercise overwhelms its detrimental effects (concerning production of free radicals), resulting in greater protection and stronger resistance against oxidative stress. We must emphasize that the effects of a single bout of exercise and regular exercise are quite different. With a single bout of exercise the adaptation is marginal, if there is any, because oxidative damage is often observed following a single bout of exercise. On the other hand, regular exercise brings about numerous beneficial effects that adapt the body to later and possibly stronger stress.

Single Bout of Exercise

Macromolecules are oxidatively modified when the extent of oxidative stress exceeds the preventive capability against oxidative stress. A single bout of exercise is a good model to study oxidative stress, because the magnitude of oxidative damage is moderate and can be repeated without jeopardizing vital physiological functions. In addition, the whole-body stress response can be studied, because exercise stresses different organs in different ways. For example, blood flow and oxygen supply increase many-fold in working skeletal muscle yet are maintained in the brain and decreased in the liver and kidney. Increasing information is available about the effect of exercise on DNA damage. However, to our knowledge, data are not available on mtDNA damage. Most human studies have identified DNA damage or consequence of the damage in urine and blood samples, primarily leukocytes. Sumida et al. (1997) studied the effect of single bouts of exercise (exhaustive treadmill running, exhaustive cycling, and 20-km running) on urine excretion of 8-OHdG and observed no significant changes in its amount, which suggested little DNA damage. The same group reported that exhaustive running did not increase 8-OHdG content in urine samples of untrained individuals, and supplementation of beta-carotene tended to decrease the steady-state level of DNA damage (Sumida et al., 1997). On the other hand, Hartmann et al. (1994, 1995) showed that a single bout of exercise induced DNA damage in white blood cells. After short-distance triathlon running, no changes in urinary 8-OHdG level were found in trained subjects. However, increased DNA migration was observed in leukocytes (Hartmann et al., 1998). Pilger et al. (1997) measured the urinary 8-OHdG content of sedentary and regularly exercising subjects and observed no difference in this kind of oxidative damage (table 6.1). Single cell electrophoresis has revealed that exhaustive treadmill running increases single-strand DNA damage, and lymphocyte apoptosis has been reported (Mars et al., 1998). This observation is very important because it could be linked to an exercise-induced immune dysfunction (Pedersen et al., 1998). In our recent study, we were interested in whether muscle soreness is accompanied by increased DNA damage. We obtained muscle biopsy samples 24 h after an eccentric exercise that caused muscle soreness. Compared with the control group, the exercised group showed about a 30% increase in the level of 8-OHdG content in the skeletal muscle samples (Radák et al., 1999a). We concluded that the 8-OHdG level 24 h after exercise is increased because free radicals are formed to a larger extent during exercise. This increase in DNA damage may be due to a secondary effect, that is, the delayed effect of unaccustomed exercise in activating macrophages and/or neutrophils, which involves massive generation of reactive oxygen species. A single bout of exercise did not significantly modify the 8-OHdG content in 13 different tissue samples of dogs (Okamura et al., 1997b), the colon being the only tissue in which DNA damage was significantly above the steady-state level. These data suggest that oxidative stress induced by a single bout of exercise does not consistently increase DNA damage.

Long-Term Regular Exercise

Much remains to be learned about the effect of long-term regular exercise on DNA damage. Data obtained after prolonged physical exercise would have merit, because regular exercise–induced adaptations might be seen in the DNA damage repair process. Some studies mentioned earlier that used trained subjects found no increase in damage after a single bout of exercise (Hartmann et al., 1988; Pilger et al., 1997; Sumida et al., 1997). Asami et al. (1998) suggested that forced treadmill running increased 8-OHdG content in tissue samples

Table 6.1   Exercise-Induced Changes in 8-Hydroxydeoxyguanosine Content

Species/tissue	Activity	Changes	Reference
Human/urine	Marathon running	Increase	Alessio (1993)
Human /urine	Exhaustive running	Increase	Poulsen et al. (1996)
Human/urine	Training camp	Increase	Okamura et al. (1997a)
Human/urine	Exhaustive running	No change	Sumida et al. (1997)
Dog/13 tissues	Exhaustive running	No change in 12 tissues Increase in colon	Okamura et al. (1997b)
Rat/heart Rat/lung Rat/liver	Spontaneous running versus forced running	Increase Increase Increase	Asami et al. (1998)
Human/urine	Habitual runners versus sedentary	No change	Pilger et al. (1997)
Human/leg muscle	Muscle soreness	Increase	Radák et al. (1999a)
Rat/brain	9 wk swimming	No change	Radák et al. (1998)
Rat/leg muscle	9 wk swimming	Decrease	Radák et al. (1999b)

of liver, cardiac, and skeletal muscle of rats, whereas voluntary running depressed the 8-OHdG level. In a recent study, we trained rats for 9 wk by 1 h of swimming per day, five times a week. The 8-OHdG content was significantly smaller in the skeletal muscle DNA of exercised rats (Radák et al., 1999a). We believe that this decrease in the 8-OHdG content is mostly due to the upregulation of the DNA repair process. We also measured the 8-OHdG content in the brain, because we have conducted some measurement for cognitive function, and observed no changes. Lipid and protein oxidation and DNA damage were different, suggesting that a specific mechanism exists for oxidative damage and repair processes for different macromolecules (figure 6.1). One hypothesis is that oxidative damage to macromolecules is repaired according to the toxicity of damage. For example, mutagenic DNA damage may be repaired first, with protein and lipid damage repaired later. This hypothesis has to be verified.

According to our understanding, DNA mutation is linked very closely to cancer, and if regular exercise up-regulates the antioxidant systems and the oxidative damage repair system, these mechanisms may be a very important tool against this deadly disease.

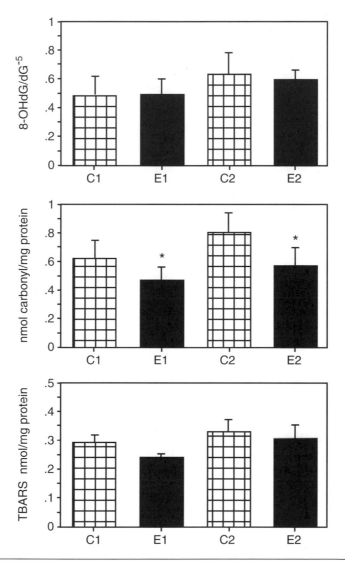

Figure 6.1 Oxidative damage markers of DNA, proteins, and lipids in rat brain were measured after 9 wk of swimming training. No significant changes were observed in 8-hydroxy-2′-deoxyguanosine (8-OHdG; DNA damage marker) and thiobarbituric acid reactive substance (TBARS; lipid damage marker). The reactive carbonyl derivatives content significantly decreased as a result of swimming in young exercised (E1) and middle-age exercised (E2) groups compared with the age-matched controls (C1 and C2). The different degree of accumulation of oxidative damage in each macromolecule could indicate the damage mediated by different mechanisms and/or that the efficiency of the repair process is different in the different macromolecules. *$P < 0.05$ versus C1.

The Effect of Aging on DNA Damage

Aging affects all individuals, and, despite its universal nature and the mounting body of scientific evidence, it remains poorly understood. Aging is unavoidable and is characterized by a general decline in physiological function that leads to morbidity and mortality in the final phase (Arking, 1998). The free radical theory of aging was first proposed by Harman (1956) and has been developed into the modern version by many investigators (Ames et al., 1993; Beckman and Ames, 1998; Ozawa, 1997; Sohal and Weindruch, 1996; Stadtman, 1992; Wallance, 1992). Harman (1956) suggested that endogenous reactive oxygen species are produced in aerobic organisms, and the interaction between these radical species and metals, especially iron, structurally alters biomolecules similar to radiation chemistry. His theory has gained considerable support but is still under debate (Yu, 1993). Positive correlation between the life span of animal species and development of an antioxidant system and oxidative damage repair system strengthened the credibility of the Harman theory (Martin, 1996). Across vertebrate species, a longer life span is associated with lower production of mitochondrial reactive oxygen species (Perez-Campo et al., 1998). It has been suggested that the production of reactive oxygen species increases and the efficiency of the antioxidant and repair systems decreases with aging, which lead to accumulation of oxidatively modified macromolecules (Mecocci et al., 1999; Orr and Sohal, 1994; Sohal and Weindruch, 1996; Stadtman, 1992). However, despite phenomenological correlation between oxidative damage and aging that may not be direct consequences of oxidative processes on cellular activities, direct evidence is still missing (Beckman and Ames, 1998).

Drawing on the free radical theory of aging, Linnane and colleagues (1989) proposed that aging could result from accumulation of mitochondrial DNA mutations. These age-associated increases in mtDNA mutations would effect mitochondrial oxidative phosphorylation and reduce ATP production. The prolonged reduction in energy supply would eventually impair functions and decrease physiological function, which increase the probability of death. Muscle and nerve are particularly prone to oxidative damage, because they consist mostly of postmitotic cells liable to accumulate oxidatively modified molecules (Radák and Goto, 1998).

Recent studies conducted to identify changes in genome stability with aging have produced evidence that DNA is oxidatively modified with aging (Fraga et al., 1990; Kaneko et al., 1996, 1997; Zhang et al., 1993). G:C to A:T transition and G:C to T:A transversion are the most commonly observed mutations resulting from oxidative damage to DNA during aging (Wang et al., 1998). The DNA repair in postmitotic cells, where the replication-induced mutation is nonexistent, mainly focuses on transcription regions of DNA, which indicates complex mechanisms for DNA damage and repair (Beckman and Ames, 1998). Many reports have shown an age-related increase in DNA damage (Ozawa, 1997; Radák and Goto, 1998; Shigenaga et al., 1994). The increase is thought to result from the decline in DNA repair capacity (Hart and Setlow, 1974). However, Higami and coworkers (1994) suggested that an increase in the vulnerability of oxidative damage to hepatocytes is more dominant than the deterioration of DNA repair capacity with age.

mtDNA is the only extrachromosomal DNA in mammalian cells, and thus mutations tend to have deleterious effects on respiratory chain function (Richter, 1988). More than 12 mutations have been identified in the mtDNA of older humans (Baumer et al., 1994). Most of these mutations are undetectable in fetal tissues and accumulate with age, but even in advanced age they are present in a relatively small amount (0.1% of the total mtDNA) (Cortopassi and Arnheim, 1990). In the healthy elderly, the decline of cytochrome c activity is a pathological hallmark of mtDNA mutation (Muller-Hocker, 1990), the detection of which is very often used as a fingerprint for oxidative damage to mtDNA. Hayakawa et al. (1991) showed that the 8-OHdG content of mtDNA in the diaphragm increased 25-fold from the age of 55 to 85, with an exponential correlation between 8-OHdG and age. The decline in oxidative metabolism associated with aging is much larger than could be calculated from the extent of mtDNA mutation and cytochrome oxidase activity. Therefore, Cooper et al. (1992) suggested that other factors, such as the decline of transcription and translation of mtDNA, play a role in the age-dependent defect of the respiratory chain. Aging in general is associated with reduced physical activity, which itself could affect the activity of respiratory chain and morphology of skeletal muscle. Indeed, physical training increases and immobilization decreases the activity of respiratory chain enzymes (Henriksson, 1992).

Immobilization has a deleterious effect on a variety of organs. For example, it increases the oxidative damage of DNA not only in the skeletal muscle but also in the brain, and it could result in related stomach bleeding and gastric ulcer due to emotional stress (Liu et al., 1996). Hence, the age-associated decline in physical activity affects the whole body, including the central nervous system. Therefore, the question arises whether the accumulation of oxidative damage of macromolecules is dominant due to aging itself or an age-dependent decrease in physical activity. One of the most striking effects of advanced age is a decline in cognitive function, which may be associated with the accumulation of reactive oxygen species–induced damage in the brain. Mecocci and coworkers (1993) reported an age-dependent accumulation of nuclear and mtDNA damage in human cerebral cortex. Cells in the central nervous system of individuals with Alzheimer's disease and Parkinson's disease accumulate massive amounts of DNA damage (Alam et al., 1997; Mukherjee and Adams, 1997). Gentle physical exercise improves mobility, prevents falls, diminishes pain, reduces mortality, and increases cerebral blood flow and cognitive function of individuals with Alzheimer's disease (Gorman, 1995). The age-dependent impairment of nerve cell function might result from significant DNA damage. However, direct evidence between the accumulation of DNA damage and the decline in cognitive performance is still missing. Moreover, the finding of our recent study suggests a relationship between the rate of damage to proteins, but not to lipid and DNA, and cognitive function in rats (Radák et al., 1999a). Movements result from complicated nerve actions that originate in the central nervous system, and, according to the "use it or lose it" concept, active neurons have a better chance to survive and cope with aging (Lucassen et al., 1998). On the other hand, massive accumulation of DNA damage might alter the synthesis of encoded proteins. Interestingly, age-dependent increases in somatic mutations in mitotic hepatic cells were observed in transgenic mice studied for detection of DNA damage, whereas no increase was observed in postmitotic nerve cells of the same animals (Dolle et al., 1997). This might indicate that accumulation of DNA damage is not closely related to neurodegeneration and that mutations are not the only major factors that influence the age-associated decline of nervous function. Leonard (1997) suggested that life span depends on two factors: the total metabolic potential over the life span and the genetic clock. As for the first, smaller animals have faster metabolic rates and generally shorter life spans. The second hypothesis is supported by the observation that nematodes, which have a mutant form of clk-1 gene, display decreased rate of embryonic development and a 50% longer life span. Both suggestions indicate that aging is set up at the cellular level.

The functional degeneration of somatic cells seems to be a necessary component of aging, which can be linked readily to the relationship between age and cance incidence (Ames et al., 1993). The incidence of cancer depends on age, and it has been suggested that the basic metabolic rate significantly influences the appearance of this disease (Ames et al., 1993; Shigenaga et al., 1994). Interestingly, physical exercise, which increases the daily metabolic rate, decreases the incidence of cancer (Shephard and Futcher, 1997).

The Preventive Effect of Regular Exercise on the Incidence of Cancer

Oxidative damage of DNA may also underpin certain cancers, decreasing the function of tumor-suppressor genes and activating tumor-promoting genes, with subsequent malignancy (Miller, 1995). The development of cancer depends partly on heredity. However, its incidence and progress probably depend on the function of health-promoting genes, which maintain the correct production of proteins that sustain the viability of cells and maintain or increase the resistance to damage and damage repair processes. Evidence is accumulating that regular physical exercise significantly reduces the incidence of cancer, and the beneficial effect can be as much as 46% for a variety of cancers (Shephard and Futcher, 1997). Chronic physical exercise may decrease the risk by affecting natural immunity, improving energy balance, inducing hormonal changes, and stimulating antioxidant defense (Kiningham, 1998). In Norway, for example, the risk of breast cancer was reduced by 52% among women whose work included walking and who regularly exercised during their leisure time (Inger and Eiliv, 1997). Another example is colon cancer, which killed 55,000 people in the United States in 1997. Regular exercise is well documented to significantly reduce the incidence of colon cancer (Ahnen, 1997). A significant inverse relationship has been observed between leisure physical activity and prostate cancer (Hartmann et al., 1998). Thompson et al. (1995) conducted a 3-mo exercise study after carcinogen treatment on rats, finding a 37% reduction in the incidence of mammary cancer. The degree of protection against cancer was proportional to the intensity but not the duration of exercise. Day (1995) reported that regular exercise might inhibit nitrosamine activation, which could play a role in the cancer-reducing effects of exercise. Whittal and Parkhouse (1996) showed that the number of nitrosomethylurea-induced mammary tumors can be significantly reduced by running exercise. Skeletal muscle is one of the main targets for cancer cachexia, and Daneryd et al. (1995) showed that exercise induces antioxidant capacity that can increase the resistance against this kind of disease. Our recent study also led us to suggest that physical exercise could be an active tool against cancer. We found that 9 wk of swimming significantly decreased ($p < .05$) the level of DNA damage, measured as 8-OHdG content, in exercised rats. The mutagenic effect of 8-OHdG is well documented. Therefore, the decrease in the steady-state level of 8-OHdG probably significantly benefits cell function (Radák et al., 1999b). Moreover, we have measured the activity of DT-diaphorase, which activates antitumor compounds and is one of the main targets of anticancer drugs. Regular exercise increases the activity of DT-diaphorase in the skeletal and cardiac muscles and the brain (Radák et al., 1998, 1999a). We suggest that the ability of regular exercise to improve the efficiency of antioxidant systems and to enhance repair mechanisms is a very important part of the cancer-reducing effect of regular exercise.

Little is known about possible biochemical pathways by which exercise reduces the incidence of cancer and promotes health. Nevertheless, a mounting body of epidemiological data suggest that a physically active lifestyle can reduce the incidence of certain types of cancer. That the incidence of cancer is much higher among the elderly than the young raises questions about the benefits of beginning or maintaining exercise by older individuals. Regular exercise has been suggested to benefit people of any age, and it is never too late to start (Peterson and Cunningham 1999).

Oxidative Modification of Proteins

Dakin reported the following in 1906: "One gram of alanine was added to the calculated amount of hydrogen peroxide solution, which had been standardized by potassium permanganate and also carefully neutralized with sodium carbonate. A few milligrams of ferrous sulphate were added and the whole allowed to stand at the ordinary temperature of the air. After a couple of minutes the solution became slightly warm and a slow evolution of carbon dioxide commenced, while the smell of acetaldehyde steadily increased" (p. 173). He had conducted a pioneering experiment on the oxidative modifications of proteins. All amino acid residues of proteins are subjected to free radical–induced oxidative modifications in the presence of the metal (table 6.2). The oxidation of tyrosine residues forms the 3,4-dihydroxy derivative (DOPA), which could further generate free radical species (Waite, 1995). For example, deamination of DOPA by monoamine oxidase generates hydrogen peroxide, and a link has been established to the etiology of Parkinson's disease (Fahn and Cohen, 1992). Moreover, oxidation of tyrosine residues could result in the formation of phenoxyl radicals and tyrosyl radicals. Another example is that oxidation of prolyl residues in proteins leads to formation of pyroglutamil residues, which could yield free glutamic acid (Amici et al., 1989). The metal-

---

Table 6.2   Oxidative Modifications of Some Amino Acid Residues

Residues	Product	Selected references
Arginine, lysine, proline	Carbonyl group	Amici et al. (1989)
Histidine	2-oxohistidine	Uchida and Kawakishi (1993)
Lysine	Lysine hydroperoxides and hydroxide	Simpson et al. (1993)
Methionine	Methionine sulphoxide	Vogt (1995)
Proline	Proline hydroperoxides and hydroxides	Simpson et al. (1993)
Threonine	2-amino-3-keto-butyric acid	Taborsky (1973)

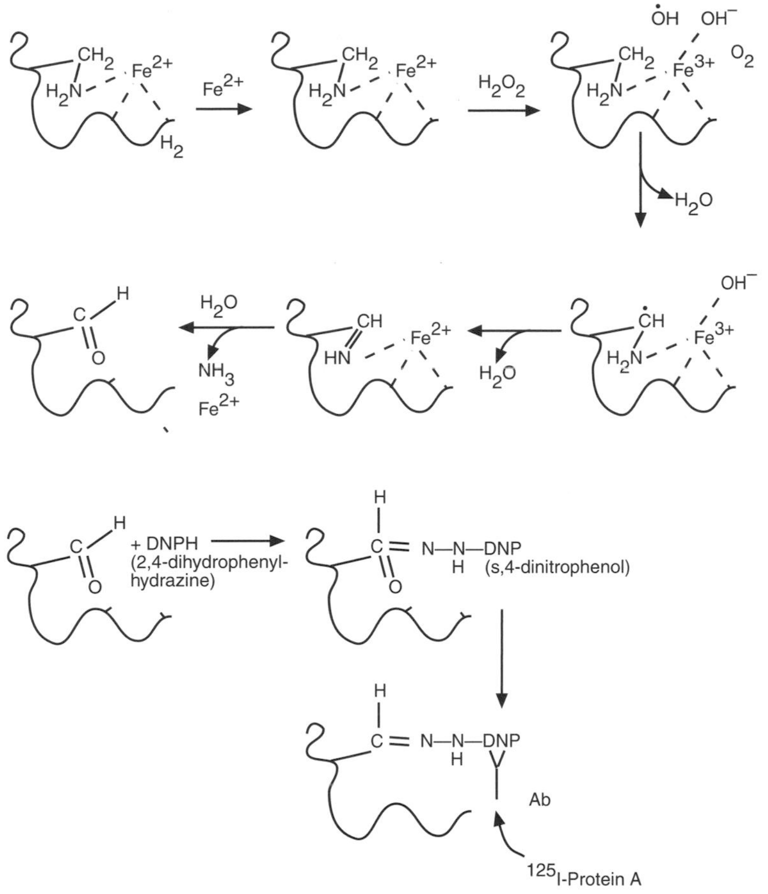

Figure 6.2   Proposed mechanisms of the reactive carbonyl derivatives formation and detection with 2,4-dinitrophenylhydrazine (DNPH) and its antibody.

catalyzed oxidation of peptide side chains of arginyl, aspartyl, glutamyl, lysyl, prolyl, and threonyl residues might result in the formation of reactive carbonyl derivatives that can be detected readily with 2,4-dinitrophenylhydrazine (DNPH) (figure 6.2). Although carbonylation of protein might be considered an unavoidable physiological consequence of aerobic organisms with a steady-state level of 0.3 to 1 nmol carbonyl per milligram protein of different tissues, it is an often-used marker of free radical–induced protein damage. Several highly specific procedures for assessing the level of carbonyl derivatives have been developed (Buss et al., 1997; Cao and Cutler, 1995; Levine et al., 1990, 1994; Nakamura and Goto, 1996; Reznick and Packer, 1994). Although the carbonyl derivatives can also be formed by reactions with unsaturated alkenals formed by lipid peroxidation or by glycation, it is generally accepted that carbonyl content is a reliable indicator of oxidative modifications of proteins. Accumulation of carbonyl derivatives is due to a variety of pathophysiological conditions such as aging (Goto and Nakamura, 1997; Radák and Goto, 1998; Stadtman, 1992; Starke-Reed and Oliver, 1989), rheumatoid arthritis (Chapman et al., 1989), Alzheimer's disease (Smith et al., 1991), smoking (Reznick et al., 1992a), and ischemia/reperfusion (Oliver et al., 1990).

Exercise and Protein Oxidation

There is little doubt that a single bout of exercise oxidatively modifies proteins. However, the question is whether these modifications represent nonspecific free radical–induced damage or specific targeted damage that could result in adaptation in the long run. The sensitivity of amino acid residues to oxidative stress varies significantly. All residues might be subjected to oxidative modifications; however, proteins in which the metal-binding sites are readily available are more prone to oxidation. Table 6.3 summarizes the findings of some studies on protein oxidation and exercise. Spectrophotometric data reveal that a single bout of exercise increases the reactive carbonyl derivatives in skeletal muscle of rats, and the administration of vitamin E attenuates the exercise-induced oxidative stress (Reznick et al., 1992a). Rajguru et al. (1994) also reported that the rate of protein oxidation measured by sulfhydryl level increases after a single bout of exercise. The authors suggested that elevation of sulfhydryl level is part of the regenerative process, because reduced glutathione is likely involved in the removal of lipid peroxides, and the increased sulfhydryl compounds in the blood act as detoxifying agents. Sen and his coworkers (1997) showed that a single bout of exercise increases reactive carbonyl derivatives in skeletal muscle, liver, and heart of rats.

However, a limitation of spectrophotometric measurement is that the modified proteins cannot be identified. We have used immunoblot technique to obtain visible information of carbonylated proteins (Radák et al., 1998a, 1998b, 1998c). A single bout of exercise greatly increased blood lactate concentration and caused significant oxidative damage in some proteins of the lung (Radák et al., 1998a; figure 6.3). Previous studies revealed that anaerobic exercise induces xanthine dehydrogenase conversion to xanthine oxidase, which is a potent reactive oxygen species generator when oxygen is available (Radák et al., 1995, 1996; Sumida and Sugawa-Katayama, 1991).

The immunoblot technique of reactive carbonyl derivatives is more specific than spectrophotometry (Goto et al., 1999) and reveals accumulation of reactive carbonyl derivatives. A particular protein with a molecular weight of 28 to 29 kDa seems to be very sensitive to exercise-induced oxidative stress. We

Table 6.3   The Effect of Exercise on Carbonyl (RCD) Formation

Muscle	Condition	RCD	Reference
Quadriceps	Exhaustive exercise	Increase	Reznick et al. (1992c)
Hindlimb	8 wk running	Increase	Witt et al. (1992)
Quadriceps	4 wk running sea level	No change	Radák et al. (1997)
	4 wk running at high altitude	Increase	
Hindlimb muscle Mitochondria Microsomes Cytoplasma	4 wk swimming	Decrease	Radák et al. (1997)
Soleus	Calorie restriction	No change	Radák et al. (unpub.)
	High altitude	No change	
Heart	8 wk swimming and 2 wk hydrogen peroxide treatment	Decrease	Radák et al. (unpub.)
Hindlimb	8 wk swimming	No change	Radák et al. (1999a)
Brain	8 wk swimming	Decrease	Radák et al. (1998c)
Lung	Exhaustive exercise	Increase	Radák et al. (1998b)
Leg, human samples	Eccentric action	No change	Saxton et al. (1994)
	Concentric contraction	Increase	
	Muscle soreness	Increase	
Liver and gastrocnemius	Single bout of exercise	No change	Sen et al. (1997)
Hindlimb	Exhaustive exercise	No change	Bejna and Ji (1999)

speculate that the protein could be carbonic anhydrase III, because it is abundant in the skeletal muscle and is sensitive to oxidation–accumulating reactive carbonyl derivatives (Cabiscol and Levine, 1995). This enzyme affects acid-base balance, CO_2 transfer, and ion exchange and thereby probably plays an important role during exercise-induced metabolic challenges. We speculate that damage to this protein is not coincidental, and we propose a physiological reason for this process. This hypothesis may be supported by the finding that a protein with the same molecular weight is carbonylated to less extent in the skeletal muscle of rats after 9 wk of swimming compared with sedentary controls (Radák et al., 1999a). The decrease in reactive carbonyl derivatives in this protein indicates faster turnover of the protein. Because oxidatively modified proteins are more sensitive to proteolytic breakdown (Stadtman, 1992), regular

Protein stain/anti-DNP immunostain

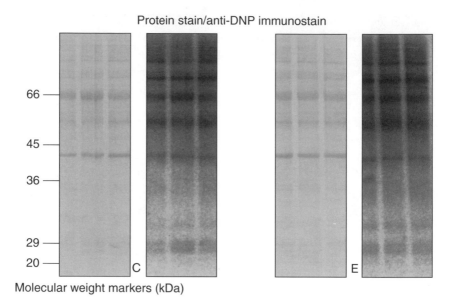

Molecular weight markers (kDa)

Figure 6.3 Anaerobic exercise increased the lactic acid concentration in blood of rats 7-fold and resulted in oxidative protein damage to the lung. Accumulation of reactive carbonyl derivatives is more significant in lungs of exercised (E) than control (C) animals. The accumulation is more visible in proteins with a molecular weight of approximately 55 kDa and 32 kDa. Molecular weight markers are shown to the left of the Coomassie blue stain panel.

exercise–induced damage of this protein is likely associated with faster turnover rate, which decreases the level of reactive carbonyl derivatives. If this hypothesis is true, it probably means that oxidative modifications to proteins are specific and that the sensitivity might depend on the function of the protein. This speculation might be true only with moderate oxidative stress.

Witt et al. (1992) conducted the first reported study on the effects of regular exercise on accumulation of reactive carbonyl derivatives. The authors noted an increase in reactive carbonyl derivatives content in hindlimbs of rats after 8 wk of moderate-intensity running training. The steady-state level of reactive carbonyl derivatives was quite high, which is probably due to limitations of the measuring technique. We have measured the reactive carbonyl derivatives of the skeletal muscle of rats after 4 wk of running training at sea level and at a simulated altitude of 4000 m. The reactive carbonyl derivatives level did not change in the red and white types of hindlimb muscle after sea-level training, whereas this level increased at high altitude (Radák et al., 1997). We used both spectrophotometric and western blot methods to appraise the reactive carbonyl derivatives level. Interestingly, the western blot data indicated that actin is likely to be the major protein that accumulates increased amounts of reactive carbonyl derivatives after regular exercise at high altitude. Our recent work (Radák et al., 1999a) also suggests no significant change in the accumulation of reactive carbonyl derivatives in skeletal muscles of rats after 9 wk of moderate training. Interestingly, the mitochondrial fraction of the hindlimb muscle of the exercised animals accumulated significantly higher levels of reactive carbonyl derivatives. Mitochondria are susceptible to oxidative damage, because the electron transport chain is one of the most significant sources of reactive oxygen species. During physical exercise, the oxygen flux into the mitochondria of working skeletal muscle might be 100-fold higher than that found in control muscle. The generation of reactive oxygen species might not increase with the same magnitude, but a large increase probably occurs (Davies et al., 1981). These findings suggest that mitochondria is a prominent site of free radical formation, and/or the oxidatively modified protein removal capability of mitochondria is very limited.

Surprisingly, a lower level of reactive carbonyl derivatives was measured in the brain of exercised rats (Radák et al., 1998c). The decreased level of reactive carbonyl derivatives in the brain was associated with enhanced cognitive functions as measured by active and passive avoidance tests (figure 6.4). Thus, regular exercise appears to improve cognitive function in rats, and this improved brain function could be due to the decreased accumulation of oxidatively modified proteins in the brain induced by the exercise (Carney et al., 1991; Forster et al., 1996). A great deal of evidence based on human epidemiological studies indicates that regular exercise delays the age-associated decline in brain functions (for review, see Chodzko-Zajko and Moore, 1994), and our findings might provide biochemical and physiological evidence of the very beneficial effects of exercise. Only limited data are available on the rate of oxidative modifications of proteins after regular physical exercise. However, regular exercise does increase the rate of protein turnover (see elsewhere in this chapter). Thus, it seems very unlikely that regular exercise induces a significant accumulation of the modified proteins. In contrast, aging is associated with increases in accumulation of reactive carbonyl derivatives and increases in the half-life of proteins (Goto et al., 1995, 1999; Reznick et al., 1981; Ishigami and Goto, 1990; Stadtman, 1990). It appears unlikely that the increased accumulation of reactive carbonyl derivatives takes place after exercise training, because decreased levels of steady-state lipid peroxidation have been reported after physical training (Kim et al., 1996). Moreover, it is well known that regular physical exercise increases the mean and maximum life span of experimental animals (Holloszy, 1993) and humans (Blair et al., 1995; Sarna et al., 1993).

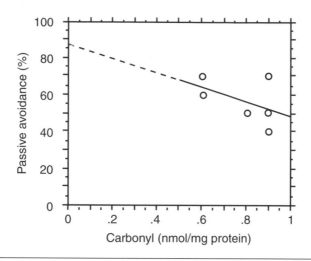

Figure 6.4 The accumulation of reactive carbonyl derivatives in rat brain is associated with decreases in passive avoidance performance. The measuring box was separated into white and black chambers, and rats were placed in the white, lighted chamber from where they entered the dark chamber within seconds. When the passive avoidance test was measured on the second day, electric shock was delivered to the rats entering the dark chamber. On the third day, the animals were also placed in the white chamber, and the time they took to enter the black chamber represented a type of memory index. The graph shows that lower levels of reactive carbonyl derivatives are associated with better memory. Data from Radák et al. (1998c).

Aging and Accumulation of Oxidatively Modified Proteins

Cell constituents undergo continuous renewal in a living organism. Metabolic turnover is fundamentally important to maintain life in all organisms. Altered molecules inevitably accumulate during aging, because the protective systems are also subject to change. Of particular importance among these systems is protein turnover. The rationale behind this thinking is that even if DNA bases are modified or membrane lipids are peroxidized, cells would not lose their function as long as enzyme proteins in the repair systems operate properly (Goto et al., 1995). Many altered proteins are known to accumulate with aging, such as β-amyloid in neural cells and modified crystallins in lens (Goto et al., 1995; Kurochkin and Goto, 1994). The age-associated changes in enzyme structure lead to the appearance of heat-labile forms of proteins, which results, at least in part, from free radical attack (Stadtman, 1986, Takahashi and Goto, 1990). Age-associated increases of o-tyrosine (a marker of oxidative damage of phenylalanine) and 3-nitrotyrosine (nitrogen radical–induced damage of tyrosine) were not observed in cardiac and skeletal muscles or the liver of old rats (Leeuwenburg et al., 1998). On the other hand, some factors similarly affect life span and the accumulation of reactive carbonyl derivatives in cells of various tissues of different species (Butterfield et al., 1997; Sohal and Dubey, 1994; Sohal et al., 1993). The amounts of reactive carbonyl derivatives increase with aging in houseflies, and the accumulation may cause senescence of the fly (Sohal and Dubey, 1994; Sohal et al., 1993). Many reports on the accumulation of reactive carbonyl derivatives in tissue proteins in aging animals and humans suggest that, despite the striking variations in metabolic rate and life span, age-associated accumulation of reactive carbonyl derivatives might be universal (Stadtman, 1992; Beckman and Ames, 1997, but also see Goto and Nakamura, 1997). Mitochondria in aged flies accumulate larger amounts of reactive carbonyl derivatives than those in the young ones (Sohal and Dubey, 1994), and protein-specific modification may also take place. For example, aconitase in housefly mitochondria appears to be particularly susceptible to oxidation (Yan et al., 1997).

Studies performed on fibroblasts and erythrocytes have revealed that reactive carbonyl derivatives are markedly increased with aging and in fibroblasts of individuals with Werner's syndrome, a disease that involves accelerated aging (Oliver et al., 1987). Starke-Reed and Oliver (1989) reported that content of the reactive carbonyl derivatives in proteins of hepatocytes isolated from 26-mo-old rats was higher than that found in younger animals. A more recent report showed that reactive carbonyl derivatives in proteins were increased in human skeletal muscle as a function of age (Mecocci et al., 1999). Immunoblot analysis of kidney proteins separated in two-dimensional polyacrylamide gel electrophoresis revealed prominent immunological signals in old rats, but these were much lower in young rats (Goto et al., 1999). The corresponding proteins are likely to be modified forms of alpha-2u-globulin as judged from their molecular weight and pI. The reason for enhanced carbonylation of this protein is not known; however, it is possible that modified forms of alpha-2u-globulin are trapped in the kidney of adult rats before or after carbonylation. Other proteins that are highly carbonylated include serum albumin trapped in this tissue. Involvement of reactive oxygen species generated by autoxidation has been implicated in the loss of dopaminergic neurons in the nigrostriatal region of the aged brain, possibly due to protein damage (Miller et al., 1996). An age-related increase in reactive carbonyl derivatives has been observed in human and animal brains (Carney and Carney 1994; Carney et al., 1991; Forster et al., 1996; Smith et al., 1991). Dubley et al. (1996) found that dietary restriction retarded the age-related decline of sensorimotor coordination of aged mice. Forster et al. (1996) found that the age-associated loss of ability to perform a spatial swim maze task was positively correlated with reactive carbonyl derivatives in mice brains. Therefore, a link has been established between cognitive function and the rate of accumulation of reactive carbonyl derivatives in the brain (Carney et al., 1991; Forster et al., 1996; Radák et al., 1998a). This indicates that age-related damage of proteins influences performance of the nervous system. These observations suggest that accumulation of reactive carbonyl derivatives in proteins not only is a marker of oxidative modification but is a cause of age-related decline of physiological capabilities (i.e., senescence) of cells.

Accumulation of oxidized proteins in at least some tissues of aged animals can be explained by either a higher rate of oxidation or a lower rate of degradation of modified proteins (Goto et al., 1995). The increases in the half-life of proteins in tissues and cells of senescent animals suggest that aging also affects the rate of proteolytic degradation (Ikeda et al., 1992; Ishigami and Goto, 1990; Reznick et al., 1981). Thus, accumulation of oxidatively modified proteins may be explained at least partly by a decrease in the protein degradation, as discussed in detail next.

Degradation of Oxidatively Modified Proteins

Abnormal proteins can arise from mutation, from errors in biosynthesis, from postsynthetic modifications such as glycation (nonenzymatic glycosylation) or deamination, or from interaction with free radical species. Accumulation of physiologically inactive, and in many cases toxic proteins, is prevented by degradation of such proteins, which is carried out by proteases (figure 6.5). Intracellular protein degradation is a tightly controlled and highly regulated physiological process. One of the major proteolytic processes is the ATP-dependent pathway by the enzymes in 26S/20S proteasome complex, which cleave a wide range of peptide bonds. This enzyme complex predominantly targets abnormal, short-lived, and ubiquitin-conjugated proteins. The 26S proteasome is made by the assembly of two seven-membered alpha and beta subunits located in the outer and inner rings of the complex, respectively (Pickart, 1997). The beta subunits contain the proteolytic active sites, presumably at the substrate entrance of the cylinder. The cylinder is approximately 17 nm in length and 12 nm in diameter with an internal cavity that makes it possible to swallow and break down substrate proteins (Grune et al., 1997). Extensively cross-linked proteins are not able to enter the substrate channel of the proteasome by themselves because of their steric hindrance. Therefore, it is believed that ATP hydrolysis takes place to provide the necessary energy to unfold the proteins (Larsen and Finley, 1997). The recognition and selection of proteins for degradation are the key elements of the proteolytic pro-

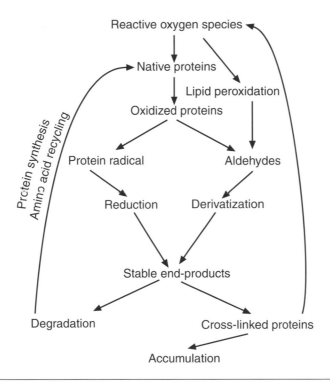

Figure 6.5   The proposed mechanisms of oxidative protein damage and accumulation. A variety of interactions are possible in each step.

cess. The covalent attachment of ubiquitin to protein substrate at the expense of ATP specially marks for proteolysis by 26S proteasome complex (Fagan and Waxman, 1992). In mammalian cells, an ATP-independent pathway also exists to degrade oxidatively modified proteins in which so-called insulin degrading enzyme is involved (Fagan et al., 1985; Kurochkin and Goto, 1994). However, little is known about how the enzyme of 26S/20S proteasome selectively recognizes oxidatively modified proteins, because degradation of proteins by the 20S core proteasome does not require the ubiquitin-ATP system (Fagan et al., 1985; Grune et al., 1997). Rivett (1985) suggested that oxidant-induced structural modifications of amino acid residues might serve as a substrate marking signal for the 20S proteasome. Proteasome appears to attach to the oxidant-exposed hydrophobic sites of damaged proteins (Giulivi and Davies, 1993). Therefore, changes in hydrophobicity might also provide signals for degradation. Proteins with exposed hydrophobic and bulky residues are especially preferred substrates for the protease complex (Davies,

1986). The removal of oxidatively modified proteins is of vital importance to the cell, because the accumulation of physiologically inactive, sometimes toxic junk, can significantly hinder cell functions (Goldberg and Boches, 1982). However, oxidative modification might lead to aggregation and massive cross-linking of proteins, resulting in giant molecular complexes resistant to degradation. In addition, Friguet and coworkers (1994) showed that cross-linked proteins may inhibit the proteasome complex and may prevent the degradation of other oxidized and short-lived proteins, resulting in accumulation of these proteins. The controlling mechanism of degradation is a very complex process that may affect a number of different pathways. IκBα (the inhibitory subunit of nuclear factor-κB transcription factor) is degraded by proteasome following polyubiquitination at specific sites (Jentsch, 1992). This process might occur without ubiquitin conjugation of IκBα (Pickart, 1997).

Physical Exercise and Proteolysis

During physiological challenges such as physical exercise cells have to cope with a variety of stresses. Regular exercise provides repeated stimuli with high metabolic demands, to which cells are forced to reply as economically as possible. Even old, still functional proteins have physiologically less function and durability for massive challenges compared with newly synthesized proteins (Smith, 1995). Moreover, accumulation of oxidatively modified proteins in cells might significantly limit physiological function. One response to metabolic challenges is to change the rate of protein synthesis or protein degradation. It seems that physical exercise in general could increase both. Ji et al. (1988) reported that, after exhaustive treadmill running, proteolysis increased in skeletal muscle of rats, as measured by the enhanced clearance of [^{3}H]leucine from pre-labeled proteins. Calpain-like (Ca^{2+}-stimulated protease) activity was also induced in rats after 1 h of running (Raj et al., 1998). We have shown that 9 wk of swimming increased chymotrypsin-like and trypsin-like activities of the proteasome complex in skeletal muscle of rats (Radák et al., 1999a), whereas protein carbonyl content did not change. Interestingly, the same exercise load increased chymotrypsin-like but not trypsin-like activity of the proteasome complex in the brain, where the accumulation of oxidized proteins decreased as a result of exercise. In a related study, we administered hydrogen peroxide for 3 wk every second day to trained and untrained rats. We found that hydrogen peroxide induced proteasome activity in the cardiac muscle, but the increase was higher in the trained group. The reactive carbonyl derivative content was larger in the untrained group than the control group and less than both in the trained group (Radák et al., 1999a). Similarly, exercise-induced proteolysis has been reported in human studies (Biolo et al., 1995; Nissen et al., 1996; Rohde et al., 1997). The evidence is convincing that exercise increases protein turnover, and therefore the half-life of proteins decreases as an adaptive response to exercise. The increased protein breakdown, primarily in the rate of protein turnover, might improve the antioxidant defense of cells (Davies, 1986). It also could be one of the beneficial effects of regular exercise that up-regulates resistance against oxidative stress.

Aging and Proteolysis

It has been widely postulated that accumulation of oxidatively modified proteins is a causative factor in aging. Oxidatively modified proteins are known to be more susceptible than native proteins to proteolytic degradation executed by proteasome or insulin degrading enzyme, or both (Grune et al., 1997). Agarwal and Sohal (1994) indicated that alkaline protease, possibly proteasome, can be induced in the housefly by the formation of reactive carbonyl derivatives. However, Agarwal and Sohal found no age-related changes in activity. On the other hand, Carney et al. (1991) reported a decrease in protease activity in the brains of old gerbils. Starke-Reed and Oliver (1989) reported similar age-associated decreases in protease activity in the livers of aged rats. The proteasome complex has at least five distinct protease activities including three types of peptidase activities (Cardozo and Michaud, 1993). Some studies focused on the behavior of proteasome complex during aging. Conconi and his coworkers (1996) showed that, among three peptidase activities of proteasome complex (trypsin-like, chymotrypsin-like, and peptidyl glutamyl peptide hydrolyzing), the peptidyl glutamyl peptide hydrolyzing activity decreased as much as 60% as a function of age. On the other hand, Shibatani and Ward (1996) reported that the basal level of chymotrypsin-like activity of proteasome complex increased with age, whereas the peptidyl glutamyl peptide hydrolyzing activity decreased. Hayashi and Goto (1998) showed that the alpha-subunit protein content of the proteasome complex did not change with aging in the livers of 25- to 28-mo-old rats as compared with that in young and middle-aged animals, whereas trypsin-like, chymotrypsin-like, and peptidyl glutamyl peptide hydrolyzing activities of the proteasome complex decreased differently with age. The age-associated decrease in 26S proteasome activity, which is known to be responsible for the degradation of ubiquitinated proteins, is in accordance with the previous findings that high–molecular weight ubiquitin protein conjugates accumulate with age (Ohtsuka et al., 1995). Thus, an age-associated increase in the ubiquitinated proteins is likely responsible in part for decreased 26S proteasome activities.

Conclusion

The continuous generation of free radicals in aerobic organisms damages DNA and proteins. A single bout of exercise might increase the generation of free radical–induced damage in DNA and proteins. However, regular exercise maintains a steady-state level of damage or decreases damage, due to the faster protein turnover rate in proteins and probably improved repair processes of DNA. The increase in proteasome complex activity as a result of regular exercise is an important factor in the exercise-induced adaptation against oxidative stress. Free radical formation increases and the antioxidant and repair processes decrease as a function of age, resulting in the accumulation of DNA and protein damage. Regular physical exercise–induced light oxidative stress might stimulate repair processes and delay the age-associated increase in DNA and protein damage accumulation and the decline in cognitive function. The cause of the cancer-reducing effects of regular exercise might be based on the down-regulation of DNA damage and increased activity of antioxidant systems, including DT-diaphorase and proteasome complex. Probable reasons for the age-related increase in protein damage include decreased protein degradation and increased half-life of proteins.

References

Agarwal, S., and R.S. Sohal. 1994. Aging and proteolysis of oxidized proteins. *Archives of Biochemistry and Biophysics* 309:24-28.

Ahnen, D.J. 1997. The genetic and biologic basis of colon cancer risk. *Proceedings of the Annual American Association of Cancer Research* 38:S663.

Alam, Z.I., A. Jenner, S.E. Daniel, A.J. Lees, N. Cairns, C.D. Marsden, P. Jenner, and B. Halliwell. 1997. Oxidative DNA damage in the Parkinsonian brain: An apparent selective increase in 8-hydroxyguanine levels in substantia nigra. *Journal of Neurochemistry* 69:1196-1203.

Alessio, H.M. 1993. Exercised-induced oxidative stress. *Medicine and Science in Sports and Exercise* 25:218-224.

Ames, B.N., M.K. Shinegawa, and T.M. Hagen. 1993. Oxidants, antioxidants, and the degenerative diseases of aging. *Proceedings of the National Academy of Sciences of the USA* 90:7915-7922.

Amici A., R. Levine, L. Tsai, and E.R. Stadtman. 1989. Conversion of amino acid residues in proteins and amino acid homopolymers to carbonyl derivatives by metal-catalysed reactions. *Journal of Biological Chemistry* 264:3341-3346.

Anderson, S., A.T. Bankler, B.G. Barrell, M.H. de Brujin, A.R. Coulson, J. Drouin, I.C. Eperson, D.P. Nierlich, B.A. Roe, F. Sanger, P.H. Schreier, A.J. Smith, R. Staden, and I.G.Young. 1981. Sequence and organization of the human mitochondrial genome. *Nature* 290: 457-465.

Arking, R. 1998. Measuring aging-related changes in individuals. In: *Biology of Aging, 2nd ed.* Sinauer Sunderland, Sunderland, Massachusetts, pp. 1-570.

Asami, S., T. Hirano, R. Yamaguchi, H. Tsurodome, and H. Kasai. 1998. Effects of forced and spontaneous exercise on 8-hydroxydeoxyguanosine levels in rat organs. *Biochemical and Biophysical Research Communications* 243:678-682.

Baumer, A., C. Zhang, A.W. Linnane, and P. Nagley. 1994. Age-related human mtDNA deletions: A heterogeneous set of deletions arising at a single pair of directly repeated sequences. *American Journal of Human Genetics* 54:618-630.

Bejna, J., L.L. Ji. 1999. Aging and acute exercise enhance free radical generation in rat skeletal muscle. *Journal of Applied Physiology* 87:465-470.

Blair, S.N., H.W. Kohl, C.E. Barlow, R.S. Paffenberger, L.W. Gillons, and C.A. Macera. 1995. Changes in physical fitness and all-cause mortality. A prospective study of healthy and unhealthy men. *Journal of the American Medical Association* 273:1093-1098.

Beckman, K.B., and B.N. Ames. 1997. Oxidative decay of DNA. *Journal of Biological Chemistry* 272:19633-19636.

Beckman, K.B., and B.N. Ames. 1998. The free radical theory of aging matures. *Physiological Reviews* 78:547-577.

Biolo, G., S.P. Maggi, B.D. Williams, K.D. Tipton, and R.R. Wolfe. 1995. Increased rates of muscle protein turnover and amino acid transport after resistance exercise in humans. *American Journal of Physiology* 268:E514-E520.

Buss, H., T.P. Chan, K.B. Sluis, N.M. Domigan, and C.C. Winterbourn. 1997. Protein carbonyl measurement by a sensitive ELISA method. *Free Radical Biology and Medicine* 23:361-366.

Butterfield, D.A., B.J. Howard, S. Yaltin, K. Allen, and J.M. Carney. 1997. Free radical oxidation of brain proteins in accelerated senescence and its modulation by N-tert-butyl-alpha-phenylnitrone. *Proceedings of the National Academy of Sciences of the USA* 94:674-678.

Cabiscol, E., and R.L. Levine. 1995. Carbonic anhydrase III. *Journal of Biological Chemistry* 270:14742-14747.

Cao, G., and R.G. Cutler. 1995. Protein oxidation and aging: I. Difficulties in measuring reactive protein carbonyls in tissues using 2,4-dinitrophenylhydrazine. *Archives of Biochemistry and Biophysics* 320:106-114.

Cardozo, M., and C. Michaud. 1993. Evidence for the presence of five distinct proteolytic components in the pituitary multicatalytic proteinase complex properties of two components cleaving bonds on the carboxyl side of branched chain and small neutral amino acids. *Biochemistry* 32:1563-1572.

Carney, J.M., and A.M. Carney. 1994. Role of protein oxidation in aging and in age-associated neurodegenerative diseases. *Life Sciences* 55:2097-2103.

Carney, J.M., P.E. Starke-Reed, C.N. Oliver, R.W. Landum, M.S. Cheng, J.F. Wu, and R.A. Floyd. 1991. Reversal of age-related increase in brain protein oxidation, decrease in enzyme activity, and loss in temporal and spatial memory by chronic administration of spin-trapping compound N-tert-butyl-alpha phenylnitrone. *Proceedings of the National Academy of Sciences of the USA* 88:3633-3636.

Chance, B., H. Siess, and A. Boveris. 1979. Hydrogen peroxide metabolism in mammalian organs. *Physiological Reviews* 59:527-605.

Chapman, M.L., B.R. Runbin, and R.W. Gracy. 1989. Increased carbonyl content of proteins in synovial fluid from patients with rheumatoid arthritis. *Journal of Rheumatology* 16:15-18.

Chodzko-Zajko, W.J., and K.A. Moore. 1994. Physical fitness and cognitive function in aging. *Exercise and Sport Sciences Reviews* 22:195-220.

Clayton, D.A., J.N. Doda, and E.C. Friedberg. 1974. The absence of a pyrimidine dimer repair mechanism in mammalian mitochondria. *Proceedings of the National Academy of Sciences USA* 71:2777-2781.

Climent I., and R.L. Levine. 1991. Oxidation of the active site of glutamine synthetase: Conversion of arginine 344 to glutamyl semialdehyde. *Archives of Biochemistry and Biophysics* 189:371-375.

Conconi, M., L.I. Szweda, R.L. Levine, E.R. Stadtman, and B. Friguet. 1996. Age-related decline of rat liver multicatalytic proteinase activity and protection from oxidative inactivation by heat-shock protein. *Archives of Biochemistry and Biophysics* 331:232-240.

Cooper, J.M., V.M. Mann, and A.H. Schapira. 1992. Analyses of mitochondrial respiratory chain function and mitochondrial DNA deletion in human skeletal muscle: Effect of ageing. *Journal of Neurological Sciences* 113:91-98.

Cortopassi, G.A., and N. Arnheim. 1990. Detection of a specific mitochondrial DNA deletion in tissues of older humans. *Nucleic Acids Research* 18:6927-6933.

Dakin, H.D. 1906. The oxidation of amino-acids with the production of substances of biological importance. *Journal of Biological Chemistry* 1:171-176.

Daneryd, P., F. Aberg, G. Dallner, L. Erstner, T. Schersten, and B. Soussi. 1995. Coenzymes Q9 and Q10 in skeletal and cardiac muscle in tumor-bearing exercise rats. *European Journal of Cancer* 31A:760-765.

Davies, K.J.A. 1986. Intracellular proteolytic system may function as secondary antioxidant defense: A hypothesis. *Free Radical Biology and Medicine* 2:155-173.

Davies, K.J.A, A.T. Quintanilha, G.A. Brooks, and L. Packer. 1981. Free radicals and tissue damage produced by exercise. *Biochemical and Biophysical Research Communications* 107:1198-1205.

Day, W.W. 1995. Exercise-induced inhibition of hepatic drug metabolism and a mechanism for decreased activity of P450IIE1 in Fisher-344 rats. *Dissertation Abstracts International* 55:2659.

Dolle, M.E., H. Giese, C.L. Hopkins, H.J. Martus, J.M. Hausdorf, and J. Vijg. 1997. Rapid accumulation of genome rearrangement in liver but not in brain of old mice. *Nature Genetics* 17:431-434.

Dubley, A., M.J. Forster, H. Lal, and R.S. Sohal. 1996. Effect of age and caloric intake on protein oxidation in different brain regions and on behavioral functions of the mouse. *Archives of Biochemistry and Biophysics* 333:189-197.

Fagan, J.M., and L. Waxman. 1992. The ATP-independent pathway in red blood cells that degrades oxidant-damaged hemoglobin. *Journal of Biological Chemistry* 267:23015-23022.

Fagan, J.M, L. Waxman, and A.L. Goldberg. 1985. Red blood cells contain a pathway for degradation of oxidant-damaged hemoglobin that does not require ATP or ubiquitin. *Journal of Biological Chemistry* 261:5705-5713.

Fahn, S., and Cohen G. 1992. The oxidant stress hypothesis in Parkinson's disease: Evidence supporting it. *Annals of Neurology* 32:804-812.

Forster, M.J., A. Dubey, K.M. Dawson, W.A. Stutts, H. Lal, and R.S. Sohal. 1996. Age-related losses of cognitive function and motor skills in ice are associated with oxidative damage in the brain. *Proceedings of the National Academy of Sciences of the USA* 93:4765-4769.

Fraga, C.G., M.K. Shigenaga, J.W. Park, P. Dagan, and B.N. Ames. 1990. Oxidative damage to DNA during aging: 8-hydroxy-2-deoxyguanosine in rat organ DNA and urine. *Proceedings of the National Academy of Sciences of the USA* 87:4533-4537.

Friguet B., L. Szweda, and E.R. Stadtman. 1994. Susceptibility of glucose-6-phosphate dehydrogenase modified by 4-hydroxy-2-nonenal and metal-catalysed oxidation to proteolysis by the multicatalytic protease. *Archives of Biochemistry and Biophysics* 311:168-173.

Giulivi, G., and K.J.A. Davies. 1993. Dityrosine and tyrosine oxidation products are endogenous markers for selective proteolysis of oxidatively modified red blood cell hemoglobin by (the 19S) proteasome. *Journal of Biological Chemistry* 268:8752-8759.

Goldberg, A.L., and F.S. Boches. 1982. Oxidized proteins in erythrocytes are rapidly degraded by the adenosine triphosphate-dependent proteolytic system. *Science* 215:1107-1108.

Gorman, W.F. 1995. Benign aging or Alzheimer's disease? *Journal of the Oklahoma State Medical Association* 89:383-391.

Goto, S., A. Hasegawa, H. Nakamoto, A. Nakamura, R. Takahashi, and I.V. Kurochkin. 1995. Age-associated changes of oxidative modification and turnover of proteins. In: *Oxidative Stress and Aging*, R.G. Cutler, L. Packer, J. Bertram, and A. Mori (eds.), pp. 151-157. Birkhauser Verlag, Basel, Switzerland.

Goto, S., and A. Nakamura. 1997. Age-associated, oxidatively modified proteins: A critical evaluation. *Age* 20:81-89.

Goto, S., A. Nakamura, Z. Radák, H. Nakamoto, R. Takahashi, K. Yasuda, Y. Sakurai, and N. Ishii. 1999. Carbonylated proteins in aging and exercise: Immunoblot approaches. *Mechanisms of Ageing and Development* 107: 245-253.

Gross, N.J., G.S. Getz, and M. Rabinowitz. 1969. Apparent turnover of mitochondrial deoxyribonucleic acid and mitochondria phospholipids in the tissues of the rat. *Journal of Biological Chemistry* 244:1552-1562.

Grune, T., T. Reinheckel, and K.J. Davies. 1997. Degradation of oxidized proteins in mammalian cells. *FASEB Journals* 11:526-534.

Harman, D. 1956. Aging: A theory based on free radical and radiation chemistry. *Journal of Gerontology* 11:298-300.

Hart, R.W., and R.B. Setlow. 1974. Correlation between deoxyribonucleic acid excision-repair and life-span in a number of mammalian species. *Proceedings of the National Academy of Sciences of the USA* 71:2169-2173.

Hartmann, A., U. Plappert, K. Raddatz, M. Grunert-Fuchs, and G. Speit. 1994. Does physical exercise induce DNA damage? *Mutagenesis* 9:269-272.

Hartmann, A., A.M. Niess, M. Grunert-Fuchs, B. Poch, and G. Speit. 1995. Vitamin E prevents exercise-induced DNA damage. *Mutation Research* 346:195-202.

Hartmann, A., S. Pfuhler, C. Dennong, D. Germadnik, A. Pilger, and G. Speit. 1998. Exercise-induced DNA effects in human leukocytes are not accompanied by increased formation of 8-hydroxy-2-deoxyguanosine or induction of micronuclei. *Free Radical Biology and Medicine* 24:245-251.

Hayakawa, M., K. Hattori, S. Sugiyama, M. Tanaka, and T. Ozawa. 1991. Age-associated accumulation of 8-hydroxydeoxyguanosine in mitochondrial DNA of human diaphragm. *Biochemical and Biophysical Research Communications* 189:979-985.

Hayashi, T., and S. Goto. 1998. Age-related changes in the 20S and 26S proteasome activities in the liver of male F344 rats. *Mechanisms of Ageing and Development* 102:55-66.

Henle, E.S., and S. Linn. 1997. Formation, prevention, and repair of DNA damage by iron/hydrogen peroxide. *Journal of Biological Chemistry* 272:19095-19098.

Henriksson, J. 1992. Effects of physical training on the metabolism of skeletal muscle. *Diabetes Care* 15:1701-1711.

Higami, Y., I. Shimokawa, T. Okimoto, and T. Ikeda. 1994. Vulnerability to oxygen radicals is more important than impaired repair in hepatocytic deoxyribonucleic acid damage in aging. *Laboratory Investigation* 71:650-656.

Holloszy, J.O. 1993. Exercise increases longevity of female rats despite increased food intake and no growth retardation. *Journal of Gerontology* 48:B97-100.

Ikeda, T., A. Ishigami, and S. Goto. 1992. Change with donor age in the degradation rate of endogenous proteins of mouse hepatocytes in primary culture. *Archives of Gerontology and Geriatrics* 15:181-188.

Inger, T., and L. Eiliv. 1997. Exercise and breast cancer. *New England Journal of Medicine* 337:708-709.

Ishigami, A., and S. Goto. 1990. Inactivation kinetics of horseradish peroxidase microinjected into hepatocytes from mice of various ages. *Mechanisms of Ageing and Development* 46:125-133.

Jentsch, S. 1992. The ubiquitin-conjugation system. *Annual Review in Genetics* 26:179-207.

Ji, L.L., F.W. Stratman, and H.A. Lardy. 1988. Enzymatic down regulation with exercise in rat skeletal muscle. *Archives of Biochemistry and Biophysics* 263:137-149.

Kaneko, T., S. Tahara, and M. Matsuo. 1996. Non-linear accumulation of 8-hydroxy-2-deoxyguanosine, a marker of oxidized DNA damage, during aging. *Mutation Research* 316:277-285.

Kaneko, T., S. Tahara, and M. Matsuo. 1997. Retarding effect of dietary restriction on the accumulation of 8-hydroxy-2'-deoxyguanosine in organs of Fisher 344 rats during aging. *Free Radical Biology and Medicine* 23:76-81.

Kim, J.D., B.P. Yu, R.J.M. McCarter, S.Y. Lee, and J.T. Herlihy. 1996. Exercise and diet modulate cardiac lipid peroxidation and antioxidant defenses. *Free Radical Biology and Medicine* 20:83-88.

Kiningham, R.B. 1998. Physical activity and primary prevention of cancer. *Primary Care* 25:515-536.

Kuchino, Y., F. Mori, H. Kasai, H. Inoue, S. Iwai, K. Miura, E. Ohtsuka, and S. Nishimura. 1987. Misreading of DNA templates containing 8-hydroxydeoxyguanosine at the modified base and at adjacent residues. *Nature* 327:77-79.

Kurochkin, I.V., and S. Goto. 1994. Alzheimer's and beta-amyloid peptide specifically interacts with and is degraded by insulin degrading enzyme. *FEBS Letters* 345:33-37.

Larsen, C.N, and D. Finley. 1997. Protein translocation channels in the proteasome and other proteases. *Cell* 91:431-434.

Leeuwenburgh, C., P. Hansen, A. Shaish, J.O. Holloszy, and J.W. Heinricke. 1998. Markers of protein oxidation by hydroxyl radical and reactive nitrogen species in tissues of aging rats. *American Journal of Physiology* 274:R453-R461.

Leonard, G. 1997. What makes us tick. *Science* 275:943-944.

Levine, R.L., D. Garland, C.N. Oliver, A. Amici, I. Climet, A. Lenz, B. Ahn, S. Shalteil, and E.R. Stadtman. 1990. Determination of carbonyl content of oxidatively modified proteins. *Methods in Enzymology* 186:464-478.

Levine, R.L., J.A. Williams, E.R. Stadtman, and E. Shacter. 1994. Carbonyl assay for determination of oxidatively modified proteins. *Methods in Enzymology* 37:346-357.

Linnane, A.W., S. Marzuki, T. Ozawa, and M. Tanaka. 1989. Mitochondrial DNA mutations as an important contributor to aging and degenerative diseases. *Lancet* 8639:642-645.

Liu, J., X. Wang, M.K. Shigenaga, H.C. Yeo, A. Mori, and B.N. Ames. 1996. Immobilization stress causes oxidative damage to lipid, protein and DNA in brain of rats. *FASEB Journal* 10:1532-1538.

Lucassen, P.J., E.J. van Someren, and D.F. Swaab. 1998. Are active neurons a better defense against aging in Alzheimer's disease? (Dutch) *Tijdschrift voor Gerontologie en Geriatrie* 29:177-184.

Mars, M., S. Govender, A. Weston, V. Naicker, and A. Chuturgoon. 1998. High intensity exercise: A cause of lymphocyte apoptosis? *Biochemical and Biophysical Research Communication* 249:366-370.

Martin, G.M. 1996. Somatic mutagenesis and antimutagenesis in aging research. *Mutation Research* 350:35-41.

Meccoci, P., G. Fano, S. Fulle, U. MacGarvey, L. Shinobu, M.C. Polidori, A. Cherubini, J. Vecchiet, U. Senin, and M.F. Beal. 1999. Age-dependent increases in oxidative damage to DNA, lipids, and proteins in human skeletal muscle. *Free Radical Biology and Medicine* 26:303-308.

Mecocci, P., U. MacGarvey, A.E. Kaufman, D. Koontz, J.M. Shoffner, D.W. Wallace, and M.F. Beal. 1993. Oxidative damage to mitochondrial DNA shows marked age-dependent increases in human brain. *Annals in Neurology* 34:609-616.

Miller, E.M., K.A. Kunugi, and T.J. Kinsella. 1995. Effect of 5-aminothymidine and leucovorin on radiosensitization by iododeoxyuridine in human colon cancer cells. *Clinical Cancer Research* 1: 407-416.

Miller, J.W., J. Selhub, and J.A. Joseph. 1996. Oxidative damage caused by free radicals produced during catacholamine autooxidation: Protective effects of O-methylation and melatonin. *Free Radical Biology and Medicine* 21:241-249.

Mukherjee, S.K., and J.D. Adams Jr. 1997. The effects of aging and neurodegeneration on apoptosis-associated DNA fragmentation and the benefits of nicotinamide. *Molecular and Chemical Neuropathology* 32:59-74.

Muller-Hocker, J. 1990. Cytocrome *c* oxidase deficient fibers in the limb muscle and diaphragm of man without muscular disease: An age-related alteration. *Journal of Neurological Sciences* 100:14-21.

Nakamura, A., and S. Goto. 1996. Analysis of protein carbonyls with 2,4-dinitrophenylhydrazine and its antibodies by immunoblot in two-dimensional gel electrophoresis. *Journal of Biochemistry* (Tokyo) 119:768-774.

Nissen, S., R. Sharp, M. Ray, J.A. Ratmacher, D. Rice, J.C. Fuller, A.S. Connelly, and N. Abumrad. 1996. Effect of leucine metabolite beta-hydroxy-beta-methylbutyrate on muscle metabolism during resistance exercise training. *Journal of Applied Physiology* 81:2095-2104.

Ohtsuka, H., R. Takahashi, and S. Goto. 1995. Age-related accumulation of high-molecular-weight ubiquitin conjugates in mouse brain. *Journal of Gerontology* 50A:B275-B281.

Okamura, K., T. Doi, K. Hamada, M. Sakurai, Y. Yoshioka, R. Mitsuzono, T. Migita, S. Sumida, and Y. Sugawa-Katayama. 1997a. Effect of repeated exercise on urinary 8-hydroxydeoxyguanosine excretion in humans. *Free Radical Research* 26:507-514.

Okamura, K., T. Doi, M. Sakurai, K. Hamada, Y. Yoshioka, S. Sumida, and Y. Sugawa-Katayama. 1997b. Effect of endurance exercise on the tissue 8-hydroxydeoxyguanosine content in dogs. *Free Radical Research* 26:523-528.

Oliver, C.N., B.-W. Ahn, E.J. Moerman, S. Goldstein, and E.R. Stadtman. 1987. Age-related changes in oxidized proteins. *Journal of Biological Chemistry* 262:5488-5491.

Oliver, C.N., P.E. Starke-Redd, E.R. Stadtman, G.J. Liu, J.M. Carney, and R.A. Floyd. 1990. Oxidative damage to brain proteins, loss of glutamine synthetase activity, and production of free radicals during ischemia/reperfusion-induced injury to gerbil brain. *Proceedings of the National Academy of Sciences of the USA* 87:5144-5147.

Orr, W.C., and R.S. Sohal. 1994. Extension of life-span by over expression of superoxide dismutase and catalase in *Drosophila melanogaster*. *Science* 263:1128-1130.

Ozawa, T. 1997. Genetic and functional changes in mitochondria associated with aging. *Physiological Reviews* 77:425-464.

Pedersen, B.K., T. Rohde, and K. Ostrowski. 1998. Recovery of immune system after exercise. *Acta Physiologica Scandinavica* 162:325-332.

Perez-Campo, R., M. Lopez-Torres, S. Cadenas, C. Rojas, and G. Barja. 1998. The rate of free radical production as a determinant of the rate of aging: Evidence from the comparative approach. *Journal of Comparative Physiology—B: Biochemical, Systemic, and Environmental Physiology* 168:149-158.

Peterson, D.H., and D.A. Cunningham. 1999. The gas transporting system: Limits and modifications with age and training. *Canadian Journal of Applied Physiology* 24:28-40.

Pickart, C.M. 1997.Targeting of substrates to the 26S proteasome. *FASEB Journal* 11:1055-1066.

Pilger, A., D. Germadnik, D. Formanek, H. Zwick, N. Winkler, and H.W. Rudiger. 1997. Habitual long-distance running does not enhance urinary excretion of 8-hydroxydeoxyguanosine. *European Journal of Applied Physiology* 75:467-469.

Poulsen, H.E., S. Loft, and K. Vistisen. 1996. Extreme exercise and DNA modification. *Journal of Sport Sciences* 14:343-346.

Radák, Z., K. Asano, Y. Fu, A. Nakamura, H. Nakamoto, and S. Goto. 1998a. The effect of high altitude and caloric restriction on carbonyl derivatives and activity of glutamine synthetase. *Life Sciences* 62:1317-1322.

Radák, Z., K. Asano, M. Inoue, T. Kizaki, S. Oh-ishi, K. Suzuki, N. Taniguchi, and H. Ohno. 1995. Superoxide dismutase derivatives reduce oxidative damage in skeletal muscle of rats during exhaustive exercise. *Journal of Applied Physiology* 79:129-135.

Radák, Z., K. Asano, M. Inoue, T. Kizaki, S. Oh-ishi, K. Suzuki, N. Taniguchi, and H. Ohno. 1996. Superoxide dismutase derivative prevents oxidative damage in liver and kidney of rats induced by exhaustive exercise. *European Journal of Applied Physiology* 72:189-194.

Radák, Z., K. Asano, K.C. Lee, H. Ohno, A. Nakamura, H. Nakamoto, and S. Goto. 1997. High altitude increases reactive carbonyl derivatives but not lipid peroxidation in skeletal muscle of rats. *Free Radical Biology and Medicine* 22:1109-1114.

Radák, Z., K. Asano, A. Nakamura, H. Nakamoto, H. Ohno, and S. Goto. 1998b. A period of anaerobic interval exercise increases accumulation of reactive carbonyl derivatives in lungs of rats. Pflugers Archiv. *European Journal of Physiology* 435:439-441.

Radák, Z., and S. Goto. 1998. The effects of exercise, ageing and caloric restriction on protein oxidation and DNA damage in skeletal muscle. In: *Oxidative Stress in Skeletal Muscle*, A.Z. Reznick, L. Packer, C.K. Sen, J.O. Holloszy, and M.J. Jackson (eds.), pp. 89-103. Birkhauser, Basel, Switzerland.

Radák, Z., T. Kaneko, S. Tahara, H. Nakamoto, H. Ohno, M. Sasvari, C. Nyakas, and S. Goto. 1999a. The effect of exercise training on oxidative damage of lipids, proteins and DNA in rat skeletal muscle: Evidence for beneficial outcomes. *Free Radical Biology and Medicine.* 26:1059-1063.

Radák, Z., C. Nyakas, T. Kaneko, and S. Goto. 1998c. Regular exercise improves cognitive function and decreases oxidative damage in rat brain. *Free Radical Biology and Medicine* 25:S77.

Radák, Z., J. Pucsok, S. Mecseki, T. Csont, and P. Ferdinandy. 1999b. Muscle soreness increases nitric oxide content and DNA damage in human skeletal muscle. *Free Radical Biology and Medicine* 27:69-74.

Raj, D.A., T.S. Booker, and A.N. Belcastro. 1998. Striated muscle calcium-stimulated cysteine protease (calpaine-like) activity promotes myeloperoxidase activity with exercise. Pflugers Archiv. *European Journal of Physiology* 435:804-809.

Rajguru, S.U., G.S. Yeargans, and N.W. Seidler. 1994. Exercise causes oxidative damage to rat skeletal muscle microsomes while increasing cellular sulfhydryls. *Life Sciences* 54:149-157.

Reznick, A.Z., C.E. Cross, M.L. Hu, Y.J. Suzuki, S. Khwaja, A. Safadi, P.A. Motchnick, L. Packer, and B. Halliwell. 1992a. Modification of plasma proteins by cigarette smoke as measured by protein carbonyl formation. *Biochemical Journal* 286:607-611.

Reznick, A.Z., V.E. Kagan, R. Ramsey, M. Tsuchiya, S. Khwaja, E.A. Serbinova, and L. Packer. 1992b. Antiradical effects in L-propionyl carnitine protection of the heart against ischemia-reperfusion injury: The possible role of iron chelation. *Archives of Biochemistry and Biophysics* 296:394-401.

Reznick, A.Z., L. Lavie, H.E. Gershon, and D. Gershon. 1981. Age-associated accumulation of altered FDP aldolase B in mice. Conditions of detection and determination of aldolase half life in young and old animals. *FEBS Letters* 128:221-224.

Reznick, A.Z., and L. Packer. 1994. Oxidative damage to proteins spectrophotometric method for carbonyl assay. *Methods in Enzymology* 233:357-363.

Reznick, A.Z., E. Witt, M. Matsumoto, and L. Packer. 1992c. Vitamin E inhibits protein oxidation in skeletal muscle of resting and exercising rats. *Biochemical and Biophysical Research Communications* 189:801-806.

Richter, C. 1988 Do mitochondrial DNA fragments promote cancer and aging? *FEBS Letters* 241:1-5.

Richter, C., J.W. Park, and B.N. Ames. 1988. Normal oxidative damage to mitochondrial and nuclear DNA is extensive. *Proceedings of the National Academy of Sciences of the USA* 85:6465-6467.

Rivett, J.A. 1985. Preferential degradation of the oxidatively modified form of glutamine synthetase by intracellular mammalian proteases. *Journal of Biological Chemistry* 260:300-305.

Rohde, T., D.A. MacLean, E.A. Richter, B. Kiens, and B.K. Pedersen. 1997. Prolonged submaximal eccentric exercise is associated with increased levels of plasma IL-6. *American Journal of Physiology* 273:E85-E91.

Sarna, S., T. Sahi, M. Koskenvuo, and J. Kaprio. 1993. Increased life expectancy of world class male athletes. *Medicine and Science in Sports and Exercise* 25:237-244.

Saxton, J.M., A.E. Donnelly, and H.P. Roper. 1994. Indices of free-radical-mediated damage following maximum voluntary eccentric and concentric muscular work. *European Journal of Applied Physiology* 68:189-193.

Sen, C.K., M. Atalay, J. Agren, D.E. Laaksonen, S. Roy, and O. Hanninen. 1997. Fish oil and vitamin E supplementation in oxidative stress at rest and after physical exercise. *Journal of Applied Physiology* 83:189-195.

Shephard, R.J., and R. Futcher. 1997. Physical activity and cancer: How may protection be maximized? *Critical Review of Oncology* 8:219-272.

Shibatani, T., and W.F. Ward. 1996. Effect of age and food restriction on alkaline protease activity in rat liver. *Journal of Gerontology* 51A:B175-B178.

Shigenaga, M.K., T.M. Hagen, and B.N. Ames. 1994. Oxidative damage and mitochondrial decay in aging. *Proceedings of the National Academy of Sciences of the USA* 91:10771-10778.

Simpson, J.A., S.P. Giesseg, and R.T. Dean. 1993. Free radical and enzymatic mechanism for the generation of protein bound reducing moieties. *Biochimica et Biophysica Acta* 1156:190-196.

Smith, J.A. 1995. Exercise, training and red blood cell turnover. *Sports Medicine* 19:9-31.

Smith, C.D., J.M. Carney, P.E. Starke-Reed, C.N. Oliver, E.R. Stadtman, R.A. Floyd, and W.R. Markesbery. 1991. Excess brain protein oxidation and enzyme dysfunction in normal aging and in Alzheimer's disease. *Proceedings of the National Academy of Sciences of the USA* 88:10540-10543.

Sohal, R.S., S. Agarwal, A. Dubey, and W.C. Orr. 1993. Protein oxidative damage is associated with life expectancy of houseflies. *Proceedings of the National Academy of Sciences of the USA* 90:7255-7259.

Sohal, R.S., and A. Dubey. 1994. Mitochondrial oxidative damage, hydrogen peroxide release, and aging. *Free Radical Biology and Medicine* 16:621-626.

Sohal, R.S., and R. Weindruch. 1996. Oxidative stress, caloric restriction, and aging. *Science* 273:59-63.

Stadtman, E.R. 1986. Oxidation of proteins by mixed-function oxidation system: Implication in protein turnover, aging and neutrophil function. *Trends in Biochemical Sciences* 11:11-12.

Stadtman, E.R. 1990. Metal ion-catalyzed oxidation of proteins: Biochemical mechanism and biological consequences. *Free Radical Biology and Medicine* 9:315-325.

Stadtman, E.R. 1992. Protein oxidation and aging. *Science* 257:1220-1224.

Stadtman, E.R., and C.N. Oliver. 1991. Metal-catalyzed oxidation of proteins. *Journal of Biological Chemistry* 266:2005-2008.

Stamler, J.S. 1994. Redox signaling: Nitrosylation and related target interactions of nitric oxide. *Cell* 78:931-936.

Starke-Reed, P.E., and C.N. Oliver. 1989. Protein oxidation and proteolysis during aging and oxidative stress. *Archives of Biochemistry and Biophysics* 275:559-567.

Sumida, S., T. Doi, M. Sakurai, Y. Yoshida, and K. Okamura. 1997. Effect of a single bout of exercise and beta-carotine supplementation on the urinary excretion of 8-hydroxy-deoxyguanosine in humans. *Free Radical Research* 27:607-618.

Sumida, S., and Y. Sugawa-Katayama. 1991. Effects of alloprinol administration and exercise on lipid peroxidation and antioxidants in rats. *Journal of Clinical Biochemistry and Nutrition* 10:127-134.

Taborsky, G. 1973. Oxidative modification of protein in the presence of ferrous iron and air. Effect of ionic constituents of the reaction medium on the nature of the oxidation products. *Biochemistry* 12:1341-1348.

Takahashi, R., and S. Goto. 1990. Alteration of aminoacyl-tRNA synthetase with aging: heat-labilization of the enzyme by oxidative damage. *Archives of Biochemistry and Biophysics* 277: 228-233.

Tanaka, M., H. Ino, K. Ohno, K. Hattori, W. Sato, and T. Ozawa. 1990. Mitochondrial mutation in fatal infatile cardiomyopathy. *Lancet* 336:1452.

Tchou, J., T. Kasai, S. Shibutani, M.H. Chung, J. Lavai, A.P. Grollman, and Nishimura, S. 1991. 8-oxoguanine (8-hydroxyguanine) DNA glycosylase and its substrate specificity. *Proceedings of the National Academy of Sciences of the USA* 88:4690-4694.

Thompson, H.J., K.C. Westerlind, J. Snedden, S. Briggs, and M. Singh. 1995. Exercise intensity dependent inhibition of methyl-1-nitrosourea induced mammary carcinogenesis in female F-344 rats. *Carcinogenesis* 16:1783-1786.

Uchida, K., and S. Kawakishi. 1993. 2-oxohistidine as a novel biological marker for oxidatively modified proteins. *FEBS Letters* 332:208-210.

Vogt, W. 1995. Oxidation of methionine residues in proteins: Tools, targets, and reversal. *Free Radical Biology and Medicine* 18:93-105.

Waite, H. 1995. Precursors of quinone tanning: DOPA-containing compounds. *Methods in Enzymology* 258:1-20.

Wallance, D.W. 1992. Mitochondrial genetics: A paradigm for aging and degenerative disease? *Science* 256:628-632.

Wang, D., D.A. Kreutzer, and J.M. Essigmann. 1998. Mutagenicity and repair of oxidative DNA damage: Insights from studies using defined lesions. *Mutation Research* 400:99-115.

Weise, A.G., R.E. Pacifici, and K.J. Davies. 1995. Transient adaptation of oxidative stress in mammalian cells. *Archives of Biochemistry and Biophysics* 318:231-240.

Whittal, K.S. and W.S. Parkhouse. 1996. Exercise during adolescence and its effects on mammary gland development, proliferation, and nitrosomethylurea (NMU) induced tumorigenesis in rats. *Breast Cancer Research Treatment* 37:21-27.

Witt, E.H., A.Z. Reznick, C.A. Viguie, P. Starke-Reed, and L. Packer. 1992. Exercise, oxidative damage and effects of antioxidant manipulation. *Journal of Nutrition* 122:766-773.

Yan, L.J., R.L. Levine, and R.S. Sohal. 1997. Oxidative damage during aging targets mitochondrial aconitase. *Proceedings of the National Academy of Sciences of the USA* 94:1168-1172.

Yu, B.P. 1993. *Free Radicals in Aging*. CRC Press, Boca Raton, FL.

Zhang, C., A.W. Linnane, and P. Nagley. 1993. Occurrence of particular base subunits (3243, A to G) in mitochondrial DNA of tissues of aging humans. *Biochemical and Biophysical Research Communications* 195:1104-1110.

Chapter 7

Role of Lipid and Lipoprotein Oxidation

Shuji Oh-ishi
5th Department of Internal Medicine, Tokyo Medical University, Inashiki, Japan

Jay W. Heinecke
Department of Medicine and Department of Molecular Biology and Pharmacology, Washington University School of Medicine, St. Louis, Missouri, USA

Tomomi Ookawara
Department of Biochemistry, Hyogo College of Medicine, Nishinomiya, Japan

Hiromi Miyazaki and Shukoh Haga
Institute of Health and Sport Sciences, University of Tsukuba, Tsukuba, Japan

Zsolt Radák
Laboratory of Exercise and Physiology, School of Sport Sciences, Semmelweis University Budapest, Budapest, Hungary

Takako Kizaki and Hideki Ohno
Department of Hygiene, Kyorin University, School of Medicine, Mitaka, Japan

Over 40 years ago, Harman first proposed that reactive oxygen species play a role in the aging process. According to Harman's theory (1956), free radicals generated from normal metabolic processes produce irreversible cellular injury. The organism's life span is then limited by its ability to cope with damaging free radical reactions. Reactive oxygen species, such as superoxide($O_2^{\cdot-}$), hydrogen peroxide (H_2O_2), and hydroxyl radical (HO•), are known to cause a wide spectrum of cellular damage due to enzyme inactivation, lipid peroxidation, nucleic acid damage, and the like (Cadenas, 1989; Clark et al., 1985; Halliwell and Gutteridge, 1984; Jenkins, 1988; Slater, 1984).

Strenuous exercise markedly increases the oxygen uptake of skeletal and cardiac muscle, probably increasing the generation of reactive oxygen species (Chance et al., 1979; Jenkins, 1988). This led to the proposal that reactive oxygen species may cause exercise-induced oxidative damage in active muscle (Davies et al., 1982; Jenkins, 1988). Davies et al. (1982), for example, reported that the levels of reactive oxygen species in muscle doubled after exhaustive exercise. In addition, Dillard et al. (1978a) observed that exercising humans increased their exhalation of pentane, which is a product of free radical–

induced lipid peroxidation. Therefore, it has been proposed that exercise may increase the rate of aging (Sohal and Allen, 1986).

Atherosclerosis is one of the leading causes of death in the aging populations of the industrialized countries. One important risk factor for this disorder is an elevated plasma level of low density lipoproteins (LDL). Many lines of evidence indicate that oxidation of LDL contributes to the early stages of atherogenesis (Luc and Fruchart, 1991; Munro and Cotran, 1988; Steinberg et al., 1989; Steinbrecher et al., 1990). On the other hand, the level of LDL oxidation in veteran endurance athletes is low (Kujala et al., 1996). Therefore, physical endurance training may enhance antioxidant defenses and prevent LDL oxidation.

In this chapter, we first briefly outline the reactions of lipid peroxidation and assays of lipid hydroperoxides. Second, we summarize data on the effect of exercise on lipid peroxidation in humans and animals, including alterations induced in antioxidant defense systems. Third, we discuss the implications of the LDL oxidation hypothesis. Fourth, we summarize recent results on the effect of exercise on the oxidation of lipoproteins.

Lipid Peroxidation

Oxidation of lipids, so-called lipid peroxidation, may be a potent feature of the injury caused by free radical attack of cells (Hochstein and Emster, 1963). Lipid peroxidation is initiated when free radicals rapidly abstract hydrogen atoms from esterified polyunsaturated fatty acids in cell membranes and lipoproteins as well as free fatty acids. One widely studied model system involves the hydroxyl radical, formed from superoxide via metal ion catalyzed reactions. Hydroxyl radical is an extremely potent oxidant that readily abstracts hydrogen atoms from bis-allylic positions in polyunsaturated fatty acids.

In vitro studies indicate that the peroxidation of polyunsaturated fatty acids usually consists of three mechanistic steps: initiation, propagation, and termination (Girotti, 1985). Because the initial interaction of molecular oxygen with unsaturated bonds of fatty acids is unfavorable because of spin-forbidden reactions, oxygen must be activated to more reactive oxygen species to initiate lipid peroxidation (Floyd, 1993).

Radical Chain Reactions

1. Initiation (figure 7.1)

When a free radical ($R\bullet$) removes a bis-allylic hydrogen atom from a polyunsaturated lipid (LH), a carbon-centered free radical ($L\bullet$) is generated and the process of lipid peroxidation is initiated. The carbon-centered $L\bullet$ reacts at a diffusion-controlled rate with oxygen to form a peroxyl radical; this reaction is

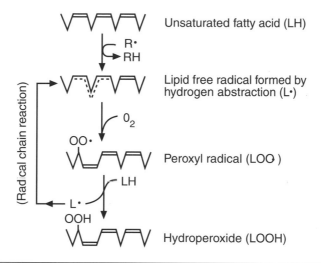

Figure 7.1   Initial process of lipid peroxidation in unsaturated fatty acids.

accompanied by rearrangement of the double bonds to produce a conjugated diene. Redox-active metals such as iron or copper can facilitate the reaction of reactive intermediates with lipids at initiation of lipid peroxidation (Aust et al., 1985).

2. Propagation (figure 7.1)

The peroxyl radical, LOO•, formed in the initial reaction may then remove a hydrogen atom from another polyunsaturated fatty acid, producing a lipid hydroperoxide (LOOH) and another L•. In the absence of chain-terminating reactions, this process can continue until polyunsaturated fatty acids are exhausted. The stable products of this process are hydroperoxides (Barber and Bemheim, 1967; Dahle et al., 1962; Pryor, 1973). In the presence of redox-active metal catalysis, hydroperoxides can be decomposed with the concomitant formation of alkoxyl (LO•) or peroxyl radicals. These reactive intermediates are capable of further reactions, and thus the propagation of lipid peroxidation continues.

3. Termination

When two free radicals meet and interact, nonradical species can be formed, terminating the lipid peroxidation process (equations 7.1-7.3):

7.1 $L\bullet + L\bullet \rightarrow LL$

7.2 $LOO\bullet + LOO\bullet \rightarrow LOOL + O_2$

7.3 $LOO\bullet + L\bullet \rightarrow LOOL$

Compounds such as vitamin E and vitamin C that yield stable free radicals, which are unable to continue the chain-propagating step, will also terminate lipid peroxidation.

Assessment of Lipid Peroxidation

Many assay methods are available for measuring lipid hydroperoxides and their degradation products (Packer, 1984; Packer and Glazer, 1990). However, each method has limitations, and no single assay method will accurately estimate the true extent of lipid peroxidation. In complex biological material, these difficulties are amplified enormously. Some of the most common methods are listed in table 7.1.

The thiobarbituric acid (TBA) assay is the method most commonly used to measure lipid peroxidation. The TBA assay method is based on the calorimetric or fluorometric assessment of the reaction of TBA with malondialdehyde (MDA), which is one of the secondary products derived from peroxidation of polyunsaturated fatty acids. However, TBA reacts not only with MDA but also with a variety of biological materials (such as carbohydrates, sialic acid, or prostaglandins), and these compounds interfere with this assay (Gutteridge, 1986). The TBA assay thus lacks specificity with biological material and cannot be applied to cultured cells, tissue, or plasma samples. Furthermore, many studies report the value of the TBA assay as the amount of MDA in the sample. However, other lipid peroxidation products react with TBA, and MDA can be generated during the assay.

Table 7.1   Assay Methods for Measuring Lipid Peroxidation

Assay method	Target molecule	Comments
TBA method (spectrophotometry or spectrofluorescence)	Hydroperoxide (LOOH), aldehyde, etc.	Nonspecific
Diene conjugation method (spectrophotometry)	Conjugated diene	An index of early events of lipid peroxidation
Analysis of expired gas (gas chromatography)	Ethane, pentane	An in vivo measurement, nonspecific
HPLC–CL method	LOOH	Highly sensitive, specific
HPLC–ECD method	LOOH	Highly sensitive, specific
GC–MS	LOOH, aldehyde, etc.	Highly sensitive, specific

Note. TBA, thiobarbituric acid; LOOH, lipid hydroperoxide; HPLC, high-performance liquid chromatography; CL, chemiluminescence; ECD, electrochemical detector; GC–MS, gas chromatography–mass spectrometry.

Another widely used method is diene conjugation (DC), which reflects the rearrangement of double bonds that occurs early in lipid peroxidation. As shown in figure 7.1, the radical formed by hydrogen abstraction is stabilized by double-bond rearrangement to yield a conjugated diene with characteristic ultraviolet absorption peak at around 233 nm. Many other methods, such as TBA assay or expired air analysis, measure end products of lipid peroxidation rather than the initial phase of the process. Vasankari et al. (1995) indicated that diene conjugation may be a more sensitive method to estimate serum lipid peroxidation induced by exercise, compared to TBA assay and fluorescent chromolipids. However, the assay has some difficulty; for example, any material that absorbs around 230 nm interferes with the assay. Moreover, conjugated dienes may be metabolized rapidly in vivo, and only a small amount of conjugated dienes exist in biological samples. Thus, diene conjugation also cannot be used to analyze tissue or plasma.

Hydrocarbons such as pentane and ethane have been measured as an index of lipid peroxidation and oxidative stress, because they are produced through peroxidation of polyunsaturated fatty acids (Dillard et al., 1977). Pentane and ethane are thought to be end products derived from ω-6 fatty acids (such as linoleic acid and arachidonic acid) and from ω-3 fatty acids (linolenic acid), respectively. However, the hydrocarbon assay cannot identify the source of pentane or ethane, and it may primarily detect hydrocarbons generated by intestinal flora. In addition, many of the methods mentioned above do not specifically detect the hydrocarbon of interest. Again, it is unclear whether the hydrocarbon method can be applied to assess lipid peroxidation in intact organisms.

Recently, more sensitive and specific assay methods, such as high-performance liquid chromatography (HPLC), HPLC-chemiluminescence (HPLC-CL), HPLC-electrochemical detection (HPLC-ECD), and gas chromatography–mass spectrometry (GC-MS), have been developed and adopted to biological materials (Packer, 1984; Packer and Glazer, 1990). Compared with the methods discussed previously, these analytical approaches are complicated and require specialized instrumentation. However, they are the only methods that yield meaningful results with biological material.

Biochemical and Biological Consequences of Lipid Peroxidation

Structures of the three types of lipids are shown in table 7.2. Polyunsaturated fatty acids possess multiple double bonds that are separated by methylene groups. The presence of two double bonds adjacent to a methylene group makes the methylene C-H bonds weaker. So the hydrogens of bis-allylic methylene

---

Table 7.2 Structures of Three Types of Lipid

---

(1) Saturated fatty acid (no double bond)

Example: palmitic acid

$$CH_3(CH_2)_{14}COOH$$

(2) Monounsaturated fatty acid (one double bond)

Example: oleic acid (18:1;9)

$$\overset{18}{C}H_3(CH_2)_7\overset{9}{C}H = \overset{1}{C}H(CH_2)_7COOH$$

(3) Polyunsaturated fatty acid (two or more double bonds)

Example: linoleic acid (18:2;9,12) (two double bonds)

$$\overset{18}{C}H_3(CH_2)_4\overset{12}{C}H = \overset{9}{C}HCH_2CH = \overset{1}{C}H(CH_2)_7COOH$$

Example: α-linolenic acid (18:3;9,12,15) (three double bonds)

$$\overset{18}{C}H_3CH_2CH = \overset{15}{C}HCH_2CH = \overset{12}{C}HCH_2CH = \overset{9}{C}H(CH_2)_7\overset{1}{C}OOH$$

Example: arachidonic acid (20:4;5,8,11,14) (four double bonds)

$$\overset{20}{C}H_3(CH_2)_4\overset{14}{C}H = \overset{}{C}HCH_2CH = \overset{11}{C}HCH_2CH = \overset{8}{C}HCH_2CH = \overset{5}{C}H(CH_2)_3\overset{1}{C}OOH$$

C-H bonds are more vulnerable to abstraction. Compared to saturated fatty acids or monounsaturated fatty acids, polyunsaturated fatty acids in lipids are more easily oxidized to hydroperoxides. The major polyunsaturated fatty acids that undergo lipid peroxidation in cell membranes are linoleic acid (18:2), linolenic acid (18:3), and arachidonic acid (20:4).

Free radicals react readily with biological membranes, because an abundance of polyunsaturated lipids provides the substrate for free radical attack (Cadenas, 1989). Transition metals such as iron and copper, also present in membranes, facilitate the rate of lipid peroxidation (Halliwell, 1984). Peroxidation by definition alters the structure of membrane lipids (Hegner, 1980; Tyler, 1975; Weddle et al., 1976; Yu et al., 1992). It may also disrupt the arrangement and structural organization of the lipid bilayer (Grinna, 1977), perhaps altering membrane fluidity and permeability.

Lipid peroxidation is also thought to be a major source of cytotoxic products, such as aldehydes produced from the decomposition of lipid hydroperoxides (Esterbauer, 1982; Esterbauer et al., 1986, 1987, 1990, 1991). Such aldehydes may be biologically active. For example, 4-hydroxynonenal, a major by-product of arachidonic acid oxidation, is both cytotoxic and mutagenic (Siu and Draper, 1982). Moreover, these aldehydes are capable of forming protein cross-linkages that inactivate many cellular constituents, including membranes and enzymes (Haberland et al., 1988; Hagihara et al., 1984). On the other hand, to protect against potential cytotoxicity (Jain, 1984), cells are endowed with mechanisms to eliminate these aldehydes (Siu and Draper, 1982).

Oxidation of MDA by mitochondrial aldehyde dehydrogenase is a good example of a detoxification process. Kim and Yu (1989) showed that mitochondrial oxidation capacity is substantially lost with age, which might explain the accumulation of MDA-modified and lipofuscin-like materials (Chio and Tappel, 1969; Dillard et al., 1977; Fletcher et al., 1973).

Lipid Peroxidation in Exercise and Aging

It is difficult to define *oxidative stress* precisely, but the term usually implies that free radical generation, lipid peroxidation by-product accumulation, or both exceed basal levels and/or antioxidant capacity to remove them. Mammalian tissues contain enzymatic and nonenzymatic antioxidant defense systems that prevent or minimize oxidative tissue damage. The process of regulating free radical generation and scavenging oxidants in biological systems is very efficient. Some conditions, however, may increase free radical generation, such as cigarette smoking (Scheffler et al., 1992), consuming a diet rich in polyunsaturated fatty acids (Kubow, 1993), and acute intense exercise (Davies et al., 1982). On the other hand, some healthy habits may reduce free radical generation and lipid peroxidation or increase antioxidant defense capacity. For example, vegetarians have higher plasma vitamin E (one of the major antioxidants) concentrations than nonvegetarians (Gey et al., 1987).

Lipid Peroxidation and Exercise in Humans

There is growing evidence that strenuous physical exercise imposes oxidative stress on the body due to oxygen free radical generation, including an increase in lipid peroxidation in various tissues (Clark et al., 1985; Jenkins, 1988; Ji, 1995; Kanter, 1995; Sen, 1995). Therefore, although physical exercise brings many health benefits, too much exercise might be harmful.

Table 7.3 displays findings on the relationship between acute exercise and lipid peroxidation in humans. A number of studies have suggested that a single bout of strenuous exercise increases an index of lipid peroxidation, although other reports have shown no change. For example, Kanter et al. (1993) demonstrated an increase in serum thiobarbituric acid reactive substance (TBARS) and in pentane in expired air after exercise at both 60% $\dot{V}O_2$max (30 min) and 90% $\dot{V}O_2$max (5 min). Laaksonen et al. (1995) reported no change in plasma TBARS after 1 hr bicycle exercise at 60% $\dot{V}O_2$max. Vasankari et al. (1995) evaluated three different assays of lipid peroxidation (TBARS, DC, and fluorescent chromolipids) to determine which method would provide the most reliable index of lipid peroxidation for an exercise-induced oxidative stress. Vasankari et al. concluded that DC might be the most sensitive of the three methods. However, Merry et al. (1991) demonstrated that acute exercise increased TBARS in synovial fluid but did not significantly change the level of synovial fluid conjugated dienes. Because the content of conjugated dienes does not indicate a specific product of lipid peroxidation or provide a conclusive result, the DC of biological samples should be interpreted carefully. The inconsistent data obtained in these studies might be due, in part, to differences in type of sample, exercise conditions (type of work, intensity, or duration), and the index of lipid peroxidation.

In human exercise studies, blood or expired air commonly has been used to assess lipid peroxidation. The production of hydrocarbon gases such as pentane and ethane in expired gas increases consistently with exercise in almost all cases, whereas the data on lipid peroxidation in blood after exercise are not consistent (see table 7.3). Leaf et al. (1997) reported that strenuous exercise increased pentane and ethane in expired breath but not TBARS in serum, which suggested that expired gas analysis may be a more sensitive method to detect exercise-induced lipid peroxidation. As previously noted, the TBA assay is nonspecific when applied to complex biological materials such as plasma. Only a few studies have directly measured exercise-induced lipid peroxidation in human skeletal muscle (Meydani et al., 1993; Ørtenblad et al., 1997; Saxton et al., 1994). Meydani et al. (1993) reported an increase in skeletal muscle conjugated dienes after running exercise (75% of maximum heart rate, 15 min × 3). Few data, however, have been collected concerning the relationship between exercising muscle and lipid peroxidation in humans, and this relationship awaits further study.

Exercise conditions used in previous studies also seemed to contribute to the inconsistent results. Dillard et al. (1978b) demonstrated that higher intensity and longer duration of exercise enhanced lipid peroxidation. In addition, Vasankari et al. (1995) reported increased conjugated dienes in serum with 10 km or 27 km maximal running but no increase in serum diene conjugation with 1 km maximal running. These results suggest that lipid peroxidation is enhanced by higher intensity of exercise and is also influenced by exercise duration.

As shown in table 7.3, three studies (Ginsburg et al., 1996; Kretzschmar et al., 1991; Rokitzki et al., 1994) found a decrease in TBARS in plasma after strenuous exercise. In all of these studies, the subjects were well-trained athletes. Among the rest of the studies in table 7.3, well-trained subjects rarely experienced increased lipid peroxidation during and after exercise. Therefore, these results suggest that endurance training may alter the response of lipid peroxidation to exercise, although a single bout of strenuous exercise may increase lipid peroxidation by increasing the generation of reactive oxygen species. On the

Table 7.3 Acute Exercise Performance and Lipid Peroxidation in Humans

Investigators	Subjects	Sample	Type of work	Marker	Change
Dillard et al. (1978b)	University students	Expired gas	1 h bicycle at 50% max	Pentane	↑
	University students	Expired gas	Graded bicycle exercise	Pentane	↑
	University students (+vit E)	Expired gas	(exhaustive)	Pentane	→
Viinikka et al. (1984)	Long-distance runners	Serum	Exhaustive cycle exercise	TBARS	↑
Lovlin et al. (1987)	Male students	Plasma	Exhaustive cycle exercise	TBARS	←
Kanter et al. (1988)	Male runners	Plasma	50 mi-road race	TBARS	←
Blake et al. (1989)	Individuals with rheumatoid arthritis	SF	Quadriceps contraction	TBARS	←
Sumida et al. (1989)	Healthy male students	Serum	Exhaustive cycle exercise	TBARS	→
	Healthy male students (+vit E)	Serum	Exhaustive cycle exercise	TBARS	↑
Maughan et al. (1989)	Healthy young males	Serum	45 min downhill running	TBARS	←
Duthie et al. (1990)	Regularly exercising males	Plasma	Half-marathon	DC	↑
	Regularly exercising males	Plasma	Half-marathon	TBARS	↑
Cannon et al. (1990)	Sedentary men (<30 yr old)	Plasma	Downhill running	TBARS	↑
	Sedentary men (>55 yr old)	Plasma	(15 min × 3)	TBARS	↑
Pincemail et al. (1990)	Healthy male volunteers	Expired gas	Submaximal cycle exercise	Pentane	←
	Healthy male volunteers (treated with propranolol)	Expired gas	Submaximal cycle exercise	Pentane	←
Merry et al. (1991)	Individuals with rheumatoid arthritis	SF	Quadriceps contraction (2 min)	TBARS	↑
	Individuals with rheumatoid arthritis	SF	or 300-yd walk	DC	→
Buczynski et al. (1991)	Healthy male volunteers	Platelet	Submaximal cycle exercise	TBARS	←
Kanaley & Ji (1991)	Eumenorrheic trained athletes	Plasma	90 min running at 60% max	TBARS	→
	Amenorrheic trained athletes	Plasma	90 min running at 60% max	TBARS	→

(continued)

219

Table 7.3 (continued)

Investigators	Subjects	Sample	Type of work	Marker	Change
Kretzschmar et al. (1991)	Untrained control subjects	Plasma	Exhaustive cycle exercise	TBARS	↑
	Well-trained runners (≤35 yr)	Plasma	Exhaustive cycle exercise	TBARS	→↓
	Well-trained runners (≤36 yr)	Plasma	Exhaustive cycle exercise	TBARS	↓↓
Kanter et al. (1993)	Young nonsmoking males	Expired gas	30 min running at 60% max	Pentane	←
	Young nonsmoking males	Serum	30 min running at 60% max	TBARS	←
	Young nonsmoking males	Expired gas	5 min running at 90% max	Pentane	↑↑
	Young nonsmoking males	Serum	30 min running at 60% max	TBARS	↑↑
	Young nonsmoking males (+vit E & C, β-carotene)	Expired gas	30 min running at 60% max	Pentane	←
	Young nonsmoking males (+vit E & C, β-carotene)	Serum	30 min running at 60% max	TBARS	←
	Young nonsmoking males (+vit E & C, β-carotene)	Expired gas	5 min running at 90% max	Pentane	↑↑
	Young nonsmoking males (+vit E & C, β-carotene)	Serum	30 min running at 60% max	TBARS	↑↑
Meydani et al. (1993)	Sedentary men (<30 yr old)	VLM	Downhill running	DC	←
	Sedentary men (<30 yr old) (+vit E)	VLM	(15 min × 3)	DC	→↑
	Sedentary men (<30 yr old)	Urine	Downhill running	TBARS	↑↑
	Sedentary men (<30 yr old) (+vit E)	Urine	(15 min × 3)	TBARS	←
	Sedentary men (>55 yr old)	Urine	(15 min × 3)	TBARS	↑↑
	Sedentary men (>55 yr old) (+vit E)	Urine	(15 min × 3)	TBARS	←

Reference	Subjects	Sample	Exercise	Measure	
Saxton et al. (1994)	Adult male volunteers	Plasma	Eccentric arm work	TBARS	↑
	Adult male volunteers	Plasma	Eccentric arm work	DC	↑
	Adult male volunteers	Plasma	Concentric arm work	TBARS	↑
	Adult male volunteers	Plasma	Concentric arm work	DC	↑
	Adult male volunteers	VLM	Eccentric knee work	MDA	↑
	Adult male volunteers	VLM	Concentric knee work	MDA	↑
Chen et al. (1994)	Individuals with hypercholesterolemia	Plasma	Exhaustive running	TBARS	←
	Healthy control subjects	Plasma	Exhaustive running	TBARS	←
Rokitzki et al. (1994)	Running-trained males	Plasma	Marathon race	TBARS	→
	Running-trained males (+ vit E & C)	Plasma	Marathon race	TBARS	→
Vasankari et al. (1995)	Well-trained endurance athletes	Serum	1 km maximal running	DC	↑
	Well-trained endurance athletes	Serum	1 km maximal running	FC	↑
	Well-trained endurance athletes	Serum	1 km maximal running	TBARS	↑
	Well-trained endurance athletes	Serum	10 km maximal running	DC	←
	Well-trained endurance athletes	Serum	10 km maximal running	FC	↑
	Well-trained endurance athletes	Serum	10 km maximal running	TBARS	↑
	Well-trained endurance athletes	Serum	27 km maximal running	DC	←
	Well-trained endurance athletes	Serum	8-d bicycle race	DC	←
	Well-trained endurance athletes	Serum	8-d bicycle race	FC	↑
	Well-trained endurance athletes	Serum	8-d bicycle race	TBARS	↑
Kedziora et al. (1995)	Healthy men	Platelet	2 watt/kg bicycle (20 min)	TBARS	↑
Hartmann et al. (1995)	Male nonsmokers	Serum	Exhaustive running	TBARS	↑
	Male nonsmokers (+vit E)	Serum	Exhaustive running	TBARS	↑
Laaksonen et al. (1995)	Young men	Plasma	1 h cycling at 60% max	TBARS	↑
	Older men	Plasma	1 h cycling at 60% max	TBARS	→

(continued)

Table 7.3 *(continued)*

Investigators	Subjects	Sample	Type of work	Marker	Change
Inayama et al. (1996)	Moderately trained male students	Plasma	Full marathon	TBARS	↑
Laaksonen et al. (1996)	Young healthy male	Plasma	40 min cycling at 60% max	TBARS	←
	Young IDDM patients	Plasma	40 min cycling at 60% max	TBARS	←
Ginsburg et al. (1996)	Well-conditioned amateur athletes	Plasma	Triathlon race	TBARS	↑
	Well-conditioned amateur athletes (+vit E)	Plasma	Triathlon race	TBARS	↑
Ørtenblad et al. (1997)	Untrained male	Plasma	Jumping	MDA	↑
	Untrained male	VLM	Jumping	MDA	↑
	Elite volleyball player	Plasma	Jumping	MDA	↑
	Elite volleyball player	VLM	Jumping	MDA	↑
Margaritis et al. (1997)	Well-trained triathletes	Plasma	Triathlon race	TBARS	↑
Leaf et al. (1997)	Healthy volunteers	Expired gas	Strenuous exercise	Ethane	←
	Healthy volunteers	Expired gas	Strenuous exercise	Pentane	←
	Healthy volunteers	Serum	Strenuous exercise	TBARS	↑
Vasankari et al. (1997a)	Male top-level skiers	Serum	Cross-country or biathlon	DC	↑
Vasankari et al. (1997b)	Male endurance runners	Serum	31 km running	DC	←
	Male keep-fit marathon runners	Serum	Marathon race	DC	↑

Note. TBARS, thiobarbituric acid reactive substance; DC, diene conjugation; MDA, malondialdehyde; FC, fluorescent chromolipids; SF, synovial fluid; VLM, vastus lateralis muscle. ↑ = increased; ↑↑ = markedly increased; → = no change; ↓ = decreased; ↓↓ = markedly decreased.

other hand, Jenkins et al. (1984) reported increased activities of superoxide dismutase (SOD) and catalase (CAT) in vastus lateralis muscles (VLM) of well-trained athletes. Ørtenblad et al. (1997) also reported increased activities of SOD and glutathione peroxidase (GPX) in VLM of elite volleyball players. Because the major antioxidant enzymes such as SOD, GPX, and CAT form the first anti-oxidant defense line against free radical attack, increasing the levels of anti-oxidant enzymes by exercise training would prevent or attenuate damaging lipid peroxidation during exercise. In addition, Vasankari et al. (1997b) reported that well-trained skiers increased not only serum antioxidant potential but also serum α-tocopherol during acute exercise. Therefore, the exercise-induced increase in serum antioxidant capacity in well-trained subjects may be due, in part, to the increase of serum α-tocopherol. However, the precise mechanism for the increase of serum antioxidant capacity or α-tocopherol remains unclear.

Kanter et al. (1993) reported that daily ingestion of an antioxidant vitamin mixture (α-tocopherol, β-carotene, and ascorbic acid) for 6 wk increased human serum vitamin concentration and significantly reduced the absolute l evels of serum TBARS and expired pentane, both at rest and after moderate or heavy exercise. However, the increased level of vitamins failed to prevent the exercise-induced increase in both serum TBARS and expired pentane. Meydani et al. (1993) reported almost the same findings as Kanter et al. Vitamin E, a peroxyl and hydroxyl scavenger, is considered the most important lipid-soluble antioxidant for protecting lipids (Bjorneboe et al., 1990) and is located in membranes and other lipids. Vitamin E can act as a lipid peroxidation chain-breaking antioxidant by providing hydrogen atoms that block lipid peroxides from further reaction. β-Carotene can also act as a chain-breaking antioxidant by quenching superoxides and other free radicals (Krinsky, 1989). Vitamin C (ascorbic acid) is an important antioxidant that directly scavenges aqueous peroxyl radicals (Frei et al., 1989). Vitamin C also regenerates oxidized vita-min E to its reduced form (Bendich et al., 1986). Many of the studies on hu-mans have indicated that supplementation of antioxidant vitamins such as vi-tamin E or vitamin C may prevent or minimize oxidative stress, lipid peroxidation, or both (Cannon et al., 1990; Dillard et al., 1978b; Hartmann et al., 1995; Kanter et al., 1993; Meydani et al., 1993; Sumida et al., 1989; Vasankari et al., 1997a).

As mentioned, chronic exercise increases the level of antioxidant enzymes in human skeletal muscle and therefore may reduce lipid peroxidation in tis-sues. Ginsburg et al. (1996) reported that chronic exercise decreased plasma TBARS. However, only a few data are available on this issue (table 7.4), and further human research is needed to clarify the effect of chronic exercise on lipid peroxidation.

Table 7.4 Chronic Exercise Performance and Lipid Peroxidation in Humans

Investigators	Subjects	Sample	Type of work	Marker	Change
Viinikka et al. (1984)	Males & females	Serum	Long-distance running	TBARS	→
Kretzschmar et al. (1991)	Males & females (≤35)	Plasma	Long-distance running	TBARS	→
	Males & females (≥36)	Plasma	Long-distance running	TBARS	→
Ginsburg et al. (1996)	Males & females	Plasma	Triathlon	TBARS	↓

Lipid Peroxidation and Exercise in Animals

In animals, a number of studies have demonstrated the direct measurement of tissue lipid peroxidation with exercise during the last two decades. We briefly summarize previous data, mainly focusing on skeletal muscles.

Acute Exercise

As shown in table 7.5, many studies have examined the effect of acute exercise on lipid peroxidation in animal tissues. Many studies indicate that a single bout of strenuous exercise induces lipid peroxidation in various tissues, although the results are sometimes conflicting. Frankiewicz-Józko et al. (1996) reported that exhaustive running increased TBARS in liver, soleus muscle, and heart of untrained rats. The authors also found that exhaustive swimming increased TBARS in liver but not in soleus muscle or heart of trained rats. Venditti and Meo (1996) showed that exhaustive swimming increased lipid peroxides in gastrocnemius muscle, heart, and liver of both of untrained and trained rats. Moreover, Koz et al. (1992) demonstrated that a longer duration of swimming induced greater lipid peroxidation in skeletal muscles. Alessio and Goldfarb (1988b) suggested that lipid peroxidation is exacerbated by higher intensities of exercise. However, other studies did not show exercise-induced lipid peroxidation (Ji et al., 1992; Leeuwenburgh and Ji, 1996). The difference among these studies might be due, in part, to type of work, training, conditions, or tissues used.

Alessio and Goldfarb (1988a) demonstrated that acute, but not exhaustive, running exercise increased the level of TBARS in white VLM but not in red VLM of untrained rats, whereas the same exercise intensity did not increase

Table 7.5 Acute Exercise Performance and Lipid Peroxidation in Animals

Investigators	Subjects	Sample	Type of work	Marker	Change
Brady et al. (1979)	Male rats (control diet)	Liver	Exhaustive swimming	TBARS	↑
	Male rats (control diet) (−Se, +vit E)	Liver	Exhaustive swimming	TBARS	↑
	Male rats (control diet) (+Se, −vit E)	Liver	Exhaustive swimming	TBARS	↑↑
	Male rats (control diet) (+Se, +vit E)	Liver	Exhaustive swimming	TBARS	↑
	Male rats (control diet) (control diet)	Hindlimb muscle	Exhaustive swimming	TBARS	←
	Male rats (control diet) (−Se, +vit E)	Hindlimb muscle	Exhaustive swimming	TBARS	←
	Male rats (control diet) (+Se, −vit E)	Hindlimb muscle	Exhaustive swimming	TBARS	←
	Male rats (control diet) (+Se, +vit E)	Hindlimb muscle	Exhaustive swimming	TBARS	←
Gee & Tappel (1981)	Male rats (standard diet)	Expired gas	Exhaustive swimming	Pentane	←
	Male rats (standard diet)	Expired gas	Exhaustive swimming	Ethane	↑
	Male rats (+vit E)	Expired gas	Exhaustive swimming	Pentane	←
	Male rats (+vit E)	Expired gas	Exhaustive swimming	Ethane	↑
	Male rats (−vit E)	Expired gas	Exhaustive swimming	Pentane	↑
	Male rats (−vit E)	Expired gas	Exhaustive swimming	Ethane	↑

(continued)

Table 7.5 *(continued)*

Investigators	Subjects	Sample	Type of work	Marker	Change
Davies et al. (1982)	Male rats (control diet)	Hindlimb muscle	Exhaustive running	TBARS	↑↑
	Male rats (control diet)	Liver	Exhaustive running	TBARS	↑↑
	Male rats (–vit E)	Hindlimb muscle	Exhaustive running	TBARS	↑
	Male rats (–vit E)	Liver	Exhaustive running	TBARS	↑
Salminen & Vihko (1983)	Male mice (6.5 mo)	Red QFM	Prolonged running (9 h)	Lipofuscin	↑
	Male mice (6.5 mo)	Red QFM	Prolonged running (9 h)	Rate of LP	←
	Male mice (6.5 mo)	Red QFM	Exhaustive running	TBARS	↑
	Male mice (6.5 mo)	Red QFM	Exhaustive running	Lipofuscin	↑
	Male mice (6.5 mo)	Red QFM	Exhaustive running	Rate of LP	←
Ji et al. (1988b)	Male untrained rats (–Se)	Liver	1 h running	TBARS	↑
	Male untrained rats (–Se)	Skeletal muscle	1 h running	TBARS	↑
	Male untrained rats (+Se)	Liver	1 h running	TBARS	←
	Male untrained rats (+Se)	Skeletal muscle	1 h running	TBARS	↑
	Male untrained rats (–Se)	Liver	1 h running	TBARS	↑
	Male untrained rats (–Se)	Skeletal muscle	1 h running	TBARS	↑
	Male untrained rats (+Se)	Liver	1 h running	TBARS	↑
	Male untrained rats (+Se)	Skeletal muscle	1 h running	TBARS	↑
Alessio & Goldfarb (1988)	Untrained rats	Red VLM	20 min running	TBARS	↑
	Untrained rats	White VLM	20 min running	TBARS	←
	Untrained rats	Liver	20 min running	TBARS	←
	Trained rats	Red VLM	20 min running	TBARS	↑
	Trained rats	White VLM	20 min running	TBARS	↑
	Trained rats	Liver	20 min running	TBARS	↑

Reference	Subject	Tissue	Exercise	Measure	Result
Alessio et al. (1988)	Male rats (6 mo)	Red VLM	20 min running	TBARS	↑
	Male rats (6 mo)	White VLM	(20 m/min)	TBAR	↑
	Male rats (6 mo)	Red VLM	1 min running	TBARS	↑↑
	Male rats (6 mo)	White VLM	(45 m/min)	TBAR	↑↑
	Male rats (6 mo)	Soleus muscle	(45 m/min)	TBAR	↑
	Male rats (6 mo)	Red VLM	(45 m/min)	LOOH	↑
	Male rats (6 mo)	White VLM	(45 m/min)	LOOH	↑
Vani et al. (1990)	Adult healthy rats	Liver	30 min swim (×1 d)	TBARS	↑
	Adult healthy rats	Liver	30 min swim (×10 d)	TBARS	↑↑
Koz et al. (1992)	Male rats	GCM	60 min swimming	TBARS	↑
	Male rats	VMM	60 min swimming	TBARS	↑
	Male rats	TBM	60 min swimming	TBARS	↑
	Male rats	Erythrocyte	60 min swimming	TBARS	↑
	Male rats	GCM	90 min swimming	TBARS	↑↑
	Male rats	VMM	90 min swimming	TBARS	↑↑
	Male rats	TBM	90 min swimming	TBARS	↑↑
	Male rats	Erythrocyte	90 min swimming	TBARS	↑
	Male rats	GCM	120 min swimming	TBARS	↑↑
	Male rats	VMM	120 min swimming	TBARS	↑↑
	Male rats	TBM	120 min swimming	TBARS	↑↑
	Male rats	Erythrocyte	120 min swimming	TBARS	↑
Reddy et al. (1992)	Female rats (control)	Lung	Exhaustive swimming	TBARS	↑
	Female rats (control) (+Se)	Lung	Exhaustive swimming	TBARS	↑
	Female rats (control) (+vit E)	Lung	Exhaustive swimming	TBARS	↑
	Female rats (control) (+Se, +vit E)	Lung	Exhaustive swimming	TBARS	↑

(continued)

Table 7.5 *(continued)*

Investigators	Subjects	Sample	Type of work	Marker	Change
Ji et al. (1992)	Male untrained rats (–Se)	Heart	1 h running	TBARS	↑
	Male untrained rats (+Se)	Heart	1 h running	TBARS	↑
	Male trained rats (–Se)	Heart	1 h running	TBARS	↑
	Male trained rats (+Se)	Heart	1 h running	TBARS	↑
Tiidus & Houston (1993)	Female rats (16 wk)	Plasma	45 min running	TBARS	↑
	Female rats (16 wk) (–vit E)	Plasma	45 min running	TBARS	↑ ←
Rajguru et al. (1993)	Male rats (7-8 wk)	Cardiac muscle	Exhaustive swimming	TBARS	←
	Male rats (7-8 wk)	Skeletal muscle	Exhaustive swimming	TBARS	←
	Male rats (7-8 wk)	Erythrocyte	Exhaustive swimming	TBARS	↑
Goldfarb et al. (1994)	Male rats (control)	Plasma	1 h running	TBARS	←
	Male rats (control)	Plasma	1 h running	LOOH	←
	Male rats (control) (+vit E)	Plasma	1 h running	TBARS	←
	Male rats (control) (+vit E)	Plasma	1 h running	LOOH	↑
	Male rats (control) (+DHEA)	Plasma	1 h running	TBARS	↑↑
	Male rats (control) (+DHEA)	Plasma	1 h running	LOOH	←
	Male rats (control) (+vit E, +DHEA)	Plasma	1 h running	TBARS	←
	Male rats (control) (+vit E, +DHEA)	Plasma	1 h running	LOOH	↑
	Male rats (control)	White QFM	1 h running	TBARS	↑↑
	Male rats (control) (+vit E)	White QFM	1 h running	TBARS	↑
	Male rats (control) (+DHEA)	White QFM	1 h running	TBARS	↑ ←

Reference	Group	Tissue	Exercise	Marker	Result
	Male rats (control) (+vit E, +DHEA)	White QFM	1 h running	TBARS	↑
	Male rats (control)	Red QFM	1 h running	TBARS	←
	Male rats (control) (+vit E)	Red QFM	1 h running	TBARS	↑
	Male rats (control) (+DHEA)	Red QFM	1 h running	TBARS	↑
	Male rats (control) (+vit E, +DHEA)	Red QFM	1 h running	TBARS	↑↑
	Male rats (control)	Soleus muscle	1 h running	TBARS	↑↑
	" (+vit E)	Soleus muscle	1 h running	TBARS	↑
	" (+DHEA)	Soleus muscle	1 h running	TBARS	↑
	" (+vit E, +DHEA)	Soleus muscle	1 h running	TBARS	↑ ←
Lawler et al. (1994)	Female rats (4 mo)	Costal diaphragm	40 min running	TBARS	↑
	Female rats (4 mo)	Crural diaphragm	(22 m/min)	TBARS	↑
	Female rats (24 mo)	Costal diaphragm	40 min running	TBARS	↑
	Female rats (24 mo)	Crural diaphragm	(14.5 m/min)	TBARS	↑
Ji & Mitchell (1994)	Female rats (control)	Heart	Exhaustive running	TBARS	↑
	Female rats (control) (treated with ADM)	Heart	Exhaustive running	TBARS	←
Mills et al. (1994)	Thoroughbred horses	Urine	Strenuous running	MDA	↑
Somani et al. (1995)	Untrained rats	Heart (cytosol)	Running at 100% max	TBARS	↑
	Trained rats	Heart (mitochondria)	Running at 100% max	TBARS	↑
Leeuwenburgh & Ji (1995)	GSH-adequate mice	Liver	Exhaustive swimming	TBARS	←
	GSH-adequate mice	Kidney	Exhaustive swimming	TBARS	↑
	GSH-adequate mice	QFM	Exhaustive swimming	TBARS	↑↑
	GSH-depleted mice	Liver	Exhaustive swimming	TBARS	↑↑
	GSH-depleted mice	Kidney	Exhaustive swimming	TBARS	↑
	GSH-depleted mice	QFM	Exhaustive swimming	TBARS	←

(continued)

Table 7.5 *(continued)*

Investigators	Subjects	Sample	Type of work	Marker	Change
Venditti & Meo (1996)	Male untrained rats	GCM	Exhaustive swimming	TBARS	↑
	Male untrained rats	GCM	Exhaustive swimming	Hydroperoxides	↑
	Male untrained rats	Heart	Exhaustive swimming	TBARS	↑
	Male untrained rats	Heart	Exhaustive swimming	Hydroperoxides	↑
	Male untrained rats	Liver	Exhaustive swimming	TBARS	↑
	Male untrained rats	Liver	Exhaustive swimming	Hydroperoxides	↑
	Male trained rats	GCM	Exhaustive swimming	TBARS	↑
	Male trained rats	GCM	Exhaustive swimming	Hydroperoxides	↑
	Male trained rats	Heart	Exhaustive swimming	TBARS	↑
	Male trained rats	Heart	Exhaustive swimming	Hydroperoxides	↑
	Male trained rats	Liver	Exhaustive swimming	TBARS	↑
	Male trained rats	Liver	Exhaustive swimming	Hydroperoxides	↑
Leeuwenburgh & Ji (1996)	48 h–unfed rats	Deep VLM	Exhaustive running	TBARS	↑
	(without refeeding)	Liver	Exhaustive running	TBARS	↑
	Refed–24 h rats	Deep VLM	Exhaustive running	TBARS	↑
	Refed–24 h rats	Liver	Exhaustive running	TBARS	↑
	Refed–48 h rats	Deep VLM	Exhaustive running	TBARS	↑
	Refed–48 h rats	Liver	Exhaustive running	TBARS	↑
Radák et al. (1996)	Male rats	Liver	Exhaustive running	TBARS	←
	Male rats	Kidney	Exhaustive running	TBARS	←
	Male rats	Liver	Exhaustive running	TBARS	↑
	(treated with SM–SOD)	Kidney	Exhaustive running	TBARS	↑

Reference	Subject	Tissue	Exercise	Marker	Effect
Hara et al. (1996)	Male rats	Liver	30 min running	MDA + HDA	↑
	Male rats	Brain	30 min running	MDA + HDA	↑
	Male rats	QFM	30 min running	MDA + HDA	←
	Male rats	Liver	30 min running	MDA + HDA	↑
	(treated with melatonin)	Brain	30 min running	MDA + HDA	↑
	(treated with melatonin)	QFM	30 min running	MDA + HDA	↑
Benderitter et al. (1996)	Male swim-trained rats	Plasma	Exhaustive swimming	TBARS	↑
	Male swim-trained rats	Heart	Exhaustive swimming	TBARS	↑
	Male swim-trained rats	GCM	Exhaustive swimming	TBARS	→
	Male swim-trained rats	Liver	Exhaustive swimming	TBARS	←
Frankiewicz-Józko et al. (1996)	Male untrained rats	Liver	Exhaustive running	TBARS	←
	Male untrained rats	Soleus muscle	Exhaustive running	TBARS	←
	Male untrained rats	Heart	Exhaustive running	TBARS	←
	Male trained rats	Liver	Exhaustive running	TBARS	←
	Male trained rats	Soleus muscle	Exhaustive running	TBARS	↑
	Male trained rats	Heart	Exhaustive running	TBARS	↑
Sen et al. (1997)	Male rats	VLM	Exhaustive running	TBARS	↑
	(soy oil + vit E)	Red GCM	Exhaustive running	TBARS	←
	(soy oil + vit E)	Liver	Exhaustive running	TBARS	←
	Male Wistar rats	VLM	Exhaustive running	TBARS	↑
	(fish oil + vit E)	Red GCM	Exhaustive running	TBARS	←
	(fish oil + vit E)	Liver	Exhaustive running	TBARS	←

(continued)

Table 7.5 *(continued)*

Investigators	Subjects	Sample	Type of work	Marker	Change
Faff & Frankiewicz-Józko (1997)	Male rats	Liver	Exhaustive running	TBARS	↑
	Male rats	Heart	Exhaustive running	TBARS	↑
	Male rats	Red GCM	Exhaustive running	TBARS	↑
	Male rats	White GCM	Exhaustive running	TBARS	↑
	Male Wistar rats	Liver	Exhaustive running	TBARS	↑
	(treated with CoQ[10])	Heart	Exhaustive running	TBARS	↑
	(treated with CoQ[10])	Red GCM	Exhaustive running	TBARS	↑↑
	(treated with CoQ[10])	White GCM	Exhaustive running	TBARS	↑

Note. TBARS, thiobarbituric acid reactive substance; QFM, quadriceps muscle; LP, lipid peroxidation; VLM, vastus lateralis muscle; LOOH, lipid hydroperoxide; GCM, gastrocnemius muscle; VMM, vastus medialis muscle; TBM, triceps brachii muscle; ADM, adriamycin; MDA, malondialdehyde; HDA, 4-hydroxyalkenals; DHEA, dehydroepiandrosterone; CoQ[10], coenzyme Q[10]; SM–SOD, poly-(styrene-co-maleic acid butyl ester)–superoxide dismutase; Se, selenium; GSH, glutathione. ↑ = increased; ↑↑ = markedly increased; → = no change; ↓ = decreased; ↓↓ = markedly decreased.

TBARS in either red or white VLM of trained rats. Previous reports have shown that slow-twitch oxidative fibers (type I) contain higher antioxidant enzymes (SOD, GPX, and CAT) than do fast-twitch fibers (type IIa or IIb) (Jenkins et al., 1984; Ji et al., 1988a; Lawler et al., 1993; Oh-ishi et al., 1995). Therefore, the level of cellular antioxidant enzymes in tissues also appears to be related to the extent of exercise-induced lipid peroxidation.

Lipid peroxidation in tissues might be influenced by antioxidant status. Studies have indicated that muscle and liver mitochondria from vitamin E–deficient rats are more sensitive to oxidative stress (Jackson et al., 1983; Quintanilha et al., 1982). Vitamin E deficiency in animals increases exercise-induced oxidative stress (Brady et al., 1979; Davies et al., 1982). Consequently, vitamin E deficiency seems to reduce endurance capacity (Davies et al., 1982; Gohil et al., 1986). In addition, Anzueto et al. (1993) showed that vitamin E deficiency enhances lipid peroxidation, activates the glutathione redox cycle, and causes early fatigue in the diaphragm muscle (indicated as the reduction of contractile function) during resistive breathing in rats. On the other hand, vitamin E supplementation diminishes lipid peroxidation in animals both at rest and after acute exercise (Goldfarb et al., 1994; Kumar et al., 1992), as mentioned before in human studies. However, these beneficial effects were not observed in all studies (Gee and Tappel, 1981; Reddy et al., 1992). Although supplementation of other antioxidants such as coenzyme Q_{10}, melatonin, and SOD derivative diminished lipid peroxidation to greater or lesser degrees (Faff and Frankiewicz-Józko, 1997; Hara et al., 1996; Radák et al., 1996), the effects of supplementation of these antioxidants on lipid peroxidation have not been studied adequately.

Chronic Exercise

As shown in table 7.6, some studies have demonstrated that exercise training reduces the level of lipid peroxidation in various tissues at rest, although other studies have not. Frankiewicz-Józko et al. (1996) showed that 4 wk of running training decreased TBARS in soleus muscle and heart but increased TBARS in liver, which suggests that the effect of chronic exercise on tissue lipid peroxidation may depend on target tissues. Pereira et al. (1994) also reported a decrease of TBARS in mesenteric lymph nodes, thymus, and spleen but no change in white gastrocnemius muscle and soleus muscle. As stated previously, the main effect of exercise training on lipid peroxidation may be not to lower the basement level of lipid peroxidation but to improve the resistance of tissues to exercise-induced lipid peroxidation.

Growing evidence suggests that physical endurance training up-regulates the level of antioxidant enzymes in tissues actively involved in exercise (Criswell et al., 1993; Higuchi et al., 1985; Ji et al., 1988b; Oh-ishi et al., 1996, 1997a, 1997b; Ohno et al., 1994), and that this may decrease susceptibility to oxidative stress. However, vitamin E concentration is known to decrease in skeletal muscle, liver, and heart in rats after endurance training (Packer et al., 1989; Quintanilha, 1984; Tiidus and Houston, 1993). These findings may support Packer's (1991) recommendation that active humans should increase daily dietary intake of vitamin E.

Lipid Peroxidation in Aging

Since Harman (1956) proposed the free radical theory of aging, evidence has accumulated concerning the relationship between aging and oxidative stress. Many studies have shown an age-related increase of lipid peroxides in various tissues such as brain, heart, liver, kidney, and skeletal muscle (Ji et al., 1991; Laganiere and Yu, 1989; Leeuwenburgh et al., 1994; Xia et al., 1995). For example, Xia et al. (1995) reported that the levels of TBARS in brain cortex, heart, liver, and kidney increased with age. The same results were found in heart muscle (Ji et al., 1991) and in skeletal muscle (Leeuwenburgh et al., 1994). Furthermore, it is well established that mitochondria isolated from the heart (Nohl and Hegner, 1978) and flight muscle (Sohal and Dubey, 1994) have a higher rate of superoxide anion production and generate more hydrogen peroxide with aging. Together, these findings suggest that oxidative stress increases with age in many important organs.

The increased oxidative stress with age could alter antioxidant enzyme levels. Although many studies have examined the effects of aging on antioxidant enzymes, the data are conflicting. Rao et al. (1990) demonstrated that the levels of one or more enzymes of SOD, GPX, and CAT decreased with age in the brain, heart, hepatocytes, intestinal mucosa, and kidney. Their results imply that tissue antioxidant capacity decreases with age, possibly resulting in an age-related increase of lipid peroxidation. Altered enzyme molecules are known to accumulate in aging animals, based on the error catastrophe theory (Orgel, 1970) or modification of the primary structure (Hardt and Rothstein, 1985), and the decreased antioxidant capacity in aged animals may be explained, in part, by this phenomenon. Another possibility is that an age-related increase of reactive oxygen species directly inactivates antioxidant enzymes with the sulfhydryl (SH) group such as SOD. On the other hand, Reiss and Gershon (1976) reported that the level of cytoplasmic SOD from rat and mouse livers decreased with age, whereas no such decrease was observed in brains and hearts. Ji et al. (1991) showed age-related decreases in copper- and zinc-containing superoxide dismutase (CuZn-SOD), GPX, and CAT of rat hearts and an age-related increase in manganese-containing superoxide dismutase (Mn-SOD) of rat hearts. In addition, Danh et al. (1983) reported that liver Mn-SOD activities decreased with age but brain Mn-SOD activities increased with age. Therefore, age-related alterations of antioxidant enzymes

Table 7.6 Chronic Exercise Performance and Lipid Peroxidation in Animals

Investigators	Subjects	Sample	Type of work	Marker	Change
Alessio & Goldfarb (1988)	Rats	Red VLM	18 wk running	TBARS	↓
	Rats	White VLM	18 wk runnin	TBARS	↑
	Rats	Liver	18 wk runnin	TBARS	↑
Starnes et al. (1989)	Male rats (12 mo)	GCM	6 mo running	TBARS	↑
	Male rats (24 mo)	GCM	6 mo running	TBARS	↑
Vani et al. (1990)	Healthy rats	Liver	60 d swimming	TBARS	↑↑
Cao & Chen (1991)	Zn-deficient mice	Hepatic homogenate	6 wk swimming	TBARS	↑
	Zn-deficient mice	Hepatic soluble fraction	6 wk swimming	TBARS	↑
	Zn-deficient mice	Hepatic mitochondria	6 wk swimming	TBARS	←
	Pair-fed mice	Hepatic homogenate	6 wk swimming	TBARS	←
	Pair-fed mice	Hepatic soluble fraction	6 wk swimming	TBARS	←
	Pair-fed mice	Hepatic mitochondria	6 wk swimming	TBARS	←
	Mice (fed ad libitum)	Hepatic homogenate	6 wk swimming	TBARS	↑
	Mice (fed ad libitum)	Hepatic soluble fraction	6 wk swimming	TBARS	↑
	Mice (fed ad libitum)	Hepatic mitochondria	6 wk swimming	TBARS	↑
Ji et al. (1992)	Male untrained rats (Se-deficient)	Heart	9 wk running	TBARS	↑
	Male untrained rats (Se-sufficient)	Heart	9 wk running	TBARS	↑

(continued)

235

Table 7.6 (continued)

Investigators	Subjects	Sample	Type of work	Marker	Change
Leeuwenburgh et al. (1994)	Rats (4.5 mo)	VLM	10 wk running	TBARS	↑
	Rats (4.5 mo)	Soleus muscle	10 wk runnin	TBARS	↑
	Rats (14.5 mo)	VLM	10 wk runnin	TBARS	↑
	Rats (14.5 mo)	Soleus muscle	10 wk runnin	TBARS	↑ →
	Rats (26.5 mo)	VLM	10 wk runnin	TBARS	→
	Rats (26.5 mo)	Soleus muscle	10 wk runnin	TBARS	→
Pereira et al. (1994)	Male rats	Mesenteric lymph nodes	8 wk swimming	TBARS	→
	Male rats	Thymus	8 wk swimming	TBARS	→
	Male rats	Spleen	8 wk swimming	TBARS	↑
	Male rats	White GCM	8 wk swimming	TBARS	↑
	Male rats	Soleus muscle	8 wk swimming	TBARS	←
Hong & Johnson (1995)	Male SHR* rats	Left ventricle	10 wk running	TBARS	→
	Male SHR rats	LDM	10 wk running	TBARS	→
	Male SHR rats	QFM	10 wk running	TBARS	→
	Male SHR rats	Liver	10 wk running	TBARS	↑
	Male SHR rats	Kidney	10 wk running	TBARS	↑
	Male WKY* rats	Left ventricle	10 wk running	TBARS	→
	Male WKY rats	LDM	10 wk running	TBARS	↑
	Male WKY rats	QFM	10 wk running	TBARS	↑
	Male WKY rats	Liver	10 wk running	TBARS	↑
	Male WKY rats	Kidney	10 wk running	TBARS	↑ ←

Study	Subjects	Tissue	Protocol	Marker	Change
Somani et al. (1995)	Rat	Heart (cytosol)	10 wk running	TBARS	→
	Rat	Heart (mitochondria)	10 wk running	TBARS	↑
Venditti & Meo (1996)	Male rats	GCM	10 wk running	TBARS	↑
	Male rats	GCM	10 wk running	LOOH	↑
	Male rats	Heart	10 wk running	TBARS	↑
	Male rats	Heart	10 wk running	LOOH	↑
	Male rats	Liver	10 wk running	TBARS	↑
	Male rats	Liver	10 wk running	LOOH	→
Kim et al. (1996)	Rats (sedentary)	Cardiac mitochondrial membrane	18.5 mo wheel running	TBARS	→
	Rats (sedentary) (food restricted)	Cardiac mitochondrial membrane	18.5 mo wheel running	TBARS	→
Frankiewicz-Józko et al. (1996)	Male rats	Liver	4 wk running	TBARS	←
	Male rats	Soleus muscle	4 wk running	TBARS	→
	Male rats	Heart	4 wk running	TBARS	→
Husain & Somani (1997)	Rats	Heart	6.5 wk running	TBARS	↑
	Male rats	White QFM	4 wk running	TBARS	↑
	Male rats	White QFM	(sea level)	LOOH	↑
	Male rats	Red QFM	(sea level)	TBARS	↑
	Male rats	Red QFM	(sea level)	LOOH	↑
	Male rats	White QFM	4 wk running	TBARS	↑
	Male rats	White QFM	(high altitude)	LOOH	↑
	Male rats	Red QFM	(high altitude)	TBARS	↑
	Male rats	Red QFM	(high altitude)	LOOH	↑

Note. VLM, vastus lateralis muscle; TBARS, thiobarbituric acid reactive substance; GCM, gastrocnemius muscle; QFM, quadriceps muscle; LOOH, lipid hydroperoxide; LDM, longissimus dorsi muscle; Se, selenium; ↑↑ = markedly increased; ↑ = increased; → = no change; ↓ = decreased; ↓↓ = markedly decreased. SHR, spontaneously hypertensive rats; WKY, normotensive rat. *SHR and WKY are symbols used in Hong & Johnson's work.

might depend on the properties of tissues or enzymes (including a type of isoenzyme and its in vivo distribution).

In skeletal muscles, some antioxidant enzymes are up-regulated with age (Ji et al., 1991; Leeuwenburgh et al., 1994; Oh-ishi et al., 1995, 1996). This might be an adaptive phenomenon to cope with an age-related increase of oxidative stress. The level of the muscle antioxidant enzymes is thought to be closely related to the oxidative capacity of the muscle cell, because the type I fibers contain higher antioxidant enzymes than type IIa and IIb fibers, as mentioned before. Therefore, the effect of aging on skeletal muscle might be different in each muscle fiber type. Indeed, our study (Oh-ishi et al., 1995) revealed an age-related increase of CuZn-SOD, GPX, and CAT in soleus muscle (rich in type I fibers) but only an increase of CuZn-SOD in extensor digitorum longus muscle (rich in type II fibers). On the other hand, physical endurance training is well established to up-regulate the level of antioxidant enzymes in exercising muscles. However, several studies have suggested that endurance training does not up-regulate antioxidant enzymes in skeletal muscle of aged animals (Oh-ishi et al., 1996; Powers et al., 1992). It is possible that the elevated basal level of the antioxidant enzyme system in aged animals prevents further training-induced increases.

Oxidative Stress and LDL Oxidation

Atherosclerotic vascular disease, the leading cause of coronary artery disease and stroke, is the major life-threatening disease of aging populations with a Western lifestyle. Although it is well established that elevated plasma concentration of LDL is associated with accelerated atherogenesis (Goldstein and Brown, 1977; Steinberg, 1983; Tyroler 1987), the underlying mechanisms remain to be clarified. The so-called oxidation hypothesis of atherosclerosis (Ross, 1993; Steinbrecher et al., 1990) states that oxidative modification of LDL is important in the pathogenesis of atherosclerosis. Growing evidence suggests that oxidative modification of LDL may make it atherogenic (Regnström et al., 1992; Salonen et al., 1992; Steinberg, 1993; Witztum, 1994). Regnström et al. (1992) also found that oxidation of LDL is associated with the severity of coronary atherosclerosis.

Oxidized LDL (Ox-LDL)

LDL consists of a single major protein (apolipoprotein B); neutral lipids; phospholipids; and lipid-soluble antioxidants, mainly vitamin E and ß-carotene. Lipoprotein oxidation seems unlikely to occur in plasma because of the presence of high concentrations of antioxidants and proteins that chelate metal ions. Oxidative modification of LDL is assumed to occur primarily in the intima of the artery, in protected microenvironments sequestered from plasma antioxidants (Witztum, 1994). Lipid peroxidation presumably starts in polyunsaturated fatty acids of phospholipids in the surface of LDL and then spreads into the core lipids of the particle. With extensive oxidation, modification of the polyunsaturated fatty acids, cholesterol and cholesteryl ester, and apolipoprotein B becomes apparent (Steinbrecher et al., 1990; Witztum, 1993). In vitro, this process generates numerous biologically active molecules, including oxidized sterols, oxidized fatty acids, and breakdown products of oxidized fatty acids such as malondialdehyde and 4-hydroxynonenal. In addition, reactive aldehydes generated during lipid peroxidation can react with lysine residues in apolipoprotein B. Consequently, modification of apolipoprotein B may generate new epitopes on apolipoprotein B, which might be recognized by the macrophage scavenger receptor.

Berliner and Heinecke (1996) summarized a wide variety of possible mechanisms for LDL oxidation in vitro. For example, reactive oxygen species generated by cells such as smooth muscle cells, endothelial cells, or macrophages promote lipid peroxidation by reactions that are metal ion–dependent, suggesting that they might cause LDL oxidation. Lipoxygenases are localized in the cytosol of smooth muscle cells, endothelial cells, and macrophages, and they can directly oxidize polyunsaturated fatty acids, although the mechanism by which an intracellular enzyme could oxidize LDL is still controversial. Nitric oxide (NO), a major regulator of vascular tone in muscular arteries generated by endothelial cells, can also contribute to LDL oxidation by interacting with superoxide and forming peroxynitrite ($ONOO^-$). Peroxynitrite is a strong oxidant that has a similar reactivity to hydroxyl radical and promotes oxidative modification of LDL in vitro. $ONOO^-$ reacts with free tyrosine to form 3-nitrotyrosine, and, indeed, Leeuwenburgh et al. (1997a) reported that 3-nitrotyrosine was found in LDL extracted from atherosclerotic lesions. However, it is not known whether any of the systems identified in vitro promote LDL oxidation in vivo.

Recent studies have demonstrated that active myeloperoxidase, a heme protein secreted by activated phagocytes, is present in human atherosclerotic lesions (Daugherty et al., 1994). The enzyme uses hydrogen peroxide to convert chloride to hypochlorous acid (Harrison and Schultz, 1976) and to convert L-tyrosine to tyrosyl radical (Heinecke et al., 1993b). Both hypochlorous acid and tyrosyl radical can oxidatively damage LDL by reactions that do not require free metal ions (Hazell and Stocker, 1993; Hazell et al., 1994; Savenkova et al., 1994). Tyrosyl radical might be a physiological catalyst for LDL oxidation, because tyrosine is present in plasma. As stated before, previous studies have suggested that myeloperoxidase may be one potent agent for lipoprotein oxidation in vivo.

Foam Cell Formation

Fatty streaks, the hallmark of the earliest lesion of atherosclerosis, are characterized by cholesterol-loaded macrophage "foam cells." Figure 7.2 shows the hypothetical contributions of oxidized lipoproteins to the formation of fatty streak, based on in vitro studies. Leukocyte entry into the vessel wall is essential for foam cell formation, and this process consists of three steps: rolling, activation of leukocytes, and leukocyte adhesion to endothelium. Treatment of endothelial cells in vitro with minimally oxidized LDL increases monocyte binding to endothelial cells (Berliner et al., 1990; Kim et al., 1994) and the production of monocyte chemotactic protein 1 (Cushing et al., 1990) as well as monocyte colony stimulating factor (Rajavasisth et al., 1990). Highly oxidized LDL also induces monocyte binding to endothelium (Lehr et al., 1993). Moreover, highly oxidized LDL attracts monocytes and inhibits macrophage migration (Quinn et al., 1988). Therefore, oxidized lipoproteins may play an important role in entry and retention of monocytes in the intima.

A series of studies by Brown and Goldstein (Brown and Goldstein, 1976, Brown et al., 1981; Goldstein and Brown, 1977) demonstrated the critical role of LDL receptor in the cellular uptake of LDL. According to in vivo studies, two-thirds or more of LDL removal from plasma is mediated by the LDL receptor (Kesaniemi et al., 1983). However, monocytes and macrophages in culture cannot be converted to foam cells by incubation with high concentrations of native LDL (Brown and Goldstein, 1983; Goldstein et al., 1979). In addition, the LDL receptors on macrophages are down-regulated in the presence of high concentrations of LDL. Nevertheless, because elevated plasma LDL concentrations promote foam cell formation in vivo, there must be another pathway to remove LDL from plasma that is independent of the LDL receptor. Goldstein et al. (1979) demonstrated that acetylated LDL converted peritoneal macrophages to foam cells. This pathway is mediated by a macrophage cell surface protein, the "scavenger receptor," which is clearly distinct from the LDL receptor. The scavenger receptor has been found on monocytes and monocyte-derived macrophages, Kupffer cells, and endothelial cells (Goldstein et al., 1979; Nagelkerke et al., 1983; Stein and Stein, 1980). Since then, a wide variety of oxidatively modified or chemically modified LDLs have been investigated, and the same receptor has been found to recognize many other modified LDLs, including acetoacetyl LDL (Mahley et al., 1979) and malondialdehyde-modified LDL (Fogelman et al., 1980).

In vascular lesions, macrophages and smooth muscle cells accumulate cholesterol in an unregulated manner, perhaps in part through the scavenger receptor to form foam cells. On the other hand, native LDL does not promote foam cell formation, because cell cholesterol content regulates the uptake of

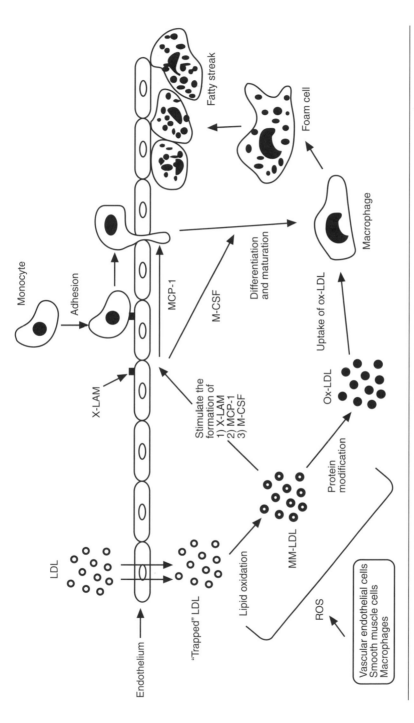

Figure 7.2 Oxidized LDL and the formation of fatty streak. LDL, low density lipoprotein; Ox-LDL, oxidized LDL; MM-LDL, minimally oxidized LDL; ROS, reactive oxygen species; X-LAM, monocyte binding molecules; M-CSF, macrophage colony stimulating factor; MCP-1, monocyte chemotactic protein-1. From Berliner and Heinecke (1996).

native LDL through the LDL receptor. The recognition of oxidized LDL by macrophage scavenger receptors is associated with loss of lysine residues in apolipoprotein B100 (Steinbrecher, 1987), and chemical modification of LDL (derivation of lysine residues with malondialdehyde or acetic anhydride) is also thought to form ligands for the scavenger receptor on the surface of LDL (Haberland and Steinbrecher, 1992). The scavenger receptor also has been shown to take up a wide variety of proteins derived with peroxidation products generated from arachidonic acid (Zhang et al., 1993). In addition, lipid removal from cholesterol-loaded macrophages might be impaired because oxidized lipoproteins are not effectively degraded by lysosomal enzymes (Lougheed et al., 1991).

The Role of High Density Lipoproteins on LDL Oxidation

Previous studies have demonstrated that high density lipoprotein (HDL) is associated with a decreased risk of atherosclerosis. For example, Hessler et al. (1979) found that HDL inhibited LDL-induced cytotoxicity in human vascular smooth muscle and endothelial cells in culture. Parthasarathy et al. (1990) showed that HDL inhibited the copper-catalyzed oxidative modification of LDL. Furthermore, HDL completely inhibited oxidation of LDL in co-cultures of human aortic wall cells (Navab et al., 1991), and HDL blocked the induction by oxidized LDL of monocyte adhesion to the endothelium (Maier et al. 1994). The mechanism for these effects of HDL, however, is unknown.

HDL binds essential fatty acids at 100-fold higher rate than serum albumin (Xu et al., 1992). Moreover, Bowry et al. (1992) demonstrated that HDL was the major carrier of oxidized lipids in plasma, whereas peroxides were much lower in plasma LDL. Therefore, HDL might be an effective carrier for oxidized lipids, and it is possible that HDL preferentially accumulates and sequesters oxidized lipids, decreasing the accumulation of oxidized lipids in LDL.

Studies related to the role of tyrosyl radical in HDL modification have provided another possible pathway for the protective role of HDL; tyrosylated HDL removed cholesterol from cultured fibroblasts and macrophage foam cells (Francis et al., 1993). Because myeloperoxidase is a component of human atherosclerotic lesions (Daugherty et al., 1994), tyrosyl radicals, which are generated by myeloperoxidase (Heinecke et al., 1993a, 1993b), may modify HDL into a form that readily takes up cholesterol even from the artery wall. Indeed, dityrosine levels are elevated in lipoprotein isolated from human atherosclerotic lesions (Leeuwenburgh et al., 1997b). Therefore, the oxidative tyrosylation of HDL by phagocytes may reverse the damaging effects of LDL oxidation.

Exercise and LDL Oxidation

Regular physical exercise helps prevent coronary heart disease (Berlin and Colditz, 1990). Enzymes related to lipoprotein metabolism are also known to be affected by physical exercise. For instance, increased activities in lipoprotein lipase and lecithin cholesterol acyl transferase have been shown in trained subjects (Tsopanakis et al., 1988; Weintraub et al., 1989). Furthermore, activity and mass concentration of cholesterol ester transfer protein have been reported to decrease in aerobically trained subjects (Serrat et al., 1993; Seip et al., 1993). Consequently, decreased concentrations of plasma triglycerides and very low density lipoprotein cholesterol, or an increased concentration of plasma HDL cholesterol, seem to be closely related to exercise (Godberg and Elliot, 1987; Nakamura, 1983; Weintraub et al., 1989). Some studies have reported a decrease in plasma LDL concentrations of trained subjects (Kujala et al., 1996), whereas others have found no significant difference in plasma LDL concentrations between trained and sedentary subjects (Sánchez-Quesada et al., 1997; Weintraub et al., 1989). Cullinane et al. (1982) observed a greater decrease after repeated, intense exercise. Moreover, Ginsburg et al. (1996) reported that well-conditioned amateur athletes exhibit decreased plasma triglyceride and LDL cholesterol concentrations but increased plasma HDL cholesterol concentrations after prolonged endurance exercise (triathlon). These findings probably suggest favorable effects of aerobic exercise on lipids and lipoprotein levels.

LDL has been operationally classified into two phenotypes: large, buoyant particles and small, dense particles (Williams et al., 1992). Large, buoyant LDL particles are less susceptible to in vitro oxidation, in contrast to small, dense LDL particles, which have increased susceptibility to oxidative stress and are more atherogenic (Chait et al., 1993; de Graaf et al., 1991; Tribble et al., 1992). Some studies have reported that predominance of large, buoyant LDL particles is an adaptation to exercise training (Slyper, 1994; Williams et al., 1986). Therefore, exercise training may improve the resistance of LDL to oxidative stress and help prevent atherosclerosis.

Electronegative LDL (LDL⁻), a subfraction of LDL that binds to positively charged resins, exists in human plasma and is more easily oxidized by copper than is buoyant LDL (Cazzolato et al., 1991; Hodis et al., 1994; Shimano et al., 1991; Tribble et al., 1995). LDL⁻ seems to be minimally oxidized and contributes to the susceptibility of LDL to oxidation (Avogaro et al., 1988; Sevanian et al., 1996), and LDL⁻ and small, dense LDL share many characteristics. Sánchez-Quesada et al. (1995) found that strenuous aerobic exercise increased the susceptibility of LDL to oxidation and the proportion of LDL⁻, which suggests that LDL⁻ is related to the susceptibility of LDL to oxidation and is generated as a consequence of exercise-induced oxidative stress. In other work, Sánchez-Quesada et al. (1997) found no significant difference in the proportion of LDL⁻ between trained and untrained subjects, although trained subjects showed increased resistance of LDL to oxidation, which added to the protective effect against atherosclerotic disease by increasing HDL and decreasing triglyceride concentrations.

Vasankari et al. (1997b) and Kujala et al. (1996) reported that acute prolonged physical exercise did not affect the baseline level of LDL diene conjugation (LDL-DC) or antioxidant potential of LDL, and that LDL-DC was significantly lower in veteran endurance atheletes than in control subjects. Moreover, the antioxidant potential of LDL tended to be higher, though not statistically significant ($p = .056$), in athletes than in controls. So the low level of LDL-DC in athletes appears to be an adaption to exercise training. LDL-DC may indicate LDL oxidation; however, LDL-DC does not appear to change during acute exercise.

Antioxidants contained in LDL particles are rapidly consumed when these particles are exposed to oxidants, and under certain conditions the propagation process of lipid peroxidation does not start until most antioxidants are exhausted (Esterbauer et al., 1987) (*propagation phase* of LDL oxidation is a descriptive term that does not correspond to the chemical term for the propagation phase of lipid peroxidation). Vitamin E (α-tocopherol) supplementation increases not only serum α-tocopherol but also the antioxidant capacity of LDL (Vasankari et al., 1997c), because a major part of α-tocopherol in plasma is transported to LDL (Kayden and Traber, 1993). Moreover, physical exercise increases serum α-tocopherol (Sánchez-Quesada et al., 1997; Vasankari et al., 1997c), possibly explaining, in part, the exercise-induced enhancement of LDL resistance to oxidation. However, several studies failed to find any correlation between the content of α-tocopherol and the LDL susceptibility to oxidation (Esterbauer et al., 1992). In addition, endurance training reportedly reduces exercise-induced extracellular reactive oxygen species generated from neutrophils during strenuous exercise (Oh-ishi et al., 1997b), which suggests that exercise training may improve LDL susceptibility to exercise-induced oxidative stress by decreasing production of reactive oxygen species.

Conclusion

1. A number of human and animal studies concerning exercise-induced lipid peroxidation have been reported, although the results of these studies are not always consistent. The inconsistency of the data may be due, in part, to differences in methodologies, that is, the intensity and duration of exercise, training conditions, and the animals and tissues used.

2. In general, higher intensity and longer duration of exercise seem to cause greater lipid peroxidation.

3. Physical training tends to prevent or minimize the degree of exercise-induced lipid peroxidation, possibly by increasing cellular antioxidant enzymes.

4. Aging causes the up-regulation of antioxidant enzymes in skeletal muscle, which may reflect the response to age-related increases of oxidative stress (expressed as increased lipid peroxides). However, aging seems to down-regulate the antioxidant enzymes in other important organs such as brain, liver, and heart.

5. The effect of endurance training on the muscle antioxidant enzyme system, as seen in skeletal muscle from young subjects, is not observed in old subjects.

6. Physical training may decrease the susceptibility of LDL to oxidative stress by acute exercise.

7. Supplemental antioxidants, especially vitamin E, may inhibit lipid and lipoprotein lipid peroxidation. It may be beneficial to increase dietary intake of vitamin E during strenuous physical exercise, because such training decreases plasma and tissue vitamin E concentrations. To date, however, no long-term, randomized, prospective trials have demonstrated that vitamin E exerts any beneficial effects on oxidative stress in vivo.

References

Alessio, H.M., and A.H. Goldfarb. 1988a. Lipid peroxidation and scavenger enzymes during exercise: Adaptive response to training. *Journal of Applied Physiology* 64:1333-1336.

Alessio, H.M., A.H. Goldfarb, and R.G. Cutler. 1988b. NOA content increases in fast- and slow-twitch skeletal muscle with intensity of exercise in a rat. *American Journal of Physiology* 255:C874-C877.

Anzueto, A., F.H. Andrade, L.C. Maxwell, S.M. Levine, R.A. Lawrence, and S.G. Jenkinson. 1993. Diaphragmatic function after resistive breathing in vitamin E deficient rats. *Journal of Applied Physiology* 74:267-271.

Aust, S.D., L.A. Morehouse, and C.E. Thomas. 1985. Role of metals in oxygen radical reactions. *Advances in Free Radical Biology and Medicine* 1:3-25.

Avogaro, P., G. Bittolo-Bon, and G. Cazzolato. 1988. Presence of a modified low density lipoprotein in humans. *Arteriosclerosis* 8:79-87.

Barber, A.A., and F. Bernheim. 1967. Lipid peroxidation: Its measurement, occurrence, and significance in animal tissue. *Advances in Gerontological Research* 2:355-403.

Benderitter, M., F. Hadj-Saaii, M. Lhuissier, V. Maupoil, J.-C. Guilland, and L. Rochette. 1996. Effects of exhaustive exercise and vitamin B6 deficiency on free radical oxidative process in male trained rats. *Free Radical Biology and Medicine* 21:541-549.

Bendich, A.L., J. MacWin, and O. Scandurra. 1986. The antioxidant role of vitamin C. *Advances in Free Radical Biology and Medicine* 2:419-444.

Berlin, J.A., and G.A. Colditz. 1990. A meta-analysis of physical activity in the prevention of coronary heart disease. *American Journal of Epidemiology* 132:612-628.

Berliner, J.A., and J.W. Heinecke. 1996. The role of oxidized lipoproteins in atherogenesis. *Free Radical Biology and Medicine* 20:707-727.

Berliner, J.A., M.C. Territo, A. Sevanian, S. Ramin, J.A. Kim, B. Bamshad, M. Esterson, and A.M. Fogelman. 1990. Minimally modified LDL stimulates monocyte endothelial interactions. *Journal of Clinical Investigation* 85:1260-1266.

Bjorneboe, A., G.E.A. Bjomeboe, and C.A. Drevon. 1990. Absorption, transport, and distribution of vitamin E. *Journal of Nutrition* 120:233-242.

Blake, D.R., J. Unsworth, J.M. Outhwaite, C.J. Morris, P. Merry, B.L. Kidd, R. Ballard, and L. Gray. 1989. Hypoxic-reperfusion injury in the inflamed human joint. *Lancet* (8633):289-293.

Bowry, V.W., K.K. Stanley, and R. Stocker. 1992. High density lipoprotein is the major carrier of lipid hydroperoxides in human blood plasma from fasting donors. *Proceedings of the National Academy of Sciences of the USA* 89:10316-10320.

Brady, P.S., L.J. Brady, and D.E. Ullrey. 1979. Selenium, vitamin E and the response to swimming stress in the rat. *Journal of Nutrition* 109:1103-1109.

Brown, M.S., and J.L. Goldstein. 1976. Receptor-mediated control of cholesterol metabolism. *Science* 191:150-154.

Brown, M.S., and J.L. Goldstein. 1983. Lipoprotein metabolism in the macrophage: Implications for cholesterol deposition in atherosclerosis. *Annual Review of Biochemistry* 52:223-261.

Brown, M.S., P.T. Kovanen, and J.L. Goldstein. 1981. Regulation of plasma cholesterol by lipoprotein receptors. *Science* 212:628-635.

Buczynski, A., J. Kedziora, W. Tkaczewski, and B. Wachowicz. 1991. Effect of submaximal physical exercise on antioxidative protection of human blood platelets. *International Journal of Sports Medicine* 12:52-54.

Cadenas, E. 1989. Biochemistry of oxygen toxicity. *Annual Review of Biochemistry* 58:79-110.

Cannon, J.G., S.F. Orencole, R.A. Fielding, M. Meydani, S.N. Meydani, M.A. Fiatarone, J.B. Blumberg, and W.J. Evans. 1990. Acute phase response in exercise: Interaction of age and vitamin E on neutrophils and muscle enzyme release. *American Journal of Physiology* 259: R1214-R1219.

Cao, G., and J. Chen. 1991. Effects of dietary zinc on free radical generation, lipid peroxidation, and superoxide dismutase in trained mice. *Archives of Biochemistry and Biophysics* 291:147-153.

Cazzolato, G., P. Avogaro, and G. Bittolo-Bon. 1991. Characterization of a more electronegatively charged LDL subfraction by ion exchange HPLC. *Free Radical Biology and Medicine* 11:247-253.

Chait, A., R.L. Brazg, D.L. Tribble, and R.M. Krauss. 1993. Susceptibility of small, dense, low-density lipoproteins to oxidative modification in subjects with the atherogenic lipoprotein phenotype, pattern B. *American Journal of Medicine* 94:350-356.

Chance, B., H. Sies, and A. Boveris. 1979. Hydroperoxide metabolism in mammalian organs. *Physiological Reviews* 59:527-605.

Chen, M-F., H-C. Hsu, and Y-T. Lee. 1994. Effects of acute exercise on the changes of lipid profiles and peroxides, prostanoids, and platelet activation in hypercholesterolemic patients before and after treatment. *Prostaglandins* 48:157-174.

Chio, K.S., and A.L. Tappel. 1969. Synthesis and characterization of the fluorescent products derived from malondialdehyde and amino acids. *Biochemistry* 8:2821-2827.

Clark, I.A., W.B. Cowden, and N.H. Hunt. 1985. Free radical-induced pathology. *Medicinal Research Reviews* 5:297-332.

Criswell, D., S. Powers, S. Dodd, J. Lawler, D. Martin, and L.L. Ji. 1993. High intensity training-induced changes in skeletal muscle antioxidant enzyme activity. *Medicine and Science in Sports and Exercise* 25:1135-1140.

Cullinane, E., S. Siconolfi, A. Saritelli, and P.D. Thompson. 1982. Acute decrease in serum triglycerides with exercise: Is there a threshold for an exercise effect? *Metabolism* 31:844-847.

Cushing, S.D., J.A. Berliner, A.J. Valente, M.C. Territo, M. Navab, F. Parhami, R. Gerrity, C.J. Schwartz, and A.M. Fogelman. 1990. Minimally modified LDL induces monocyte chemotactic proteins I in human endothelial and smooth muscle cells. *Proceedings of the National Academy of Sciences of the USA* 87:5134-5138.

Dahle, L.K., E.G. Hill, and R.T. Holman. 1962. The thiobarbituric acid reaction and the autoxidations of polyunsaturated fatty acid methyl esters. *Archives of Biochemistry and Biophysics* 98:253-261.

Danh, H.C., M.S. Benedetti, and P. Dostert. 1983. Differential changes in superoxide dismutase activity in brain and liver of old rats and mice. *Journal of Neurochemistry* 40:1003-1007.

Daugherty, A., D.L. Rateri, J.L. Dunn, and J.W. Heinecke. 1994. Myeloperoxidase, a catalyst for lipoprotein oxidation, is expressed in human atherosclerotic lesions. *Journal of Clinical Investigation* 94:437-444.

Davies, K.J.A., A.T. Quintanilha, G.A. Brooks, and L. Packer. 1982. Free radicals and tissue damage produced by exercise. *Biochemical and Biophysical Research Communications* 107:1198-1205.

de Graaf, J., H.L.M. Hak-Lenimers, M.P.C. Hectors, P.N.M. Demacker, J.C.M. Hendriks, and A.F.H. Stalenhoef. 1991. Enhanced susceptibility to in vitro oxidation of the dense low density lipoprotein subtraction in healthy subjects. *Arteriosclerosis and Thrombosis* 11:298-306.

Dillard, C.J., E.E. Dumelin, and A.L. Tappel. 1977. Effect of dietary vitamin E on expiration of pentane and ethane by the rat. *Lipids* 12:109-114.

Dillard, C.J., R.E. Litov, W.M. Savin, E.E. Dumelin, and A.L. Tappel. 1978b. Effects of exercise, vitamin E, and ozone on pulmonary function and lipid peroxidation. *Journal of Applied Physiology* 45:133-138.

Dillard, C.J., T.L. Litov, and A.L. Tappel. 1978a. Effects of dietary vitamin E, selenium, and polyunsaturated fats on in vivo lipid peroxidation. *Journal of Applied Physiology* 45:927-932.

Duthie, G.G., J.D. Robertson, R.J. Maughan, and P.C. Morrice. 1990. Blood antioxidant status and erythrocyte lipid peroxidation following distance running. *Archives of Biochemistry and Biophysics* 282:78-83.

Esterbauer, H. 1982. Aldehydic products of lipid peroxidation. In: *Free Radicals, Lipid Peroxidation, and Cancer*, D.C.H. McBrien and T.F. Slater (eds.), pp. 101-128. Academic Press, London.

Esterbauer, H., J. Gebicki, H. Puhl, and G. Jurgens. 1992. The role of lipid peroxidation and antioxidants in oxidative modification of LDL. *Free Radical Biology and Medicine* 13:341-390.

Esterbauer, H., G. Jurgens, O. Quehenberger, and E. Koller. 1987. Autoxidation of human low density lipoprotein loss of polyunsaturated fatty acids and vitamin E and generation of aldehydes. *Journal of Lipid Research* 28:495-509.

Esterbauer, H., E. Koller, R.G. Slee, and J.F. Koster. 1986. Possible involvement of the lipid-peroxidation product 4-hydroxynonenal in the formation of fluorescent chromolipids. *Biochemical Journal* 239:405-409.

Esterbauer, H., R.J. Schaur, and H. Zollner. 1991. Chemistry and biochemistry of 4-hydroxynonenal, malondialdehyde and related aldehydes. *Free Radical Biology and Medicine* 11:81-128.

Esterbauer, H., H. Zollner, and R.J. Schaur. 1990. Aldehydes formed by lipid peroxidation: Mechanisms of formation, occurrence, and determination. In: *Membrane Lipid Oxidation*, C. Vigo-Pelfrey (ed.), pp. 240-268. CRC Press, Boca Raton, FL.

Faff, J., and A. Frankiewicz-Józko. 1997. Effect of ubiquinone on exercise induced lipid peroxidation in rat tissues. *European Journal of Applied Physiology* 75:413-417.

Fletcher, B.L., C.J. Dillard, and A.L. Tappel. 1973. Measurement of fluorescent lipid peroxidation products in biological systems and tissues. *Analytical Biochemistry* 52:1-9.

Floyd, R.A. 1993. Basic free radical biochemistry. In: *Free Radicals in Aging*, B.P. Yu (ed.), pp. 39-55. CRC Press, Boca Raton, FL.

Fogelman, A.M., I. Shechter, J. Seager, M. Hokom, J.S. Child, and P.A. Edwards. 1980. Malondialdehyde alteration of low density lipoprotein leads to cholesteryl ester accumulation in human monocyte-macrophages. *Proceedings of the National Academy of Sciences of the USA* 77:2214-2218.

Francis, G.A., A.J. Mendez, E.L. Bierman, and J.W. Heinecke. 1993. Oxidative tyrosylation of high density lipoprotein by peroxidase enhances cholesterol removal from cultured fibroblasts and macrophage foam cells. *Proceedings of the National Academy of Sciences of the USA* 90:6631-6635.

Frankiewicz-Józko, A., J. Faff, and B. Sieradzan-Gabelska. 1996. Changes in concentrations of tissue free radical marker and serum creatine kinase during the postexercise period in rats. *European Journal of Applied Physiology* 74:470-474.

Frei, B., L. England, and B.N. Ames. 1989. Ascorbate is an outstanding antioxidant in human blood plasma. *Proceedings of the National Academy of Sciences of the USA* 86:6377-6381.

Gee, D.L., and A.L. Tappel. 1981. The effect of exhaustive exercise on expired pentane as a measure of in vivo lipid peroxidation in the rat. *Life Sciences* 28:2425-2429.

Gey, K.F., G.B. Brubacher, and H.B. Stahelin. 1987. Plasma levels of antioxidant vitamins in relation to ischemic heart disease and cancer. *American Journal of Clinical Nutrition* 45:1368-1377.

Ginsburg, G.S., A. Agil, M. O'Toole, E. Rimm, P.S. Douglas, and N. Rifai. 1996. Effects of a single bout of ultraendurance exercise on lipid levels and susceptibility of lipids to peroxidation in triathletes. *Journal of the American Medical Association* 276:221-225.

Girotti, A.W. 1985. Mechanisms of lipid peroxidation. *Free Radical Biology and Medicine* 1:87-95.

Godberg, L., and D.C. Elliot. 1987. The effect of exercise on lipid metabolism in men and women. *Sports Medicine* 4:307-321.

Gohil, K., L. Packer, B. DeLumen, G.A. Brooks, and S.E. Terblanche. 1986. Vitamin E deficiency and vitamin C supplements: Exercise and mitochondrial oxidation. *Journal of Applied Physiology* 60:1986-1991.

Goldfarb, A.H., M.K. McIntosh, and J. Fatouros. 1994. Vitamin E effects on indexes of lipid peroxidation in muscle from DHEA-treated and exercised rats. *Journal of Applied Physiology* 76:1630-1635.

Goldstein, J.L., and M.S. Brown. 1977. The low-density lipoprotein pathway and its relation to atherosclerosis. *Annual Review of Biochemistry* 46:897-930.

Goldstein, J.L., Y.K. Ho, S.K. Basu, and M.S. Brown. 1979. Binding site on macrophages that mediates uptake and degradation of acetylated low density lipoprotein, producing massive cholesterol deposition. *Proceedings of the National Academy of Sciences of the USA* 76:333-337.

Grinna, L.S. 1977. Age-related changes in the lipids of the microsomal and the mitochondrial membranes of rat liver and kidney. *Mechanisms of Ageing and Development* 6:197-205.

Gutteridge, J.M. 1986. Aspects to consider when detecting and measuring lipid peroxidation. *Free Radical Research Communications* 1:173-184.

Haberland, M.E., D. Fong, and L. Chena. 1988. Malondialdehyde-altered protein occurs in atheroma of Watanabe heritable hyperlipidemic rabbits. *Science* 241:215-218.

Haberland, M.E., and U.P. Steinbrecher. 1992. Modified low density lipoproteins: Diversity and biological relevance in atherogenesis. In: *Molecular Genetics of Coronary Artery Disease, vol. 14*, A.J. Lusis, J.I. Rotter, and R.S. Sparkes (eds.), pp. 35-61. Karger, Basel, Switzerland.

Hagihara, M., I. Nishigaki, M. Maseki, and K. Yagi. 1984. Age-dependent changes in lipid peroxide levels in the lipoprotein fractions of human serum. *Journal of Gerontology* 39:269-272.

Halliwell, B., and J.M.C. Gutteridge. 1984. Oxygen toxicity, oxygen radicals, transition metals and disease. *Biochemical Journal* 219:1-14.

Hara, M., M. Abe, T. Suzuki, and R.J. Reiter. 1996. Tissue changes in glutathione metabolism and lipid peroxidation induced by swimming are partially prevented by melatonin. *Pharmacology and Toxicology* 78:308-312.

Hardt, H., and M. Rothstein. 1985. Altered phosphoglycerate kinase from old rat muscle shows no change in primary structure. *Biochimica et Biophysica Acta* 831:13-21.

Harman, D. 1956. Aging: A theory based on free radical and radiation chemistry. *Journal of Gerontology* 11:298-300.

Harrison, J.E., and J. Schultz. 1976. Studies on the chlorinating activity of myeloperoxidase. *Journal of Biological Chemistry* 251:1371-1374.

Hartmann, A., A.M. Nieb, M. Grünert-Fuchs, B. Poch, and G. Speit. 1995. Vitamin E prevents exercise-induced DNA damage. *Mutation Research* 346:195-202.

Hazell, L.J., and R. Stocker. 1993. Oxidation of low-density lipoprotein with hypochlorite causes transformation of the lipoprotein into a high-uptake form for macrophages. *Biochemical Journal* 290:165-172.

Hazell, L.J., J.J. van den Berg, and R. Stocker. 1994. Oxidation of low-density lipoprotein by hypochlorite causes aggregation that is mediated by modification of lysine residues rather than by lipid oxidation. *Biochemical Journal* 302:297-304.

Hegner, D. 1980. Age-dependence of molecular and functional changes in biological membrane properties. *Mechanisms of Ageing and Development* 14:101-118.

Heinecke, J.W., W. Li, H.L. Daehnke, and J.A. Goldstein. 1993a. Dityrosine, a specific marker of oxidation, is synthesized by the myeloperoxidase. *Journal of Biological Chemistry* 268:4069-4077.

Heinecke, J.W., W. Li, H.G.A. Francis, and J.A. Goldstein. 1993b. Tyrosyl radical generated by myeloperoxidase catalyzes the oxidative cross-linking of proteins. *Journal of Clinical Investigation* 91:2866-2872.

Hessler, J.R., A.L. Robertson, and G.M. Chisolm. 1979. LDL-induced cytotoxicity and its inhibition by HDL in human vascular smooth muscle and endothelial cells in culture. *Arteriosclerosis* 32:213-229.

Higuchi, M., L.J. Cartier, M. Chen, and J.O. Holloszy. 1985. Superoxide dismutase and catalase in skeletal muscle: Adaptive response to exercise. *Journal of Gerontology* 40:281-286.

Hochstein, P., and L. Emster. 1963. ADP-activated lipid peroxidation coupled on TPNH oxidase system of microsomes. *Biochemical and Biophysical Research Communications* 12:388-394.

Hodis, H.N., D.M. Kramsch, P. Avogaro, G. Bittolo-Bon, G. Cazzolato, J. Hwang, H. Peterson, and A. Sevanian. 1994. Biochemical and cytotoxic characteristics of an in vivo circulating oxidized low density lipoprotein (LDL⁻). *Journal of Lipid Research* 31:1382-1398.

Hong, H., and P. Johnson. 1995. Antioxidant enzyme activities and lipid peroxidation levels in exercised and hypertensive rat tissues. *International Journal of Biochemistry and Cell Biology* 27:923-931.

Husain, K., and S.M. Somani. 1997. Response of cardiac antioxidant system to alcohol and exercise training in the rat. *Alcohol* 14:301-307.

Inayama, T., Y. Kumagai, M. Sakane, M. Saito, and M. Matsuda. 1996. Plasma protein-bound sulfhydryl group oxidation in humans following a full marathon race. *Life Sciences* 59:573-578.

Jackson, M.J., D.A. Jones, and R.H.T. Edwards. 1983. Vitamin E and skeletal muscle. In: *Ciba Foundation Symposium 1: Biology of Vitamin E*, pp. 224-239. Pitman Books, London.

Jain, S.K. 1984. The accumulation of malondialdehyde, a product of fatty acid peroxidation, can disturb aminophospholipid organization in the membrane bilayer of human erythrocytes. *Journal of Biological Chemistry* 269:3391-3394.

Jenkins, R.R., R. Friedland, and H. Howald. 1984. The relationship of oxygen uptake to superoxide dismutase and catalase activity in human skeletal muscle. *International Journal of Sports Medicine* 5:11-14.

Jenkins, R.R. 1988. Free radical chemistry: Relationship to exercise. Sports Medicine 5:156-170.

Ji, L.L. 1995. Oxidative stress during exercise: Implications of antioxidant nutrients. *Free Radical Biology and Medicine* 18:1079-1086.

Ji, L.L., D. Dillon, and E. Wu. 1991. Myocardial aging: Antioxidant enzyme systems and related biochemical properties. *American Journal of Physiology* 261:R386-R392.

Ji, L.L., and E.W. Mitchell. 1994. Effects of adriamycin on heart mitochondrial function in rested and exercised rats. *Biochemical Pharmacology* 47:877-885.

Ji, L.L., F.W. Stratman, and H.A. Lardy. 1988a. Enzymatic down regulation with exercise in rat skeletal muscle. *Archives of Biochemistry and Biophysics* 263:137-149.

Ji, L.L., F.W. Stratman, and H.A. Lardy. 1988b. Antioxidant enzyme systems in rat liver and skeletal muscle. *Archives of Biochemistry and Biophysics* 263:150-160.

Ji, L.L., F.W. Stratman, and H.A. Lardy. 1992. Antioxidant enzyme response to selenium deficiency in rat myocardium. *Journal of the American College of Nutrition* 11:79-86.

Kanaley, J.A., and L.L. Ji. 1991. Antioxidant enzyme activity during prolonged exercise in amenorrheic and eumenorrheic athletes. *Metabolism* 40:88-92.

Kanter, M. 1995. Free radicals and exercise: Effects of nutritional antioxidant supplementation. *Exercise and Sport Sciences Reviews* 23:375-397.

Kanter, M.M., G.R. Lesmes, L.A. Kaminsky, J.L. Ham-Saeger, and N.D. Nequin. 1988. Serum creatine kinase and lactate dehydrogenase changes following an eighty kilometers race. *European Journal of Applied Physiology* 57:60-63.

Kanter, M.M., L.A. Nolte, and J.O. Holloszy. 1993. Effects of an antioxidant vitamin mixture on lipid peroxidation at rest and postexercise. *Journal of Applied Physiology* 74:965-969.

Kayden, H.J., and M.G. Traber. 1993. Absorption, lipoprotein transport, and regulation of plasma concentrations of vitamin E in humans. *Journal of Lipid Research* 34:343-358.

Kedziora, J., A. Buczynski, and K. Kedziora-Komatowska. 1995. Effect of physical exercise on antioxidative enzymatic defense in blood platelets from healthy men. *International Journal of Occupational Medicine and Environmental Health* 8:33-39.

Kesaniemi, Y.A., J.L. Witztum, and U.P. Steinbrecher. 1983. Receptor-mediated catabolism of low density lipoprotein in man: Quantitation using glucosylated low density lipoproteins. *Journal of Clinical Investigation* 71:950-959.

Kim, J.A., M.C. Territo, E. Wayner, T.M. Carlos, F. Parhami, C.W. Smith, M.E. Haberland, A.M. Fogelman, and J.A. Berliner. 1994. Partial characterization of leukocyte binding molecules on endothelial cells induced by minimally oxidized LDL. *Arteriosclerosis and Thrombosis* 14:427-433.

Kim, J.D., B.P. Yu, R.J.M. McCarter, S.Y. Lee, and J.T. Herlihy. 1996. Exercise and diet modulate cardiac lipid peroxidation and antioxidant defenses. *Free Radical Biology and Medicine* 20:83-88.

Kim, J.W., and B.P. Yu. 1989. Characterization of age-related malondialdehyde oxidation: The effect of modulation by food restriction. *Mechanisms of Ageing and Development* 50:277-287.

Koz, M., D. Erbas, A. Bilgihan, and A. Erbas. 1992. Effects of acute swimming exercise on muscle and erythrocyte malondialdehyde, serum myoglobin, and plasma ascorbic acid concentrations. *Canadian Journal of Physiology and Pharmacology* 70:1392-1395.

Kretzschmar, M., D. Muller, J. Hubscher, E. Marin, and W. Klinger. 1991. Influence of aging, training and acute physical exercise on plasma glutathione and lipid peroxides in man. *International Journal of Sports Medicine* 12:218-222.

Krinsky, N.I. 1989. Antioxidant functions of carotenoids. *Free Radical Biology and Medicine* 7:617-635.

Kubow, S. 1993. Lipid oxidation products in food and atherogenesis. *Nutrition Reviews* 51:33-40.

Kujala, U.M., M. Ahotupa, T. Vasankari, J. Kaprio, and M.J. Tikkanen. 1996. Low LDL oxidation in veteran endurance athletes. *Scandinavian Journal of Medicine and Science in Sports* 6:303-308.

Kumar, C.T., V.K. Reddy, M. Prasad, K. Thyagaraju, and P. Reddanna. 1992. Dietary supplementation of vitamin E protects heart tissue from exercise-induced oxidative stress. *Molecular and Cellular Biochemistry* 111:109-115.

Laaksonen, D.E., M. Atalay, L. Niskanen, M. Uusitupa, O. Hänninen, and C.K. Sen. 1996. Increased resting and exercise-induced oxidative stress in young IDDM men. *Diabetes Care* 19:569-574.

Laaksonen, R., M. Fogelholm, J.-J. Himberg, J. Laakso, and Y. Salorinne. 1995. Ubiquinone supplementation and exercise capacity in trained young and older men. *European Journal of Applied Physiology* 72:95-100.

Laganiere, S., and B.P. Yu. 1989. Effect of chronic food restriction in aging rats: I. Liver subcellular membranes. *Mechanisms of Ageing and Development* 48:207-219.

Lawler, J.M., S.K. Powers, H.V. Dijk, T. Visser, M.J. Kordus, and L.L. Ji. 1994. Metabolic and antioxidant enzyme activities in the diaphragm: Effects of acute exercise. *Respiration Physiology* 96:139-149.

Lawler, J.M., S.K. Powers, T. Visser, H.V. Dijk, M.J. Kordus, and L.L. Ji. 1993. Acute exercise and skeletal muscle antioxidant and metabolic enzymes: Effects of fiber type and age. *American Journal of Physiology* 265:R1344-R1350.

Leaf, D.A., M.T. Kleinman, M. Hamilton, and T.J. Barstow. 1997. The effect of exercise intensity on lipid peroxidation. *Medicine and Science in Sports and Exercise* 29:1036-1038.

Leeuwenburgh, C., R. Fiegib, R. Chandwaney, and L.L. Ji. 1994. Aging and exercise training in skeletal muscle: Responses of glutathione and antioxidant enzyme systems. *American Journal of Physiology* 267:R439-R445.

Leeuwenburgh, C., M.M. Hardy, S.L. Hazen, P. Wagner, S. Oh-ishi, U.P. Steinbrecher, and J.W. Heinecke. 1997a. Reactive nitrogen intermediates promote low density lipoprotein oxidation in human atherosclerotic intima. *Journal of Biological Chemistry* 272:1433-1436.

Leeuwenburgh, C., and L.L. Ji. 1995. Glutathione depletion in rested and exercised mice: Biochemical consequence and adaptation. *Archives of Biochemistry and Biophysics* 316:941-949.

Leeuwenburgh, C., and L.L. Ji. 1996. Alteration of glutathione and antioxidant status with exercise in unfed and refed rats. *Journal of Nutrition* 126:1833-1843.

Leeuwenburgh, C., J.E. Rasmussen, F.F. Hsu, D.M. Mueller, S. Pennathur, and J.W. Heinecke. 1997b. Mass spectrometric quantification of markers for protein oxidation by tyrosyl radical, copper, and hydroxyl radical in low density lipoprotein isolated from human atherosclerotic plaques. *Journal of Biological Chemistry* 272:3520-3526.

Lehr, H.A., J. Seemuller, C. Hubner, M.D. Menger, and K. Messmer. 1993. Oxidized LDL induced leucocyte-endothelium interaction in vivo involves the receptor for platelet-activating factor. *Arteriosclerosis and Thrombosis* 13:1013-1018.

Lougheed, M., H. Zhang, and U.P. Steinbrecher. 1991. Oxidized low density lipoprotein is resistant to cathepsins and accumulates within macrophages. *Journal of Biological Chemistry* 266:14519-14525.

Lovlin, R., W. Cottle, I. Pyke, M. Kavanagh, and A.N. Belcastro. 1987. Are indices of free radical damage related to exercise intensity? *European Journal of Applied Physiology* 56:313-316.

Luc, G., and J.C. Fruchart. 1991. Oxidation of lipoproteins and atherosclerosis. *American Journal of Clinical Nutrition* 53:206S-209S.

Mahley, R.W., T.L. Innerarity, K.H. Weisgraber, and S.Y. Oh. 1979. Altered metabolism (in vivo and in vitro) of plasma lipoprotein after selective chemical modification of lysine residues of the apoproteins. *Journal of Clinical Investigation* 64:743-750.

Maier, J.A., L. Barenghi, F. Pagani, S. Bradamante, P. Comi, and G. Ragnotti. 1994. The protective role of HDL on oxidized LDL induced U937 endothelial cell interaction. *European Journal of Biochemistry* 221:35-41.

Margaritis, I., F. Tessier, M.-J. Richard, and P. Marconnet. 1997. No evidence of oxidative stress after a triathlon race in highly trained competitors. *International Journal of Sports Medicine* 18:186-190.

Maughan, R.J., A.E. Donnelly, M. Gleeson, P.H. Whiting, K.A. Walker, and P.J. Clough. 1989. Delayed-onset muscle damage and lipid peroxidation in man after a downhill run. *Muscle and Nerve* 12:332-336.

Merry, P., M. Grootveld, J. Lunec, and D.R. Blake. 1991. Oxidative damage to lipids within the inflamed human joint provides evidence of radical-mediated hypoxic-reperfusion injury. *American Journal of Clinical Nutrition* 53:362S-369S.

Meydani, R.J., W.J. Evans, G. Handelman, L. Biddle, R.A. Fielding, S.N. Meydani, J. Burrill, M.A. Fiatarone, J.B. Blumberg, and J.G. Cannon. 1993. Protective effect of vitamin E on exercise-induced oxidative damage in young and older adults. *American Journal of Physiology* 264:R992-R998.

Mills, P.C., J.C. Ng, J. Thornton, A.A. Seawright, and D.E. Auer. 1994. Exercised-induced connective tissue turnover and lipid peroxidation in horses. *British Veterinary Journal* 150:53-63.

Munro, J.M., and S. Cotran. 1988. The pathogenesis of atherosclerosis: Atherogenesis and inflammation. *Laboratory Investigation* 58:249-261.

Nagelkerke, J.F., K.P. Barto, and T.J. van Berkel. 1983. In vivo and in vitro uptake and degradation of acetylated low density lipoprotein by rat liver endothelial, Kupffer, and parenchymal cells. *Journal of Biological Chemistry* 258:12221-12227.

Nakamura, N., H. Uzawa, H. Haeda, and T. Inomoto. 1983. Physical fitness, its contribution to serum high density lipoprotein. *Atherosclerosis* 48:173-181.

Navab, M., S.S. Imes, S.Y. Hama, G.P. Hough, L.A. Ross, R.W. Bork, A.J. Valente, J.A. Berliner, D.C. Drinkwater, H. Laks, and A.M. Fogelman. 1991. Monocyte transmigration induced by modification of LDL in co-culture of human aortic wall cells is due to induction of MCP-1 synthesis and is abolished by HDL. *Journal of Clinical Investigation* 88:2039-2046.

Nohl, N., and D. Hegner. 1978. Do mitochondria produce oxygen radicals in vivo? *European Journal of Biochemistry* 82:563-567.

Oh-ishi, S., T. Kizaki, J. Nagasawa, T. Izawa, T. Komabayashi, N. Nagata, K. Suzuki, N. Taniguchi, and H. Ohno. 1997a. Effects of endurance training on superoxide dismutase activity, content and mRNA expression in rat muscle. *Clinical and Experimental Pharmacology and Physiology* 24:326-332.

Oh-ishi, S., T. Kizaki, T. Ookawara, T. Sakurai, T. Izawa, N. Nagata, and H. Ohno. 1997b. Endurance training improves the resistance of rat diaphragm to exercise induced oxidative stress. *American Journal of Respiratory and Critical Care Medicine* 156:1579-1585.

Oh-ishi, S., T. Kizaki, H. Yamashita, N. Nagata, K. Suzuki, N. Taniguchi, and H. Ohno. 1995. Alterations of superoxide dismutase iso-enzyme activity, content, and mRNA expression with aging in rat skeletal muscle. *Mechanisms of Ageing and Development* 84:65-76.

Oh-ishi, S., K. Toshinai, T. Kizaki, S. Haga, K. Fukuda, N. Nagata, and H. Ohno. 1996. Effects of aging and/or training on antioxidant enzyme system in diaphragm of mice. *Respiration Physiology* 105:195-202.

Ohno, H., K. Suzuki, J. Fujii, H. Yamashita, T. Kizaki, S. Oh-ishi, and N. Taniguchi. 1994. Superoxide dismutases in exercise and disease. In: *Exercise and Oxygen Toxicity*, C.K. Sen, L. Packer, and O. Hänninen (eds.), pp. 127-161. Elsevier Science, Amsterdam.

Orgel, L.E. 1970. The maintenance of the accuracy of protein synthesis and its relevance to aging: A correction. *Proceedings of the National Academy of Sciences of the USA* 67:1476.

Ørtenblad, N., K. Madsen, and M.S. Djurhuus. 1997. Antioxidant status and lipid peroxidation after short-term maximal exercise in trained and untrained humans. *American Journal of Physiology* 272:R1258-R1263.

Packer, L., ed. 1984. *Methods in Enzymology, vol. 105*. Academic Press, San Diego, CA.

Packer, L. 1991. Protective role of vitamin E in biological systems. *American Journal of Clinical Nutrition* 53:1050s-1055s.

Packer, L., and A.L. Glazer, eds. 1990. *Methods in Enzymology, vol. 186*. Academic Press, San Diego, CA.

Packer, L., T.F. Slater, A.L. Almada, L.M. Rothfuss, and D.S. Wilson. 1989. Modulation of tissue vitamin E levels by physical activity. *Annals of the New York Academy of Sciences* 570:311-321.

Parthasarathy, S., J. Barnett, and L.J. Fong. 1990. High-density lipoprotein inhibits the oxidative modification of low-density lipoprotein. *Biochimica et Biophysica Acta* 1044:275-283.

Pereira, B., L.F.B.C. Rosa, D.A. Safi, M.H.G. Medeiros, R. Curi, and E.J.H. Bechara. 1994. Superoxide dismutase, catalase, and glutathione peroxidase activities in muscle and lymphoid organs of sedentary and exercise-trained rats. *Physiology and Behavior* 56:1095-1099.

Pincemail, J., G. Camus, A. Roesgen, E. Dreezen, Y. Bertrand, M. Lismonde, G. Deby-Dupont, and C. Deby. 1990. Exercise induces pentane production and neutrophil activation in humans. Effect of propranolol. *European Journal of Applied Physiology* 61:319-322.

Powers, S.K., J. Lawler, D. Criswell, F.-K. Lieu, and D. Martin. 1992. Aging and respiratory muscle metabolic plasticity: Effects of endurance training. *Journal of Applied Physiology* 72:1068-1073.

Pryor, W.A. 1973. Free radical reactions and their importance in biochemical systems. *Federation Proceedings* 32:1862-1869.

Quinn, M.T., S. Parthasarathy, and D. Steinberg. 1988. Lysophosphatidyl choline: A chemotactic factor for human monocytes and its potential role in atherogenesis. *Proceedings of the National Academy of Sciences of the USA* 85:2805-2809.

Quintanilha, A.T. 1984. Effects of physical exercise and/or vitamin E on tissue oxidative metabolism. *Biochemical Society Transactions* 12:403-404.

Quintanilha, A.T., L. Packer, J.M. Szyszlo-Davies, T.L. Racanelli, and K.J.A. Davies. 1982. Membrane effects of vitamin E deficiency: Bioenergetic and surface charge density studies of skeletal muscle and liver mitochondria. *Annals of the New York Academy of Sciences* 393:32-47.

Radák, Z., K. Asano, M. Inoue, T. Kizaki, S. Oh-ishi, K. Suzuki, N. Taniguchi, and H. Ohno. 1996. Superoxide dismutase derivative prevents oxidative damage in liver and kidney of rats induced by exhausting exercise. *European Journal of Applied Physiology* 72:189-194.

Radák, Z., K. Asano, K-C. Lee, H. Ohno, A. Nakamura, H. Nakamoto, and S. Goto. 1997. High altitude training increases reactive carbonyl derivatives but not lipid peroxidation in skeletal muscle of rats. *Free Radical Biology and Medicine* 22:1109-1114.

Rajavasisth, T.B., A. Andalibi, M.C. Territo, J.A. Berliner, M. Navab, A.M. Fogelman, and A.J. Lusis. 1990. Modified LDL induce endothelial cell expression of granulocyte and macrophage colony stimulating factors. *Nature* 344:254-257.

Rajguru, S., G.S. Yeargans, and N.W. Seidler. 1993. Exercise causes oxidative damage to rat skeletal muscle microsomes while increasing cellular sulfhydryls. *Life Sciences* 54:149-157.

Rao, G., E. Xia, and A. Richardson. 1990. Effect of age on the expression of antioxidant enzymes in male Fischer 344 rats. *Mechanisms of Ageing and Development* 53:49-60.

Reddy, V.K., C.T. Kumar, M. Prasad, and P. Reddanna. 1992. Exercise-induced oxidant stress in the lung tissue: Role of dietary supplementation of vitamin E and selenium. *Biochemistry International* 26:863-871.

Regnström, J., J. Nilsson, P. Tomvall, C. Landou, and A. Hamsten. 1992. Susceptibility to low-density lipoprotein oxidation and atherosclerosis in man. *Lancet* 339:1183-1186.

Reiss, U., and D. Gershon. 1976. Comparison of cytoplasmic superoxide dismutase in liver, heart, and brain of aging rats and mice. *Biochemical and Biophysical Research Communications* 73:255-262.

Rokitzki, L., E. Logemann, A.N. Sagredos, M. Murphy, W. Wetzel-Roth, and J. Keul. 1994. Lipid peroxidation and antioxidative vitamins under extreme endurance stress. *Acta Physiologica Scandinavica* 151:149-158.

Ross, R. 1993. The pathogenesis of atherosclerosis: A perspective for the 1990s. *Nature* 362:801-809.

Salminen, A., and V. Vihko. 1983. Lipid peroxidation in exercise myopathy. *Experimental and Molecular Pathology* 38:380-388.

Salonen, J.T., S. Ylä-Herttuala, R. Yamamoto, S. Butler, H. Korpela, R. Salonen, K. Nyyssönen, W. Palinski, and J.L. Witztum. 1992. Autoantibody against oxidized LDL and progression of carotid atherosclerosis. *Lancet* 339:883-888.

Sánchez-Quesada, J.L., R. Homs-Serradesanferm, J. Serrat-Serrat, J.R. Serra-Grima, F. González-Sastre, and J. Ordóñez-Llanos. 1995. Increase of LDL susceptibility to oxidation occurring after intense, long duration aerobic exercise. *Atherosclerosis* 118:297-305.

Sánchez-Quesada, J.L., H. Ortega, A. Payés-Romero, J. Serrat-Serrat, F. González-Sastre, M.A. Lasunción, and J. Ordónez-Llanos. 1997. LDL from aerobically-trained subjects shows higher resistance to oxidative modification than LDL from sedentary subjects. *Atherosclerosis* 132:207-213.

Savenkova, M.I., D.M. Mueller, and J.W. Heinecke. 1994. Tyrosyl radical generated by myeloperoxidase is a physiological catalyst for initiation of lipid peroxidation in low density lipoprotein. *Journal of Biological Chemistry* 269:20394-20400.

Saxton, J.M., A.E. Donnelly, and H.P. Roper. 1994. Indices of free-radical mediated damage following maximum voluntary eccentric and concentric muscular work. *European Journal of Applied Physiology* 68:189-193.

Scheffler, E., E. Wiest, J. Woehrle, I. Otto, I. Schulz, C. Huber, R. Ziegler, and H.A. Dressel. 1992. Smoking influences the atherogenic potential of low-density lipoprotein. *Clinical Investigator* 70:263-268.

Seip, R.L., P. Moulin, T. Cocke, A. Tall, W.M. Kohrt, K. Mankowitz, C.F. Semenkovich, R. Ostlund, and G. Schonfeld. 1993. Exercise training decreases plasma cholesteryl ester transfer protein. *Arteriosclerosis and Thrombosis* 13:1359-1367.

Sen, C.K. 1995. Oxidants and antioxidants in exercise. *Journal of Applied Physiology* 79:675-686.

Sen, C.K., M. Atalay, J. Agren, D.E. Laaksonen, S. Roy, and O. Hänninen. 1997. Fish oil and vitamin E supplementation in oxidative stress at rest and after physical exercise. *Journal of Applied Physiology* 83:189-195.

Serrat, J., J. Ordónez, R. Serra, J.A. Gómez-Gerique, E. Pellicer, A. Payés, and F. González-Sastre. 1993. Marathon runners presented lower serum cholesterol activity than sedentary people. *Atherosclerosis* 101:43-49.

Sevanian, A., J. Hwang, H. Hodis, G. Cazzolato, P. Avogaro, and G. Bittolo-Bon. 1996. Contribution of an in vivo oxidized LDL to LDL oxidation and its association with dense LDL subpopulations. *Arteriosclerosis, Thrombosis and Vascular Biology* 16:784-793.

Shimano, H., N. Yamada, S. Ishibashi, H. Mokuno, Y. Mori, and F. Takaku. 1991. Oxidation-labile subtraction of human plasma low density lipoprotein isolated by ion-exchange chromatography. *Journal of Lipid Research* 32:763-773.

Siu, G.M., and H.H. Draper. 1982. Metabolism of malondialdehyde in vivo and in vitro. *Lipids* 17:349-355.

Slater, T.F. 1984. Free-radical mechanisms in tissue injury. *Biochemical Journal* 222:1-15.

Slyper, A.H. 1994. Low-density lipoprotein density and atherosclerosis. Unraveling the connection. *Journal of the American Medical Association* 272:305-308.

Sohal, R.S., and R.G. Allen. 1986. Relationship between oxygen metabolism, aging and development. *Advances in Free Radical Biology and Medicine* 2:117-160.

Sohal, R.S., and A. Dubey. 1994. Mitochondrial oxidative damage, hydrogen peroxide release, and aging. *Free Radical Biology and Medicine* 16:621-626.

Somani, S.M., S. Frank, and L.P. Rybak. 1995. Responses of antioxidant system to acute and trained exercise in rat heart subcellular fractions. *Pharmacology Biochemistry and Behavior* 51:627-634.

Starnes, J.W., G. Cantu, R.P. Farrar, and J.P. Kehrer. 1989. Skeletal muscle lipid peroxidation in exercised and food-restricted rats during aging. *Journal of Applied Physiology* 67:69-75.

Stein, O., and Y. Stein. 1980. Bovine aortic endothelial cells display macrophagelike properties towards acetylated 125I-labelled low density lipoprotein. *Biochimica et Biophysica Acta* 620:631-635.

Steinberg, D. 1983. Lipoproteins and atherosclerosis: A look back and a look ahead. *Arteriosclerosis* 3:283-301.

Steinberg, D. 1993. Antioxidant vitamins and coronary heart disease. *New England Journal of Medicine* 328:1487-1489.

Steinberg, D., S. Parthasarathy, T.E. Carew, J.C. Khoo, and J.L. Witztum. 1989. Beyond cholesterol: Modification of low-density lipoprotein that increases its atherogenicity. *New England Journal of Medicine* 320:915-924.

Steinbrecher, U.P. 1987. Oxidation of human low density lipoprotein results in derivatization of lysine residues of apolipoprotein B by lipid peroxide decomposition products. *Journal of Biological Chemistry* 262:3603-3608.

Steinbrecher, U.P., H.F. Zhang, and M. Lougheed. 1990. Role of oxidatively modified LDL in atherosclerosis. *Free Radical Biology and Medicine* 9:155-168.

Sumida, S., K. Tanaka, H. Kitao, and F. Nakadomo. 1989. Exercise-induced lipid peroxidation and leakage of enzymes before and after vitamin E supplementation. *International Journal of Biochemistry* 21:835-838.

Tiidus, P., and M.E. Houston. 1993. Vitamin E status does not affect the responses to exercise training and acute exercise in female rats. *Journal of Nutrition* 123:834-840.

Tribble, D.L., L.G. Holl, P.D. Wood, and R.M. Krauss. 1992. Variations in oxidative susceptibility among six low density lipoprotein subtractions of differing density and particle size. *Atherosclerosis* 93:189-199.

Tribble, D.L., R.M. Krauss, M.G. Lansberg, P.M. Thiel, and J.J.M. van den Bero. 1995. Greater oxidative susceptibility of the surface monolayer in small dense LDL may contribute to differences in copper-induced oxidation among LDL density subtractions. *Journal of Lipid Research* 36:662-671.

Tsopanakis, C., D. Kotsarellis, and A. Tsopanakis. 1988. Plasma lecithin:cholesterol acyltransferase in elite athletes from selected sports. *European Journal of Applied Physiology* 58:262-265.

Tyler, D.D. 1975. Role of superoxide radicals in the lipid peroxidation of intracellular membranes. *FEBS Letters* 51:180-183.

Tyroler, H.A. 1987. Lowering plasma cholesterol levels decreases risk of coronary heart disease: An overview of clinical trials. In: *Hypercholesterolemia and Atherosclerosis*, D. Steinberg and J.M. Olefsky (eds.), pp. 99-116. Churchill Livingstone, New York.

Vani, M., G.P. Reddy, G.R. Reddy, K. Thyagaraju, and P. Reddanna. 1990. Glutathione-*S*-transferase, superoxide dismutase, xanthine oxidase, catalase, glutathione peroxidase and lipid peroxidation in the liver of exercised rats. *Biochemistry International* 21:17-26.

Vasankari, T., U. Kujala, O. Hcinoncn, J. Kapanen, and M. Ahotupa. 1995. Measurement of serum lipid peroxidation during exercise using three different methods: diene conjugation, thiobarbituric acid reactive material and fluorescent chromolipids. *Clinica Chimica Acta* 234:63-69.

Vasankari, T.J., U.M. Kujala, II. Rusko, S. Sarma, and M. Ahotupa. 1997a. The effect of endurance exercise at moderate altitude on serum lipid peroxidation and antioxidative functions in humans. *European Journal of Applied Physiology* 75:396-399.

Vasankari, T.J., U.-M. Kujala, T.M. Vasankari, T. Vuorimaa, and M. Ahotupa. 1997b. Effects of acute prolonged exercise on serum and LDL oxidation and antioxidant defenses. *Free Radical Biology and Medicine* 22:509-513.

Vasankari, T.J., U.M. Kujala, T.M. Vasankari, T. Vuorimaa, and M. Ahotupa. 1997c. Increased serum and low-density-lipoprotein antioxidant potential after antioxidant supplementation in endurance athletes. *American Journal of Clinical Nutrition* 65:1052-1056.

Venditti, P., and S.D. Meo. 1996. Antioxidants, tissue damage, and endurance in trained and untrained young male rats. *Archives of Biochemistry and Biophysics* 331:63-68.

Viinikka, L., J. Vuori, and O. Ylikorkala. 1984. Lipid peroxides, prostacyclin, and thromboxane A2 in runners during acute exercise. *Medicine and Science in Sports and Exercise* 16:275-277.

Weddle, C.C., K.R. Hornbrook, and P.B. McCay. 1976. Lipid peroxidation and alteration of membrane lipids in isolated hepatocytes exposed to carbon tetrachloride. *Journal of Biological Chemistry* 251:4973-4978.

Weintraub, M.S., Y. Rosen, R. Otto, S. Eisenberg, and J.L. Breslow. 1989. Physical exercise conditioning in the absence of weight loss reduces fasting and postprandial triglyceride rich lipoprotein levels. *Circulation* 79:1007-1014.

Williams, P.T., R.M. Krauss, P.D. Wood, F.T. Lindcren, C. Giotas, and K.M. Vranizan. 1986. Lipoprotein subfractions of runners and sedentary men. *Metabolism* 35:45-52.

Williams, P., K. Vranizan, and K. Krauss. 1992. Correlations of plasma lipoproteins with LDL subfractions by particle size in men and women. *Journal of Lipid Research* 33:765-774.

Witztum, J.L. 1993. Role of oxidized low density lipoprotein in atherogenesis. *British Heart Journal* 69: S12-S18.

Witztum, J.L. 1994. The oxidation hypothesis of atherosclerosis. *Lancet* 344:793-795.

Xia, E., G. Rao, H.V. Remmen, A.R. Heydari, and A. Richardson. 1995. Activities of anti-oxidant enzymes in various tissues of male Fischer 344 rats are altered by food restriction. *Journal of Nutrition* 125:195-201.

Xu, Q., E. Buhler, A. Steinmetz, D. Schonitzer, G. Bock, G. Jurgens, and G. Wick. 1992. A high-density-lipoprotein receptor appears to mediate the transfer of essential fatty acids from high-density lipoprotein to lymphocytes. *Biochemical Journal* 287:395-401.

Yu, B.P., E.A. Suescun, and S.Y. Yang. 1992. Effect of age-related lipid peroxidation on membrane fluidity and phospholipase A2: Modulation by dietary restriction. *Mechanisms of Ageing and Development* 65:17-33.

Zhang, H.F., Y.H. Yang, and U.P. Steinbrecher. 1993. Structural requirements for the binding of modified proteins to the scavenger receptor of macrophages. *Journal of Biological Chemistry* 268:5535-5542.

Index

About the Editor

Zsolt Radák is an Associate Professor and Head of the Course of Human Kinesiology at in the School of Sport Science at Semmelweis University Budapest. He has worked in the fields of sport biomechanics and training theory, and he formerly coached the Hungarian national track and field team. In the last seven years he has become interested in free radical physiology, focusing on exercise and aging.

Radák received his PhDs from the Hungarian University of Physical Education in 1990 and from the University Tsukuba in Japan in 1996. In 1998 he received a Senior Research Fellowship from the Japanese Foundation of Aging and Health, and in 1999 he received Hungary's Bolyai Research Fellowship in Medicine. He belongs to the American College of Sports Medicine and the Oxygen Society, and he has served as Secretary of the Sport Science Committee of the Hungarian Academy of Sciences. His articles have been published in the *Journal of Applied Physiology* and *Free Radical Biological Medicine*.

Radák lives in Budapest with his wife Kalmar Zsuzsa and his children Fanni and Bence. In his leisure time he enjoys tennis, skiing, travel, and reading books published by Human Kinetics.

*You'll find
other outstanding
exercise psychology resources at*

www.humankinetics.com

In the U.S. call

1-800-747-4457

Australia ....(08) 8277-1555
Canada..............................(800) 465-7301
Europe .... +44 (0) 113-278-1708
New Zealand ....(09) 309-1890

HUMAN KINETICS
The Information Leader in Physical Activity
P.O. Box 5076 • Champaign, IL 61825-5076 USA